Lecture Notes in Mechanical Engineering

Lecture Notes in Mechanical Engineering (LNME) publishes the latest developments in Mechanical Engineering—quickly, informally and with high quality. Original research or contributions reported in proceedings and post-proceedings represents the core of LNME. Volumes published in LNME embrace all aspects, subfields and new challenges of mechanical engineering.

To submit a proposal or request further information, please contact the Springer Editor of your location:

Europe, USA, Africa: Leontina Di Cecco at Leontina.dicecco@springer.com
China: Ella Zhang at ella.zhang@springer.com
India: Priya Vyas at priya.vyas@springer.com
Rest of Asia, Australia, New Zealand: Swati Meherishi at swati.meherishi@springer.com

Topics in the series include:

- Engineering Design
- Machinery and Machine Elements
- Mechanical Structures and Stress Analysis
- Automotive Engineering
- Engine Technology
- Aerospace Technology and Astronautics
- Nanotechnology and Microengineering
- Control, Robotics, Mechatronics
- MEMS
- Theoretical and Applied Mechanics
- Dynamical Systems, Control
- Fluid Mechanics
- Engineering Thermodynamics, Heat and Mass Transfer
- Manufacturing Engineering and Smart Manufacturing
- Precision Engineering, Instrumentation, Measurement
- Materials Engineering
- Tribology and Surface Technology

Indexed by SCOPUS, EI Compendex, and INSPEC.

All books published in the series are evaluated by Web of Science for the Conference Proceedings Citation Index (CPCI).

To submit a proposal for a monograph, please check our Springer Tracts in Mechanical Engineering at https://link.springer.com/bookseries/11693.

Niroj Maharjan · Wei He
Editors

Proceedings of the 3rd International Conference on Advanced Surface Enhancement (INCASE) 2023

Surface Engineering for Sustainability

Springer

Editors
Niroj Maharjan
Advanced Remanufacturing
and Technology Centre (ARTC)
Singapore, Singapore

School of Engineering
Swinburne University of Technology
Hawthorn, Australia

Wei He
Singapore Institute of Manufacturing
Technology (SIMTech)
Singapore, Singapore

ISSN 2195-4356 ISSN 2195-4364 (electronic)
Lecture Notes in Mechanical Engineering
ISBN 978-981-99-8642-2 ISBN 978-981-99-8643-9 (eBook)
https://doi.org/10.1007/978-981-99-8643-9

This Springer imprint is published by the registered company Springer Nature Singapore Pte Ltd.
The registered company address is: 152 Beach Road, #21-01/04 Gateway East, Singapore 189721, Singapore

Contents

Advanced Techniques for Surface Engineering Towards Enhanced Performance

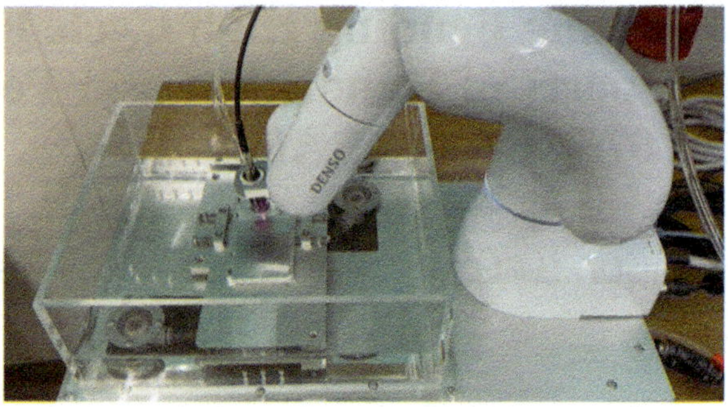

Fig. 1 The newly developed handheld pulse laser oscillator on a robot arm

1 Introduction

Laser peening (LP) processing is one of surface modification technologies. In this process, a pulse laser is irradiated to a metal material placed in water, and the metal is peened by the generated shock wave [1]. At that time, since conventional LP treatment requires large-scale and expensive pulse laser oscillator, smaller and cheaper pulse laser oscillators have been desired for industrial applications [2, 3]. Recently, new handheld pulse laser oscillators as shown in Fig. 1 have been developed by the ImPACT program in Japan [4, 5]. If it turns out that this device can be used as an LP treatment device, there is a possibility that LP treatment can be applied to various industrial fields. In this study, in order to investigate the effectiveness of the newly developed handheld pulse laser oscillator for LP treatment, rotating bending fatigue tests were conducted on Al–Zn–Mg–Cu system aluminum alloys treated by LP.

2 Experimental Methods

2.1 Material

The material used in this study is A7075BE-T6511 aluminum alloy. After extruding into a bar with a diameter of 15 mm, T6511 heat treatment was applied, namely, solution heat treatment, residual stress relief treatment and artificial aging treatment. It is a commercially available material, and detail of the heat treatment conditions are unknown. The chemical composition and the mechanical properties of the material are shown in Table 1 and 2. The microstructure is shown Fig. 2. In this material, an oriented texture is well developed in the axial direction.

Table 1 Chemical composition of the material (%)

Si	Fe	Cu	Mn	Cr	Zn	Ti	Ti + Zr	Al
0.08	0.19	1.6	0.04	2.6	5.8	0.01	0.02	Bal

Table 2 Mechanical properties of the material (experimental data)

Tensile strength σ_B MPa	0.2% proof stress $\sigma_{0.2}$ MPa	Elongation δ %	Reduction of area Ψ %	Hardness Hv
685	631	12.6	23.4	189

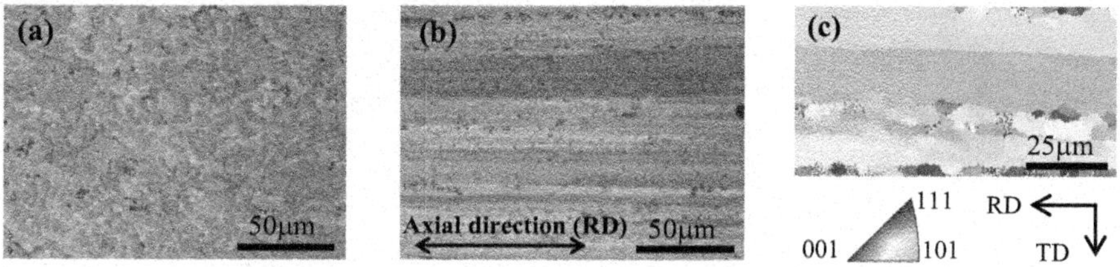

Fig. 2 Microstructure of material **a** cross section **b** longitudinal section **c** EBSD analysis result (IP Map)

2.2 *Specimen and Fatigue Test*

The shape of the test specimens in this study is shown in Fig. 3. The stress concentration factor of this shape is 1.074 [6]. The LP treatment was applied only to the notch area. The laser device used in this study was a prototype handheld pulse laser oscillator provided by OPTOQUEST Co., Ltd. The pulse width of this oscillator is 1.3 ns. An over-view of the laser peening system is shown in Fig. 4. The LP treatment was performed under the following conditions; an irradiation energy of 1.68 mJ, a spot diameter of 0.3 mm, a repetition rate of 50 Hz, and laser irradiation densities of 400 and 800 pulse/mm^2. The LP-treated specimen with under the condition of 400pulse/mm^2 is called 400pls, and the one with under the condition of 800pulse/mm^2 is called 800pls. Cantilever type rotating bending fatigue test was carried out with a rotating speed of 4000 rpm in the air at room temperature.

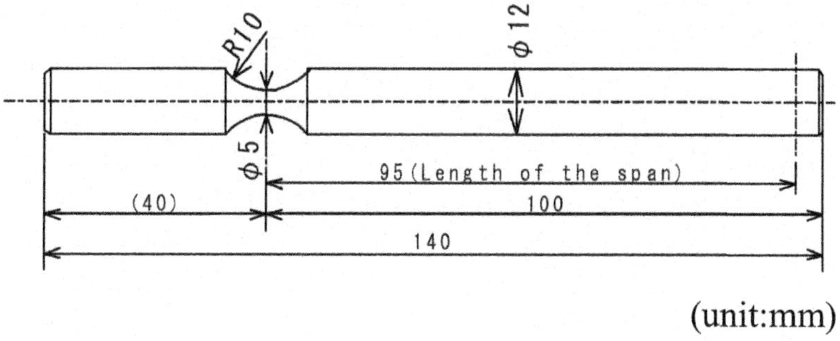

Fig. 3 The shape of fatigue test specimen

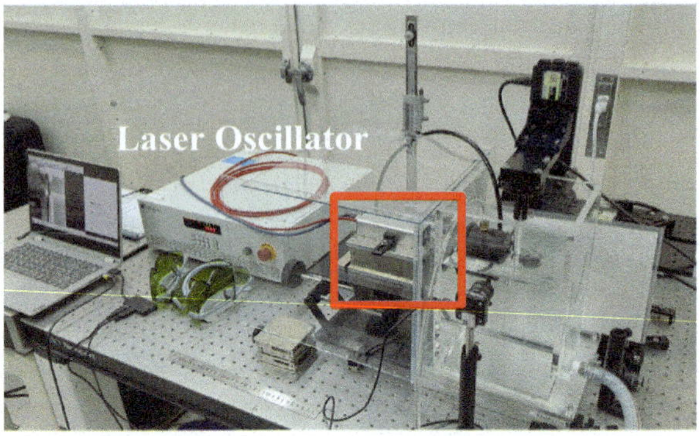

Fig. 4 Over-view of laser peening system

3 Experimental Results and Discussion

3.1 Fatigue Results

The S–N curves of the rotating bending fatigue testing result are shown in Fig. 5. The fatigue properties of A7075BE-T6511 were significantly improved by the handheld LP treatment. The difference in fatigue properties between 400 and 800pls was not significant. The handheld LP treatment improved the fatigue strength at 10^7 cycles by about 1.5 times and extended the fatigue life by about 100 times compared to those of the non-peened materials. These fatigue results indicate that the newly developed handheld pulse laser oscillator can be used for LP treatment. Typical fatigue fracture surface models are shown in Fig. 6. These surface modeling images were obtained from 3D models constructed by X-ray CT observations. There was a big difference between the non-peening specimens and the LP specimens. The fracture surfaces of all the non-peening specimens were on a flat plane perpendicular to the axial direction as shown in Fig. 6a. On the other hand, most of the fracture surfaces of the LP specimens have a slant part as shown in Fig. 6b, c. From the fracture surface observation results by optical microscope and scanning electron microscope, it was suggested that this slant part was not the final fracture site but the fatigue crack initiation site. The angle of the slant part to the cross section of the specimen was about 55–60°. From the results of crystal orientation analysis (EBSD analysis), it was clear that this material had an oriented texture in the axial direction in which the (001) plane of the fcc crystal structure coincides with the cross section of the specimen. The angle of the slant part agrees with 54.7 degrees which is the angle of the slip system $\{111\} < 110 >$ for the cross section of specimen.

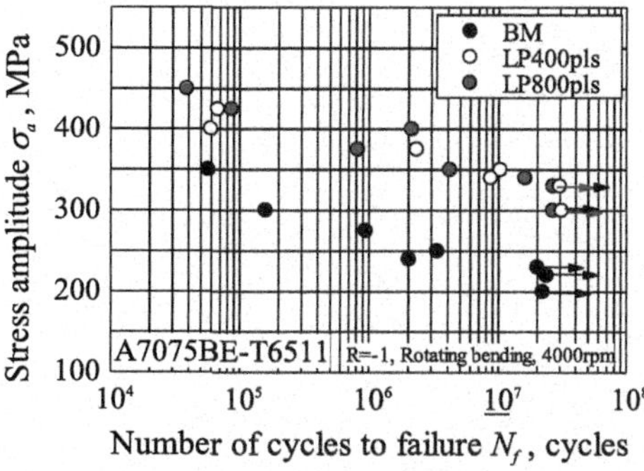

Fig. 5 S–N curves obtained by fatigue testin

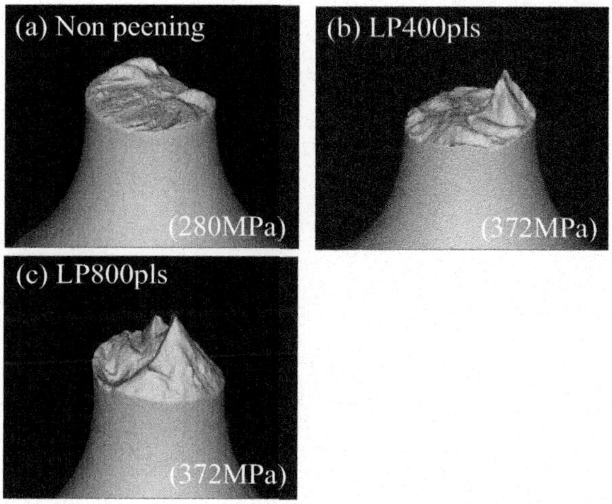

Fig. 6 Surface modeling images of fracture surfaces

3.2 Peening Effect and Fracture Mechanism

Hardness and residual stress, which were the main peening effects of fatigue property improvement, were investigated. The hardness distributions of the LP specimens are shown in Fig. 7. Strain hardening due to the handheld LP treatment is slight. Only the condition of LP800pls shows a slight increase in the surface hardness value. The thickness of the hardened layer is about 50 μm and the increase value is about 20Hv. The residual stress distributions of the LP specimens are shown in Fig. 8. The LP specimens have a large compressive residual stress near the surface layer despite a small increase in hardness. This phenomenon is one of the characteristics of handheld LP treatment. Although it is not clear from the hardness distribution, it is thought that the microstructure at only the specimen surface is in a significantly hardened state or in a highly strained state. It is considered that the generation and propagation of surface cracks are prevented by the high compressive residual stress and special surface state. The reason for the formation of this special surface state is thought to

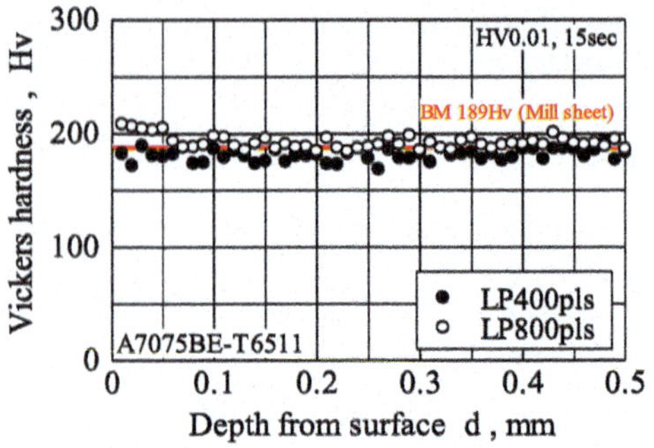

Fig. 7 Hardness distributions of LP specimens

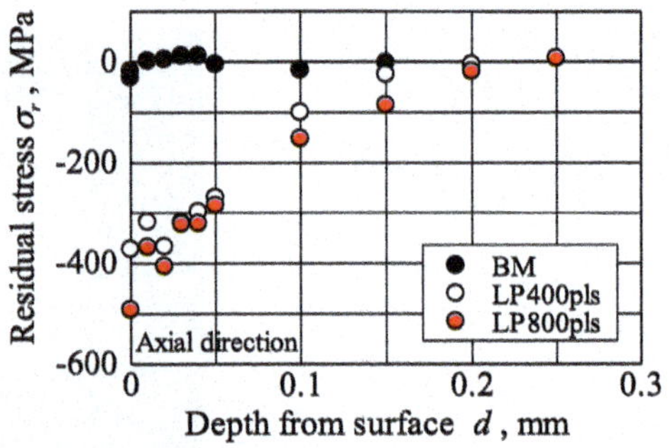

Fig. 8 Residual stress distributions of specimens

be related to the fact that the pulse width of the handheld pulse laser oscillator is less than 1/10 that of conventional pulse laser oscillators. We plan to investigate the details in the future. Even if the fatigue crack can be generated at the surface, the fatigue crack cannot grow in mode I, which is the crack open mode. More than that, the large-scale glide plane decohesion occurred inside the test specimen due to the influence of the developed texture in a triaxial stress state.

4 Conclusion

In order to investigate the effectiveness of the newly developed handheld pulse laser oscillator as a laser source of LP treatment, rotating bending high cycle fatigue tests were conducted on A7075BE-T6511 material used for aircraft components. The fatigue strength at 10^7 cycles is improved by about 1.5 times, and the fatigue life is extended by about 100 times than those of the non-peening material. It is clarified that the newly developed handheld pulse laser oscillator can be used for LP treatment and that it can be applied in a wide variety of industrial fields.

Acknowledgements This work was partially supported by the JST-MIRAI Program (grant number JPMJMI17A1), the Supporting Industry Program of the Small and Medium Enterprise Agency (grant number 202041907014), the Amada Foundation (grant number AF-2020239-C2), and the

Impulsing Paradigm Change through Disruptive Technologies Program (ImPACT) of the Council for Science, Technology and Innovation.

References

1. Sano Y, Mukai N, Okazaki K, Obata M (1997) Residual stress improvement in metal surface by underwater laser irradiation. Nucl Instrum Methods Phys Res B 121:432–436
2. Clauer AH (2019) Laser shock peening, the path to production. Metals 9:626
3. Sano Y (2020) Quarter century development of laser peening without coating. Metals 10:152
4. Sano Y, Kato T, Mizuta Y, Tamaki S, Yokofujita K, Taira T, Hosokai T, Sakino Y (2022) Development of a portable laser peening device and its effect on the fatigue properties of HT780 butt-welded joints. Forces in Mech 7:100080
5. Sano Y, Masaki K, Mizuta Y, Tamaki S, Hosokai T, Taira T (2021) Effects of laser peening with a pulse energy of 1.7 mJ on the residual stress and fatigue properties of A7075 aluminum alloy. Metals 11:1716
6. Walter DP, Deborah FP (2007) Third edition—Peterson's stress concentration factors. Wiley

Developing Domeless, Circular Vibratory Finishing for Aerospace Applications

Jeremy Weng Keong Ho, Kai Liang Tan, and Swee Hock Yeo

Abstract This paper proposed a vibrofinisher with a unique design—domeless, non-elevating and circular bowl vibrofinisher to be used for the aerospace industry. Compared to conventional circular dome vibrofinishers, this new setup could be more effective for material removal and surface roughness improvement than current industrial practices because of its better accessibility to annular geometries. There was limited literature to support this statement. In this paper, a model of a cuboid-body vibration system was proposed and a numerical analysis of this design with reference to past papers of standard design was conducted. The theoretical analysis showed a high potential for faster material removal rate due to an increase of 80.5% in the excitation force and excitation moment of up to 144.6%. Both displacement and rotational angle increased by 68% and 47% respectively, which suggests a higher amplitude. Using the theoretical results, the law of abrasive law and a roughness model was used to substantiate the effectiveness of the domeless bowl as a potential replacement to both the trough and dome bowl.

Keywords Mass finishing · Vibratory finishing · Aerospace

J. W. K. Ho (✉) · S. H. Yeo
Nanyang Technological University, 50 Nanyang Ave, Singapore 639798, Singapore
e-mail: jeremy-ho@artc.a-star.edu.sg

S. H. Yeo
e-mail: mshyeo@ntu.edu.sg

J. W. K. Ho · K. L. Tan
Advanced Remanufacturing & Technology Centre (ARTC), Agency for Science, Technology and Research (A*STAR), 3 CleanTech Loop, #01-01 CleanTech Two, Singapore 637143, Republic of Singapore
e-mail: tan_kai_liang@artc.a-star.edu.sg

N. Maharjan and W. He (eds.), *Proceedings of the 3rd International Conference on Advanced Surface Enhancement (INCASE) 2023*, Lecture Notes in Mechanical Engineering, https://doi.org/10.1007/978-981-99-8643-9_2

1 Introduction

Aerospace components, such as turbine blades, fans, vanes, stator rings, and blisk or integrally bladed rotors, play a critical role in aircraft engines, ensuring aerodynamic efficiency and service life. Surface integrity, both in terms of surface and sub-surface characteristics, is vital for their performance and safety. Vibratory finishing, a popular mass finishing process, is widely used in the aerospace industry due to its versatility in handling components of various sizes and meeting production schedules. Different machine variations and shapes, including domeless round vibrofinishers, are employed to target specific geometries and enhance the interaction between the media and the components. However, there is a lack of research on the application of domeless round vibrofinishers for annular aerospace components, warranting further investigation. This paper aims to conduct a theoretical study on a domeless, circular vibrofinisher and compare its performance to a conventional dome round vibrofinisher using numerical and economic analysis.

2 Theoretical Analysis Methodology

The domeless, circular vibratory finishing machine can be modelled using a spring-dampening-load dynamic system in x–y–z axes as seen in Fig. 1.

With reference to the model of the dome vibratory finishing machine by Hashimoto and Johnson [1], we can simplify the bowl as a charged cuboid-body with mass, M, representing the total mass which includes the mass of the bowl, media, component,

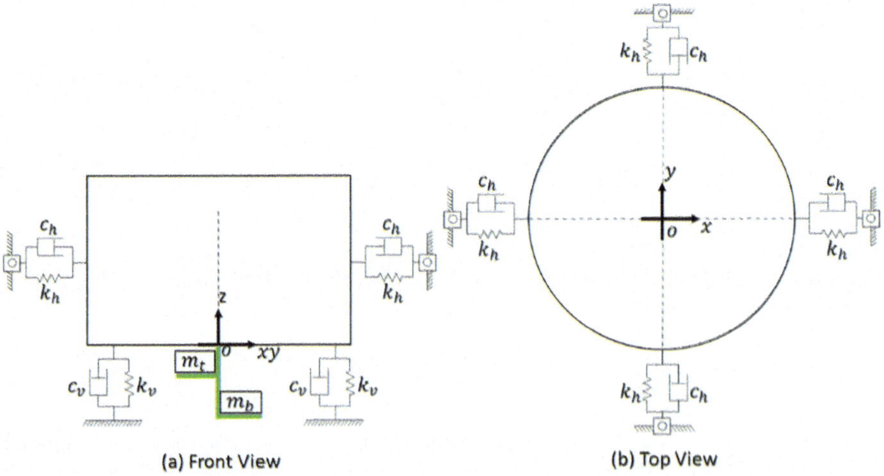

Fig. 1 Dynamic Model of domeless vibratory finishing bowl with indicative position of eccentric weights attached under the bowl

fixture, liquid compound, and water. This mass M is induced with forced vibration by two eccentric plates with attached flyweights along two ends of a motor shaft. The motor rotates and generates a centrifugal force to trigger the forced vibration. At the same time, a moment of inertia J is generated from the centrifugal force. The two eccentric plates are labelled top and bottom plates and are respectively labelled as mt, mb, with their rotational movement along the motor shaft as angular velocity, Ω. The alignment angle between the two masses will be labelled as the phase angle, φ. The distance from the centroid which is the motor axial shaft to the centroid of the flyweights is labelled as r.

Similar to Fig. 1, the springs supporting the mass, M, are consolidated into 4 spring/dampener at the corners on the body vertically (z-axis) while horizontally (x–y-axes) supported by 4 spring/dampener as well. To which the springs are labelled as kh and kv in the horizontal and vertical direction respectively, and dampeners labelled as ch and cv in the horizontal and vertical direction respectively as well.

The distance of the vertical springs from the centroid of M is labelled as L. Finally, mass M is understood to rotate along the x and y axes but not along the z axis. Therefore, there is a rotational angle along x and y axes, θx and θy respectively. The full representation is drawn in Fig. 2.

A set of parametric equations can be used to explain the model, as represented in Fig. 4 Schematic drawings of estimated positions and dimensions of Lt and Lb of (left) dome and (right) domeless Fig. 3. By using Newton's Second Law, the forces in a mechanical vibration system in the x–y-z coordinate can be established as follows:

$$F_x = M\ddot{x} + C_h\dot{x} + K_h x \tag{1}$$

$$F_y = M\ddot{y} + C_h\dot{y} + K_h y \tag{2}$$

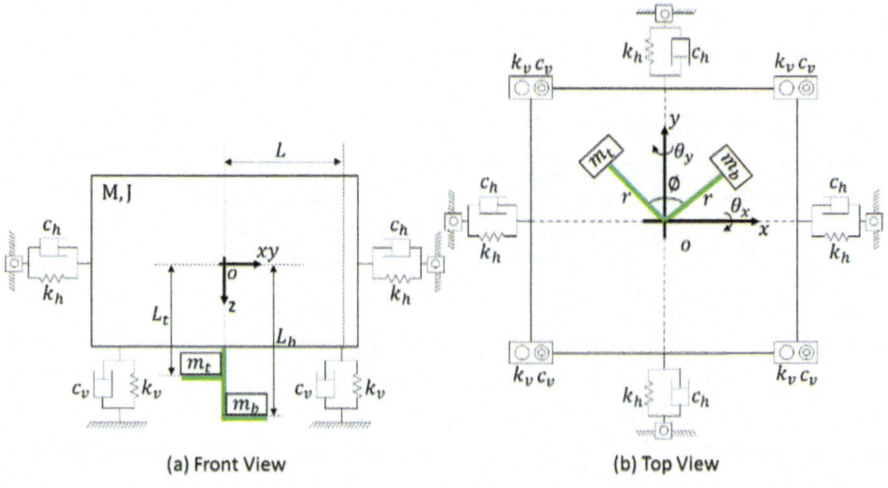

Fig. 2 Domeless vibratory finishing machine modelled as a mechanical vibration system

$$F_z = M\ddot{z} + C_h z + K_h z \tag{3}$$

With the nature of the bowl vibration, we can assume small values for θx and θy which approximated as $\sin\theta_x \approx \theta_x$, $\cos\theta_x \approx 1$, $\sin\theta_y \approx \theta_y$ and $\cos\theta_y \approx 1$. Therefore, F_z can be equated as:

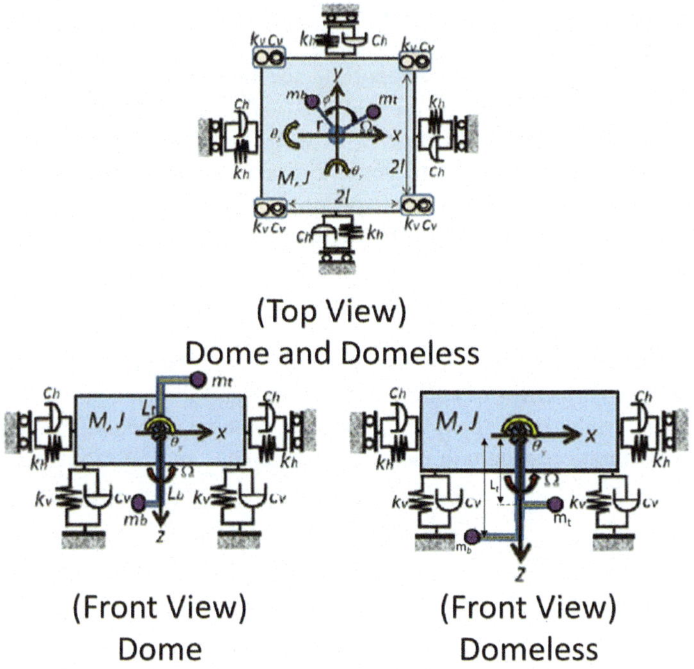

Fig. 3 Schematic drawing of dome and domeless, top view represented similar for both designs, while front view vary in terms of positioning of eccentric plates

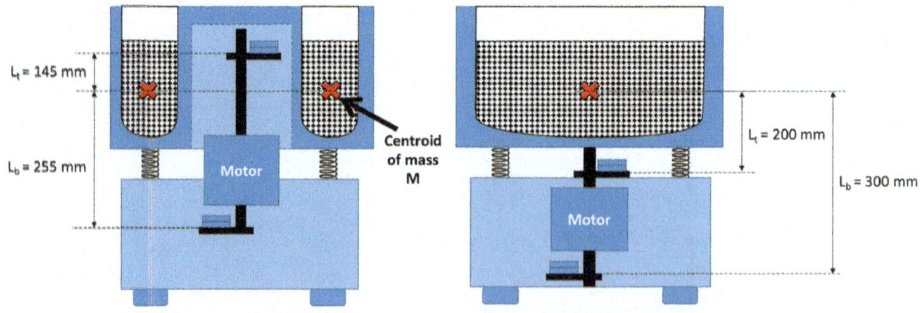

Fig. 4 Schematic drawings of estimated positions and dimensions of L_t and L_b of (left) dome and (right) domeless

$$F_z = F_x\theta_x + F_y\theta_y \tag{4}$$

The angular motion equations are established as follows:

$$P_X = J\ddot{\theta}_x + \left(C_v L^2\right)\dot{\theta}_x + \left(K_v L^2\right)\theta_x \tag{5}$$

$$P_y = J\ddot{\theta}_y + \left(C_v L^2\right)\dot{\theta}_y + \left(K_v L^2\right)\theta_y \tag{6}$$

$$P_z = J\ddot{\theta}_z = 0 \tag{7}$$

where $C_v = 4\,c_v$, $K_v = 4k_v$, $C_h = 2c_h$, $K_h = 2k_h$.

Using Fig. 2 to establish the dynamic positioning of top and bottom eccentric masses mt and mb during processing, we can establish the linear and rotational forces along x–y axes as:

$$F_x = r\Omega^2[m_t \cos\Omega t + m_b \cos(\Omega t + \varphi)] \tag{8}$$

$$F_y = r\Omega^2[m_t \sin\Omega t + m_b \sin(\Omega t + \varphi)] \tag{9}$$

$$P_X = r\Omega^2[m_t L_t \sin\Omega t + m_b L_b \sin(\Omega t + \varphi)] \tag{10}$$

$$P_y = r\Omega^2[m_t L_t \cos\Omega t + m_b L_b \cos(\Omega t + \varphi)] \tag{11}$$

Finally, the motion Eqs. (1)–(3), (5) and (6) can be solved to yield free and forced vibrations. A general solution of the second order differential equation of each of the Eqs. (1)–(3), (5) and (6) can be summated by the auxiliary solution with subscript c to label the free vibrations equations and particular solution subscript p to label the forced vibration equations. This is to follow the model of Hashimoto and Johnson [1] to enable comparison in the next chapter. Therefore, $x = x_c + x_p$, $y = y_c + y_p$, $z = z_c + z_p$, $\theta_x = \theta_{xc} + \theta_{xp}$ and $\theta_y = \theta_{yc} + \theta_{yp}$.

It is also noted that is a slight variation to Eqs. (10) and (11) for a domeless-type due to the change in direction of L_t as compared to a dome-type. Hence, the free vibrations in the x–y–z coordinates and their rotational axes, with the exception of rotation along z axis, are represented by the following equations:

$$x_c = A_1 e^{-\zeta_h \omega_{nh} t} \sin(\omega_{dh} t - \psi_1) \tag{12}$$

$$y_c = A_2 e^{-\zeta_h \omega_{nh} t} \sin(\omega_{dh} t - \psi_2) \tag{13}$$

$$z_c = A_3 e^{-\zeta_v \omega_{nh} t} \sin(\omega_{dv} t - \psi_3) \tag{14}$$

$$\theta_{xc} = A_4 e^{-\zeta_h \omega_{nh} t} \sin(\omega_{dh} t - \psi_4) \tag{15}$$

$$\theta_{yc} = A_5 e^{-\zeta_h \omega_{nh} t} \sin(\omega_{dh} t - \psi_5) \tag{16}$$

where A1—A5 are arbitrary constants, ψ_1 to ψ_5 are alignment (phase) angles determined by arbitrary constants. ζ represents the damping factors, ωn represents natural frequency and ωd represent damped natural frequency, with all three having subscripts to represent their directions—v for vertical, h for horizontal and θ for rotational. Each of them can be calculated with the following equations:

$$\zeta_h = \frac{C_h}{2\sqrt{MK_h}} \tag{17}$$

$$\omega_{nh} = \sqrt{\frac{K_h}{M}} \tag{18}$$

$$\omega_{dh} = \left(\sqrt{1 - \zeta_h^2}\right)\omega_{nh} \tag{19}$$

$$\zeta_v = \frac{C_v}{2\sqrt{MK_v}} \tag{20}$$

$$\omega_{nh} = \sqrt{\frac{K_v}{M}} \tag{21}$$

$$\omega_{dv} = \left(\sqrt{1 - \zeta_v^2}\right)\omega_{nv} \tag{22}$$

$$\zeta_\theta = \frac{C_v L}{2\sqrt{MK_v}} \tag{23}$$

$$\omega_{n\theta} = \sqrt{\frac{K_v L^2}{J}} \tag{24}$$

$$\omega_{d\theta} = \left(\sqrt{1 - \zeta_\theta^2}\right)\omega_{n\theta} \tag{25}$$

The forced vibrations generated by the rotational angular velocity Ω of eccentric masses mt and mb are represented by the auxiliary equations of Eqs. (1)–(5) in the x–y-z coordinates and their rotational axes, with the exception of rotation along z axis, by the following equations:

$$x_p = G_x \cos(\Omega t - \Phi_1) \tag{26}$$

$$y_p = G_y \sin(\Omega t - \Phi_1) \tag{27}$$

$$z_p = constant \tag{28}$$

$$\theta_{xp} = E_x \sin(\Omega t - \Phi_2) \tag{29}$$

$$\theta_{yp} = E_y \cos(\Omega t - \Phi_2) \tag{30}$$

where Φ is the alignment or phase angle and variables G, E are variables, with their subscripts to represent their directional axis, are determined by the following equations:

$$G_x = \frac{F_x}{\sqrt{(C_h\Omega)^2 + (K_h - M\Omega)^2)^2}} \tag{31}$$

$$G_y = \frac{F_y}{\sqrt{(C_h\Omega)^2 + (K_h - M\Omega)^2)^2}} \tag{32}$$

$$E_x = \frac{P_x}{\sqrt{(C_v L^2\Omega)^2 + (K_v L^2 - J\Omega)^2)^2}} \tag{33}$$

$$E_y = \frac{P_y}{\sqrt{(C_v L^2\Omega)^2 + (K_v L^2 - J\Omega)^2)^2}} \tag{34}$$

$$\Phi_1 = \tan^{-1} \frac{C_h\Omega}{(K_h - M\Omega^2)} \tag{35}$$

$$\Phi_2 = \tan^{-1} \frac{C_h L^2\Omega}{(K_h L^2 - J\Omega^2)} \tag{36}$$

The Eqs. (1)–(36) will be used as reference with Hashimoto and Johnson's [1] dome model to be described and comparatively plotted.

2.1 Comparison to Dome, Circular Vibratory Finishing

According to Hashimoto and Johnson's mechanical vibration model [1], the domeless machine can be drawn as another re-representation from Fig. 2 to create a direct comparison of between a dome and domeless schematic as shown in Fig. 3.

The key different difference between a dome and domeless vibratory finishing machine is the position of the eccentric plates. Parameters are taken from the paper from Hashimoto and Johnson [1] and scaled accordingly to have a fair comparison between dome and domeless type. A parameter table is established in Table 1.

Table 1 Parameters to be used for comparison

Parameter	Symbol	Dome	Domeless
Phase angle (alignment angle between eccentric plates)	φ	$90° \approx$ 1.5708 rad/s	
Angular velocity	Ω	146.9 rad/s	
Spring constant—vertical	k_v	39.2 N/mm	
Spring constant—horizontal	k_h	15.9 N/mm	
Damping constant—vertical	c_v	0.0055 N s/mm	
Damping constant—horizontal	c_h	0.0222 N s/mm	
Damping factors—vertical	ζ_v	0.0031	
Damping factors—horizontal	ζ_h	0.0014	
Damping factors—angular	ζ_θ	0.0031	
Natural frequency—vertical	ω_{nv}	44.6 rad/s	
Natural frequency—horizontal	ω_{nh}	20.1 rad/s	
Natural frequency—angular	$\omega_{n\theta}$	45.1 rad/s	
Damping frequency—horizontal	ω_{dv}	44.6 rad/s	
Damping frequency—vertical	ω_{dh}	20.1 rad/s	
Damping frequency—angular	$\omega_{d\theta}$	45.1 rad/s	
Horizontal length of vertical springs to mass M centroid	L	180 mm	
Horizontal length of eccentric plate to mass M centroid	r	50 mm	
Vertical length of top eccentric plate to mass M centroid	L_t	145 mm	200 mm
Vertical length of bottom eccentric plate to mass M centroid	L_b	255 mm	300 mm
Top eccentric plate and flyweight (pre-assigned) weight	$m_t \cdot g$	8.4 N	8.4 N (2 plates*)
Bottom eccentric plate and flyweight (pre-assigned) weight	$m_b \cdot g$	9.3 N	21.0 N (5 plates*)
Ratio of top eccentric plate and flyweight and bottom eccentric plate and flyweight	m_t / m_b	1 (round up)	0.4
Body weight	$M \cdot g$	772.3 N	816.1 N

*Based on weight referred from study Hashimoto & Johnson [1] and industrial practice for top/bottom flyweights' setup to be 2:5

To ensure a proper baseline comparison, machine setups that are similar in nature for both types such as springs, and dampeners will be kept constant. The process setup such as alignment (phase) angle and angular velocity will also be set similar. Additionally, natural constants and factors will be set similar. Therefore, the similar values are as follows: φ, Ω, k_v, k_h, c_v, c_h, ζ_h, ζ_v, ζ_θ, ω_{nh}, ω_{nv}, $\omega_{n\theta}$, ω_{dh}, ω_{dv}, $\omega_{d\theta}$, L, and r.

The key difference between the two bowl types will be on the masses and their relative locations. For eccentric masses m_t and m_b, a baseline qualitative ratio of m_t and m_b was used depending on the number of flyweights are attached to them.

Additionally, the flyweight quantity will be selected based on the dome and domeless current best practice to achieve an equal frequency and amplitude. This is done by using reference of ERBA EVP-250 dome and ERBA EVP-250 CL domeless equipment. For a similar setup between dome and domeless-type, a 4.5 mm amplitude and 70° lead angle for the feed frequency will require to have two flyweight at the top and bottom plates respectively for a dome-type. This makes the ratio of m_t/m_b 1:1. In a domeless-type setup, it is two at the top and five at the bottom, making the ratio 2:5. To summarise, the values of 1.0 for dome and 0.4 for domeless is used.

The difference between dome and domeless types for positions L_t and L_b are mainly that direction of L_t is under the bowl and therefore opposite in direction as opposed to dome type, and that L_b is further from the centroid M as compared to dome type due to the lowered placement of the motor. These observations of domeless type with comparison were based on the difference in EVP-250 CL domeless and EVP-250 dome type machines, respectively. Based on these pointers, for domeless type eccentric weight positions, L_t and L_b, will be estimated based on the study's machine specification [1] and also with industrial best practices, are placed at the values 200 mm and 300 mm as drawn in Fig. 4.

For body mass M of dome machine will be calculated based on similar media (Al_2O_3 spherical dia. 5 mm) of density 1.43 g/cm^3 and component mass (hardened steel cylindrical roller of dia. 15 mm x L22 mm, HRC62) from the paper. As the mass M of dome type will account for the additional volume due to the lack of a dome by a calculated approximation of 3.123L, which is the volume of the dome. Using this additional volume to calculate the weight from the additional media, the weight of the bowl (assume weight is similar for domeless-type as dome is hollow therefore weight is negligible), and weight of the component, will amount 816.1 N. The detailed calculations are appended in Appendix A.

3 Theoretical Comparison Results

Figure 5 show a comparison of dome and domeless type of the external forces in x and y axes, which are the sources of the vibration motion during machine operation with an amplitude of 1400 RPM. This is calculated by Eqs. (8) and (9). The results show that the domeless type has its excitation force amplitude peak higher than dome type at 2.44 kN, compared to domeless type at 1.352 kN, a difference of around 80.5%.

Figure 6 shows a comparison of dome and domeless type of the external moments in x and y axes. This is calculated by Eqs. (8) and (9) for domeless type and cited from the study for dome type [1]. The results show that the domeless type has its excitation moment amplitude peak higher than dome type at 703.5kN, compared to domeless type at 287.7kN, a difference of 144.6%. It is also noted that there is a shift in frequency by almost 50%.

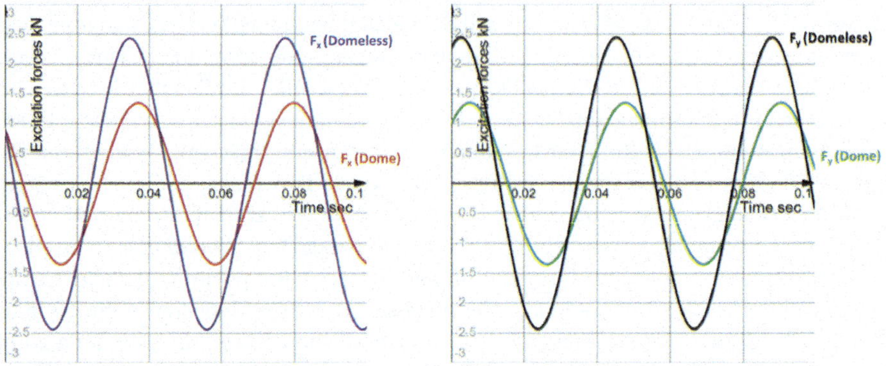

(a) Excitation Force (F_x) in the horizontal direction between dome and domeless bowl

(b) Excitation Force (F_y) in the vertical direction between dome and domeless bowl

Fig. 5 Theoretical comparison between dome and domeless-type by the external forces causing the forced vibration in **a** horizontal and **b** vertical direction

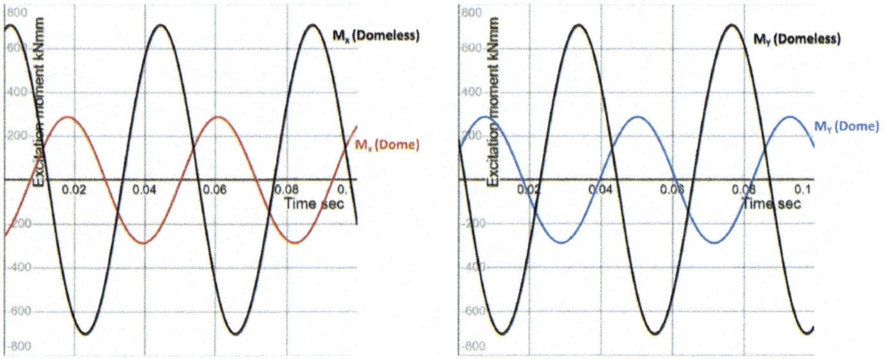

(a) Excitation Moment (M_x) in the horizontal direction between dome and domeless bowl

(b) Excitation Moment (M_y) in the vertical direction between dome and domeless bowl

Fig. 6 Theoretical comparison between dome and domeless-type by the external moment causing the forced vibration in rotational directions in horizontal axes along **a** X-axis and **b** Y-axis

4 Discussion

Based on results shown in Figs. 5 and 6, there is an agreement to qualitative reasoning of a higher processing capability. With the absence of the dome, there is a larger radial roll which allow longer displacement of component-media contact. This in turn provides a longer exposure for material removal. To validate this statement, the law of abrasive wear [2] can be used:

$$V_s = \frac{k_s N L}{H} \tag{37}$$

V_s wear volume

k_s wear coefficient
N normal load
H hardness of worn surface
L media sliding distance

In this equation, the normal load (N) is the perpendicular force acting on the component surface and thus is directly proportionate to the wear volume (V_s). From an energy standpoint, the larger radial roll caused by an increased excitation forces from the bowl vibration will cause an increased energy transfer from the bowl vibration to media and finally onto the component surface. Hence, this implies a higher material removal due to higher excitation forces from the bowl.

Illustrated in Fig. 7, a micro view of a component-media contact point can be used to show a conversion of the law of abrasive wear into a steady-state abrasive wear per unit area. Thus, wear volume (V_s) is converted into wear depth per unit time, $\delta h_s \delta t$ and normal load (N) is converted into normal specific load (p_g) per unit area, $p_g \delta A$. This allows a roughness model to be derived as done by Wan et al. [3]):

$$R_a = (R_i - R_E) \cdot e^{-\frac{k_T p_g v_s}{H} t} + R_E \tag{38}$$

R_a Current Roughness
R_E Roughness at saturation
R_i Initial Roughness
v_s Velocity along unit area
k_T wear coefficient

By making time (t) the subject:

$$\mathrm{t} = \frac{-\mathrm{H} \cdot \ln\left(\frac{R_a - R_E}{R_i - R_E}\right)}{k_T \, p_g v_s} \tag{39}$$

The Eq. (39) shows that the normal specific load (p_g) is inversely proportionate to time (t). As earlier described, the higher excitation forces caused by a domeless-type as compared to a dome-type will cause an increase in energy transfer from the

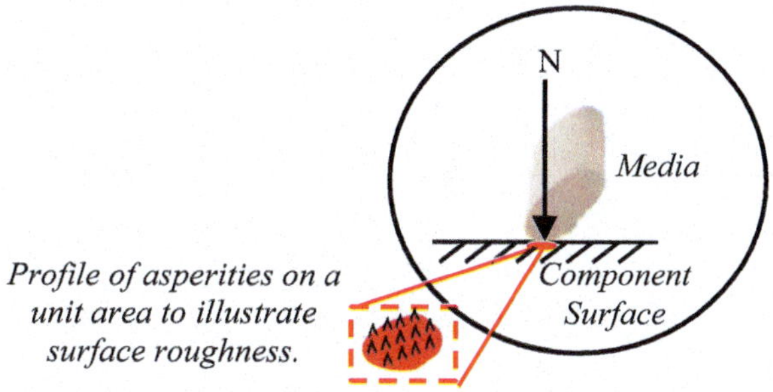

Fig. 7 Illustration of media contact to component surface with a micro illustration of surface roughness per unit area

bowl to component-media interaction via the point contact. This point contact is referred as the normal specific load (p_g), which essentially proves that time taken for surface finishing will reduce. Hence, a potentially faster surface finishing by using domeless-type.

5 Conclusion

This paper proposed a new vibratory finishing machine with a unique design—domeless, non-elevating and circular vibratory finishing bowl—to be used for the aerospace industry. A study on current industrial best practices such as machine-component pairing was conducted, and it was determined that this proposed machine could be more effective than current industrial practices.

A review of existing vibratory finishing models was conducted, and a model of a cuboid-body vibration system was adapted. Through a theoretical comparison with a dome-type, the domeless-type showed a higher processing potential than dome type vibratory finishing due to the increase in excitation forces by 80.5%. Additionally, the increased distance between the motor and two eccentric plates to the centroid of the bowl mass augmented the forced vibration component of the bowl and increased the excitation moment by up to 144.6%. This suggests that a domeless-type allows a larger roll radius due to the absence of the dome. The displacement value showed an increase of 68% and rotational angle increase of 47%. This suggests that a domeless-type could result in a higher bowl amplitude. Overall, these results showed higher processing potential and were substantiated through an inference to the law of abrasive wear and a roughness model by Wan et al. [3].

References

1. Hashimoto F, Johnson S (2015) Modeling of vibratory finishing machines. CIRP Ann Manuf Technol 64:345–348
2. Hutchings IM (1992) Tribology: friction and wear of engineering.Materials, London
3. Wan S, Liu Y, Woon K (2016) A simple general process model for vibratory finishing. Int J Adv Manuf Technol 86:2393–2400

Improvement of Fatigue Strength of 3D-Metal by Combined Process of Blasting and Cavitation Peening

Hitoshi Soyama

Abstract Three dimensional additively manufactured metallic materials, i.e., 3D-Metal, are attractive materials. However, the fatigue strength of as-built 3D-metal is nearly half of bulk metals. In order to enhance the fatigue strength of 3D-metal, a combined process of blasting and cavitation peening (CP) was proposed. In the process, the blasting makes smoother surface by collisions of abrasives, and introduces compressive residuals stress in the sub-surface region. After the blasting, CP using a submerged water jet introduces compressive residual stress into the deeper region. In the present paper, titanium alloy Ti6Al4V manufactured by direct metal laser sintering (DMLS) was treated by the combined process of blasting and CP. The fatigue strength was evaluated by torsion fatigue testers of a torsion-controlled type and an angle-controlled type. To make clear the mechanism of the improvement of fatigue strength, surface roughness and surface residual stress were measured. It was revealed that the blasting introduced the compressive residual stress while decreasing the surface roughness, and the fatigue strength was improved by about 33% compared with as-built specimen. When the specimen was treated by CP after blasting, the fatigue strength was improved by 58%, which was larger than that of CP only.

Keywords Cavitation peening · Blasting · Additive manufacturing · Fatigue · Torsion

1 Introduction

Three dimensional metallic materials produced by additive manufacturing, i.e., 3D-Metal, are attractive materials for medical implants and aviation components, as products are formed from CAD/CAM data directly. However, the fatigue strength of 3D-Metal is nearly half of wrought metals [1, 2], and it is obstacle for application

H. Soyama (✉)
Department of Finemechanics, Tohoku University, 6-6-01 Aoba-Ku, Sendai 980-8579, Japan
e-mail: soyama@mm.mech.tohoku.ac.jp

N. Maharjan and W. He (eds.), *Proceedings of the 3rd International Conference on Advanced Surface Enhancement (INCASE) 2023*, Lecture Notes in Mechanical Engineering, https://doi.org/10.1007/978-981-99-8643-9_3

of 3D-Metal. As mechanical surface treatment such as shot peening SP can improve the fatigue strength of 3D-Metal, it is worthwhile to develop mechanical surface treatment for 3D-Metal.

Hot isostatic pressing (HIP) can reduces crack propagation in 3D-Metal [3], and it enhances the fatigue strength [4]. However, HIP cannot improve the fatigue properties of 3D-Metal with surface defects [5]. The improvements of fatigue properties of 3D-Metal have been reported by using SP [1, 2, 6–10], laser peening LP [6–11], cavitation peening CP [9–12], etc. As the main reason of the weakness is large surface roughness due to un-melted particles on the surface, cavitation abrasive surface finishing CASF [13–15] was proposed. At CASF, a high-speed water jet with cavitation, i.e., a cavitating jet, is injected into a water filled chamber, in which abrasives are installed. The abrasives, which are accelerated by the cavitating jet, smoothen the surface of 3D-Metal and the cavitation impacts introduce compressive residual stress at the same time [13], then the fatigue strength is increased. As the optimum injection pressures for smoothening the surface and for the introduction of compressive residual stress are different, the fatigue strength of CP after CASF was larger than that of CASF [15]. The surface modification of 3D-Metal by abrasive finishing using ultrasonic cavitation and hydrodynamic cavitation was also developed [16, 17]. As both of smoothening the surface of 3D-Metal and the introduction of compressive residual stress are important to enhance fatigue properties of 3D-Metal.

In the present paper, in order to demonstrate the improvement of fatigue strength of 3D-Metal, titanium alloy Ti6Al4V manufactured by direct metal laser sintering DMLS was treated by blasting and CP to smoothen the surface and to introduce the compressive residual stress, then tested by torsion fatigue tests.

2 Experimental Apparatus and Procedures

Figure 1 shows a photograph of as-built specimen manufactured by DMLS and dimension of the specimen. The specimen was built by EOS M290 with 400 W laser. The diameter of used metallic particle was 40 μm, the diameter of laser spot was 100 μm, and the stacking pitch was 60 μm. After DMLS, the specimens were annealed at 923 K for 3 h to release the residual stress, and subjected to solution heat treatment at 1223 K under vacuum for 105 min before cooling under argon gas. Finally, each specimen was aged at 978 K under vacuum for 2 h before cooling under argon gas. After heat treatment, the surface of chucking part of the specimen was finished by a lathe and a metal file as shown in Fig. 1a.

At blasting, the garnet of #150 size was injected perpendicularly to the specimen surface through a nozzle 2.6 mm in diameter by air of 0.7 MPa. The distance between the nozzle and the specimen surface was 11 mm. At blasting, specimen was moved in axis-direction with rotation using a jig with a screw pitch of 2 mm.

Figure 2 illustrates a schematic diagram of test section of CP system. The cavitating jet of 30 MPa was injected into a water filled chamber, whose pressure was 0.42

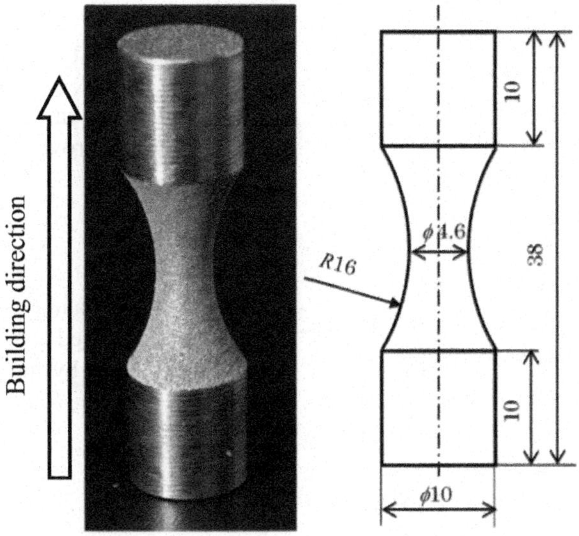

(a) Aspect of as-built specimen (b) Dimensions of specimen (unit: mm)

Fig. 1 Fatigue specimen

MPa in absolute pressure. The nozzle throat diameter was 2 mm. A specimen holder with a screw pitch of 2 mm was used to move the specimen in the axial direction. The standoff distance from the nozzle to the specimen surface was 80 mm.

For both blasting and CP, the processing time per unit length was defined by following Eq. (1).

$$t_p = \frac{n}{v} \tag{1}$$

Here, n is number of scan and v is speed in axial direction.

3 Results

Figure 3 reveals the surface of specimens of Fig. 3a as-built, Fig. 3b blasting at $t_p =$ 2s/mm, Fig. 3c CP at $t_p = 10$s/mm and Fig. 3d CP at $t_p = 10$ s/mm after blasting at $t_p = 2$ s/mm. At as-built specimen, a lot of unmelted particles were observed on the surface. In the case of CP, some unmelted particles were removed by the cavitation impacts, however, the deep valleys, where unmelted particles were removed, were recognized on the surface. At Fig. 3a as-built and Fig. 3d CP after blasting, unmelted

Fig. 2 Schematic geometry of test section of cavitation peening system

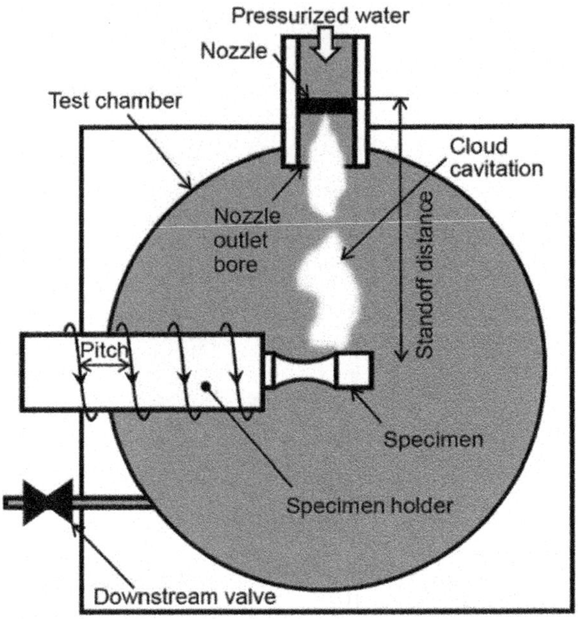

particles and deep valleys were not observed. Namely, blasting removed the unmelted particles and deep valleys by smoothening of abrasive impacts.

Figure 4 reveals fatigue life N_{FL}' at $\tau_a = 460$ MPa as a function of processing time per unit length t_p normalized by the fatigue life of as-built specimen, for blasting, CP

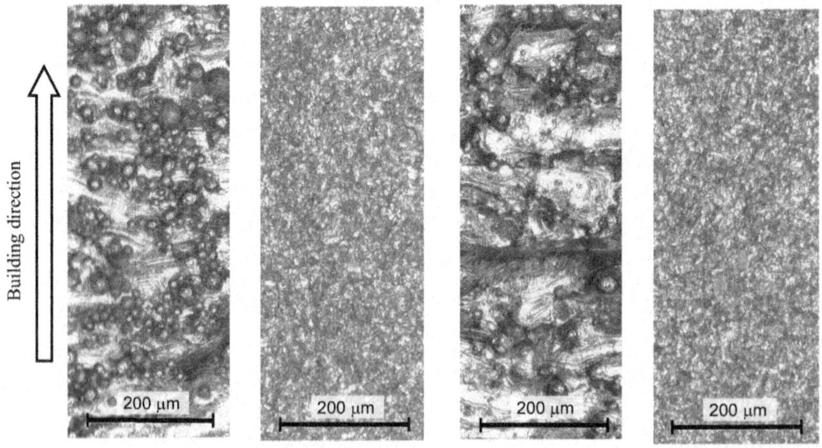

(a) As-built (b) Blasting (t_p = 2s/mm) (c) CP (t_p = 10s/mm) (d) Blasting (t_p = 2s/mm) + CP (t_p = 10s/mm)

Fig. 3 Aspect of specimen surface

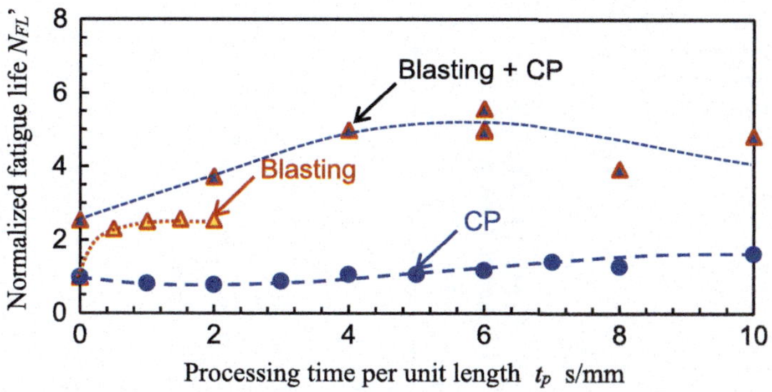

Fig. 4 Fatigue life of treated specimens as a function of processing time per unit length

and blasting + CP, respectively. The fatigue life was evaluated by using a torsion-controlled fatigue tester [10]. The fatigue strength of as-built, i.e., $(2.15 \pm 2.1) \times 10^4$, was calculated from the average value of three times. The N_{FL}' of CP was slightly decreased with t_p, and it was increased at $t_p > 2$ s/mm, then it was about 1.6 at $t_p =$ 10 s/mm. Thus, $t_p = 10$ s/mm was chosen as optimum for CP. At blasting, N_{FL}' was significantly increased from t_p, = 0 to $t_p = 0.5$ s/mm, and nearly saturated. Then, t_p = 2 s/mm was chosen as optimum for blasting. In the case of blasting + CP in Fig. 4, the specimen was treated by blasting $t_p = 2$ s/mm, then treated by CP changing with t_p. The N_{FL}' of blasting + CP was increased with t_p of CP, and it had a maximum at $t_p = 6$ s/mm, then it was slightly decreased. Thus, $t_p = 6$ s/mm for CP was chosen as optimum for blasting + CP.

Figure 5 shows the result of torsion fatigue test using an angle-controlled fatigue tester. As mentioned above, t_p was chosen as 10 s/mm for CP, 2 s/mm for blasting. At blasting + CP, t_p was 2 s/mm for blasting and 6 s/mm for CP. For all three surface treatment, i.e., CP, blasting and blasting + CP, the fatigue life and fatigue strength were improved comparing with as-built. When the fatigue strength at $N_f = 10^7$ was calculated by Little's method [18], it was 216.9 ± 6.9 MPa for as-built, 288.0 ± 27.2 MPa for blasting, 312.8 ± 11.3 MPa for CP and 342.0 ± 10.8 MPa for blasting + CP, respectively. The improvement ratio for as-built specimen was 33% for blasting, 44% for CP and 58% for blasting + CP. The fatigue strength of CP after blasting was larger than that of CP and blasting. Namely, the combined process of blasting and CP was effective to improve fatigue strength of 3D-Metal. The surface roughness was $Ra = 18.1 \pm 1.4$ μm, $Rz = 156.3 \pm 4.2$ μm for as-built, $Ra = 10.3 \pm 1.1$ μm, $Rz = 84.0 \pm 3.9$ μm for blasting, $Ra = 16.6 \pm 2.3$ μm, $Rz = 150.7 \pm 17.4$ μm for CP and $Ra = 14.2 \pm 4.2$ μm, $Rz = 96.5 \pm 17.3$ μm for blasting + CP. On the other hand, the residual stress on the surface measured by 2D-XRD method was 142 ± 23 MPa for as-built, -604 ± 57 MPa for blasting, – 150 ± 22 MPa for CP and – 582 ± 45 MPa for blasting + CP. As the surface of the specimen was considerably rough and the penetration depth of X-ray was shallow, the obtained value of residual stress only shows the tendency. It can be concluded

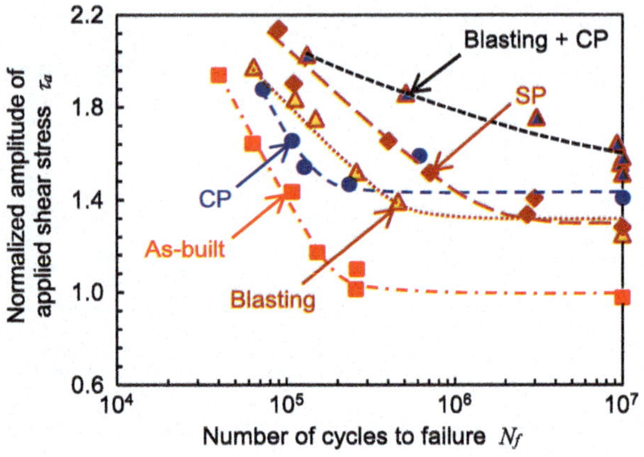

Fig. 5 Improvement of fatigue properties of Ti6Al4V manufactured by DMLS by surface treatments

that the smoothing by blasting and introduction of compressive residual stress by CP were effective for the improvement of fatigue strength. Note that the fatigue life of blasting at relatively high shear stress was larger than that of CP, however, the fatigue strength of CP was larger than that of blasting.

In order to compare present result with conventional shot peening, the result of the specimen treated by shot peening SP [19] was plotted in Fig. 5. The processing time per unit length of SP was optimized in the same way of present study [19]. The fatigue strength of SP at 10^7 was 285.0 ± 10.4 MPa. As mentioned above, the fatigue strength of blast was slightly better than that of SP. Note that the fatigue strength of CP and blasting + CP were remarkably better than that of conventional SP. Namely, the proposed combined process of blasting and cavitation peening CP is worthwhile to investigate. Note that the fatigue life of SP at low cycle fatigue was longer than that of CP, however, the fatigue strength of CP was better than that of SP as same as stainless steel [20].

4 Conclusions

In order to demonstrate the improvement of fatigue strength of 3D-Metal by combined process of blasting and cavitation peening CP, as-built specimens made of titanium alloy Ti6Al4V manufactured by direct metal laser sintering DMLS were treated by blasting, CP and CP after blasting (blasting + CP), and the fatigue properties were evaluated by torsion fatigue tests. It was concluded that the fatigue strength at 10^7 of blasting + CP was greater than that of blasting and CP. The improvement ratio of blasting + CP comparing with as-built specimen was 58%.

Acknowledgements This research was partly supported by JSPS KAKENHI grant numbers 22KK0050 and 23H01292.

References

1. Edwards P, O'Conner A, Ramulu M (2013) Electron beam additive manufacturing of titanium components: properties and performance. J Manuf Sci Eng 135:1–7
2. Edwards P, Ramulu M (2014) Fatigue performance evaluation of selective laser melted Ti-6Al-4V. Mater Sci Eng A 598:327–337
3. Lenders S, Thone M, Riemer A et al (2013) On the mechanical behaviour of titanium alloy TiAl6V4 manufactured by selective laser melting: fatigue resistance and crack growth performance. Int J Fatigue 48:300–307
4. Sun YY, Lu SL, Gulizia S et al (2020) Fatigue performance of additively manufactured Ti-6Al-4V: surface condition vs. Internal Defects JOM 72:1022–1030
5. Chern AH, Nandwana P, Yuan T et al (2019) A review on the fatigue behavior of Ti-6Al-4V fabricated by electron beam melting additive manufacturing. Int J Fatigue 119:173–184
6. Hackel L, Rankin JR, Rubenchik A et al (2018) Laser peening: A tool for additive manufacturing post-processing. Addit Manuf 24:67–75
7. Carlos Navarro P, Jesús V, Jaime D et al (2020) Effect of surface treatment on the fatigue strength of additive manufactured Ti6Al4V alloy. Frattura ed Integrità Strutturale 14:337–344
8. Kahlin M, Ansell H, Kerwin A et al (2021) Variable amplitude loading of additively manufactured Ti6Al4V subjected to surface post processes. Int J Fatigue 142:12
9. Soyama H, Korsunsky AM (2022) A critical comparative review of cavitation peening and other surface peening methods. J Mater Process Technol 305:117586
10. Soyama H, Kuji C (2022) Improving effects of cavitation peening, using a pulsed laser or a cavitating jet, and shot peening on the fatigue properties of additively manufactured titanium alloy Ti6Al4V. Surf Coat Technol 451
11. Soyama H, Okura Y (2018) The use of various peening methods to improve the fatigue strength of titanium alloy Ti6Al4V manufactured by electron beam melting. AIMS Mater Sci 5:1000–1015
12. Okura Y, Sasaki H, Soyama H (2020) Effect of mechanical properties on fatigue life enhancement of additive manufactured titanium alloy treated by various peening methods. In: Advanced surface enhancement, INCASE 2019, lecture notes in mechanical engineering, pp 88–96
13. Soyama H, Sanders D (2019) Use of an abrasive water cavitating jet and peening process to improve the fatigue strength of titanium alloy 6Al-4V manufactured by the electron beam powder bed melting (EBPB) additive manufacturing method JOM 71:4311–4318
14. Mojib M, Soyama H, Sanders D et al (2021) The high cycle fatigue behavior of surface treated electron beam melted titanium ti6al4v. In: ASME 2021 international mechanical engineering congress and exposition, IMECE 2021 2A-2021, V02AT02A018
15. Sanders D, Soyama H, De Silva C (2021) Use of cavitation abrasive surface finishing to improve the fatigue properties of additive manufactured titanium alloy Ti6Al4V. SAE Technical Papers 2021-2001-0024
16. Tan KL, Yeo SH (2017) Surface modification of additive manufactured components by ultrasonic cavitation abrasive finishing. Wear 378–379:90–95
17. Nagalingam AP, Santhanam V, Dachepally NKG et al (2021) Multiphase hydrodynamic flow characterization for surface finishing the laser powder bed fused alsi10mg conformal cooling channels. J Manuf Process 68:277–292
18. Little RE (1972) Estimating the median fatigue limit for very small up-and-down quantal response tests and for S-N data with runouts. ASTM STP 511:29–42
19. Soyama H, Kuji C (2023) Comparison on fatigue strength of additive manufactured titanium alloy by various peening methods. Proc Annual Meeting Soc Mater Sci, Japan 72:201
20. Soyama H, Chighizola CR, Hill MR (2021) Effect of compressive residual stress introduced by cavitation peening and shot peening on the improvement of fatigue strength of stainless steel. J Mater Process Technol 288:116877

Conventionally in LSP, the laser source is directed onto the specimen perpendicularly, to ensure planar shockwave propagation in the component. With rapid commercialization of LSP process, there is an increased demand to utilize LSP for surface enhancement of complex components (such as aero-blade disks, slots, cylindrical walls) [2]. In such cases, it is challenging for the laser to irradiate the component at a normal angle all the time. Therefore, laser peening at different incident angle is unavoidable. When laser peening is performed at an incident angle, it might lead to a non-uniform peening effect. This, in turn, results in non-uniform stress distribution and lowered compressive residual stresses.

Few studies have explored the effect of incident angle during LSP. Shen et al. [3] explored oblique laser shock processing for circular bead of aero-engine fan shaft peening with an oblique angle of 45°. They reported an 11% increase in microhardness and 160% in rotary bending fatigue life of the fan shafts. Similarly, Sundar et al. [4] demonstrated a significant decrease in compressive residual stresses beyond 45° incident angle for 304 stainless steel cylindrical pipes. Hong et al. [5] observed an obvious decrease in residual stresses beyond incident angle of 30° through finite element modelling.

Most of the previous studies investigated energy loss during oblique angle peening whilst evaluating the detrimental effects of energy loss on peening effect, particularly residual stress, and hardness. There is still a lack of systematic study on how oblique angle peening affects the overall residual stress distribution of the components. In addition, little attention is paid to develop methods to mitigate the energy losses during oblique angle peening in components and achieve similar residual stress distribution to normal incident angle laser shock peening.

In this study, we perform a detailed investigation on the effect of incident angle during LSP on the residual stress distribution and surface topography using Ti-6Al-4V alloy as use case. In addition, a fluence compensation approach is proposed to achieve similar peening effect as in normal incident angle. Results exhibit that increasing incident angle reduces the fluence due to increase in beam footprint area, which can be recovered by altering the laser processing conditions.

2 Methodology

2.1 Material Used

AMS 4911 standard, Grade 5, Ti-6Al-4V alloy samples were used for the study. The chemical composition of the alloy is shown in Table 1. Samples of dimensions 40 × 40 × 9.5 mm^3 were water jet cut into a cuboid geometry with all faces milled to meet the surface roughness requirements of Ra < 1 μm. The samples were then annealed in a vacuum furn0061ce at 704.4 °C for 65 min, followed by nitrogen/argon cooling to room temperature.

Table 1 Chemical composition of Ti-6Al-4V alloy used

Elements	Al	V	Fe	C	N	O	Remainder
Weight percentage (%)	6.05	4.10	0.17	0.0013	0.003	0.152	< 0.001

2.2 *Laser Shock Peening*

Laser processing trials were performed with a high-power Nd: YAG laser (Tyrida YS120-R200A), operating at 1064 nm wavelength. The laser delivers flat top pulses at 18 ns full width at half maximum with a frequency of 4 Hz. Laser output energy used was varied from 7 to 10 J and spot diameter of either 2 or 3 mm was used. The samples were held on a fixture mounted on a 6-axis robot arm, which was manipulated to mimic different oblique angles during laser peening. The laser peen pattern was designed to be registered in four steps to achieve a total coverage of 318%. Figure 1 below shows the schematic of the peen pattern used. A protective black ablative overlay was used to protect the surface from thermal effects of the laser beam. On top of the black ablative overlay, a laminar water flow was maintained in order to confine the plasma and increase shock wave intensity.

2.2.1 Single Spot Study

The laser peening was done in two configurations, p-polarized and s-polarized, to investigate the effect of laser light polarization across different incident angles. Nd: YAG laser source used in the current setup is p-polarized. Therefore, the main robot setup was manipulated to achieve different states of polarization. Figure 2 shows the typical setup when peened at p-polarized and s-polarized conditions.

When peening a region at an angle, the stand-off distance between focal lens and the component surface also changes according to the position on the surface

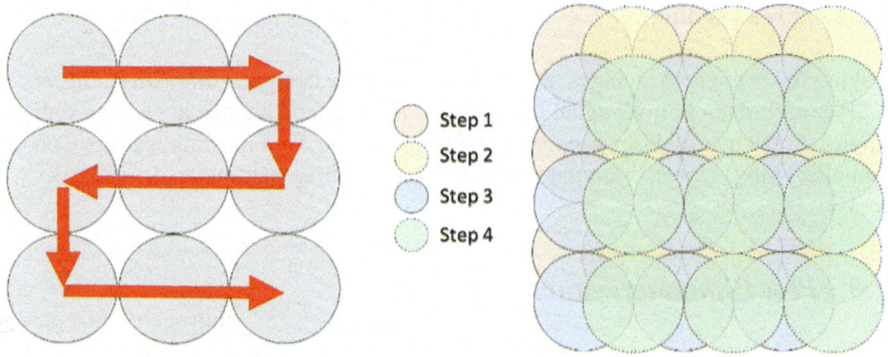

Fig. 1 Schematic of peening flow and peen pattern

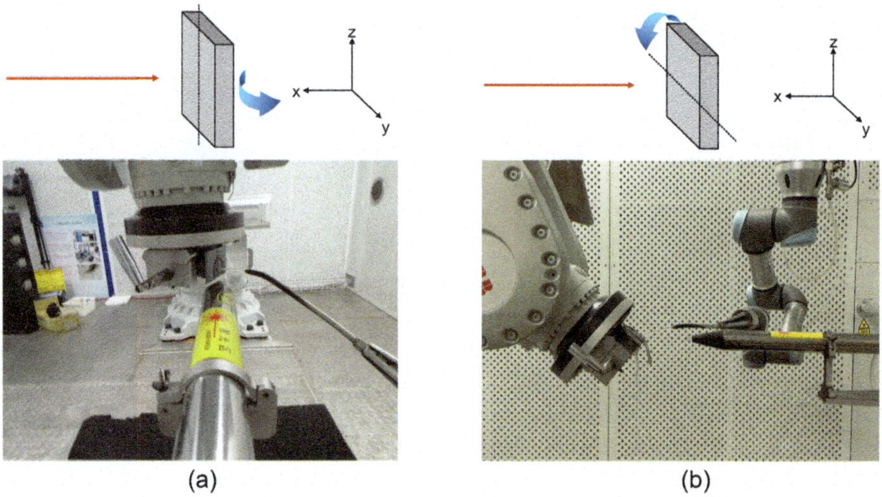

Fig. 2 Robot manipulation setup for laser shock peening at different polarizations, **a** p-polarised setup and **b** s-polarised

Table 2 Changes in beam spot size and beam footprint area as oblique angle changes

Oblique angle, θ	0°	20°	37.5°	50°
Major axis of ellipse, mm	3	3.19	3.78	4.67
Footprint area, cm^2	0.071	0.075	0.089	0.110

being peened. This means that the robot needs to be manipulated carefully in order to achieve same overlap and coverage. This was taken into consideration when designing the toolpath in the program. Table 2 shows the change in beam spot area with change in incident angle.

2.2.2 Area Peening

Taking the beam spot changes into consideration, we designed three different sets of LSP experiments to investigate the effect of beam shape change at varying oblique angles on the overall laser peening process outcomes. Area peening was done on the samples with 2 repeats with the following laser peening parameters in Table 3.

2.3 Post Characterization

All surface residual stress (depth = 0 μm) was measured by XRD, using the Stresstech Xstress Robot System, set at a voltage of 30.0 kV and current of 6.6 mA.

Table 3 Laser parameters used in experiments

Parameters used/Oblique angle	Constant energy				Constant fluence				Increasing fluence			
	0°	20°	37.5°	50°	0°	20°	37.5°	50°	0°	20°	37.5°	50°
Beam spot size/mm	3				3	3	2	2	3	3	2	2
Laser pulse energy/J	10				10	10.6	5.6	6.9	10	11.2	7	10.8
Power density, GW/cm^2	7.86	7.39	6.24	5.05	7.86				7.86	8.27	9.82	12.28
Fluence, J/cm^2	141.47	132.94	112.24	90.94	141.47				141.47	148.89	176.77	220.97

The measurement system was first being validated using a Ti-6Al-4V reference material of known stress value, before being used to assess the residual stress of the samples in two directions, namely longitudinal and transverse directions. Copper (Cu) X-rays were collimated and directed onto the samples at a single point location of about 4 mm in diameter. The intensity of the reflected X-rays was recorded and processed by the detectors rotated at angles between – 38 and 38°.

For sub-surface residual stress (depth > 0 μm), hole drilling method was employed to collect and analyse residual stress information. Stresstech PRISM electronic speckle pattern interferometry (ESPI) hole drilling system was used for depth-wise residual stress analysis. The hole drilling was carried out at a drill depth increment of 10 μm, recording the relaxed strain until it reached a total depth of 1.2 mm.

Similarly, Taylor Hobson CCI HD system was used to analyse the peened surface. A 5 × 5 mm^2 area was analysed to obtain surface topographical information. Green light laser was used in this measurement method to scan the peened surface and stitching was employed.

3 Results

3.1 Single Spot Study

Residual stress measurement on the single spots showed that the magnitude of compressive residual stress decreased with increase in incident angle for both p-polarisation and s-polarisation as shown in Fig. 3. This is expected as the fluence decreases with increase in beam footprint area at the surface at higher incident angles. However, it is to be noted that the residual stress values are lower than the typical values reported in literature for Ti-6Al-4V alloy [6]. Studies report compressive stress values almost double that of the maximum compressive residual stress recorded from our experiment [6]. This can be attributed to the fact that the spots were peened only once (single spot). Generally, a higher coverage of around 300% is used for peening of actual components in industry. An interesting observation is seen in p-polarised, at incident angle 37.5° where the residual stress value (– 281.5 MPa) is lower than that of the normal incident angle single spot peen (– 218.7 MPa). The reason for such result is inconclusive and may be just an experimental error. Apart from that, polarisation of the beam did not seem to have any significant effect on residual stresses.

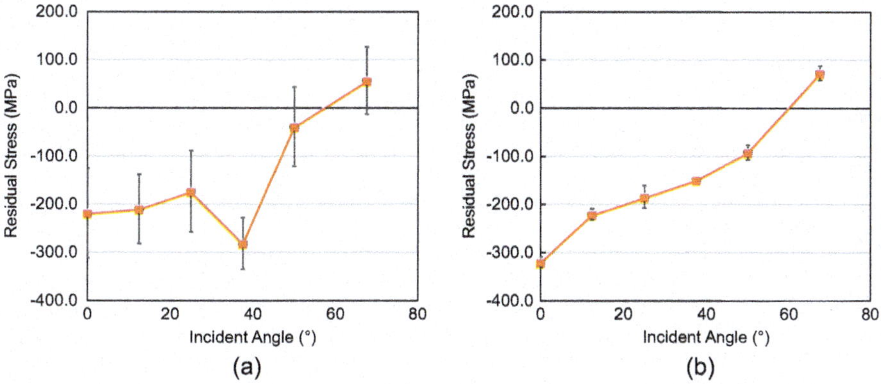

Fig. 3 Single spot XRD analysis results for **a** p-polarized peening and **b** s-polarized peening

3.2 *Area Peening*

Figure 4 shows residual stress profiles at different incident angles. We can observe an upward shift in residual stress profiles with increase in incident angle. This indicates lower magnitudes of compressive residual stresses in samples peened at larger incident angles. The decreased compressive residual stress values reflect attenuation of laser fluence at larger incident angles.

To combat attenuation of laser fluence, focal beam spot and laser energy at the surface were modified to maintain constant fluence when peened at increasing incident angles. Figure 5 shows residual stresses distributions at different incident angles peened using a constant laser fluence of 141.47 J/cm^2. The stress profiles

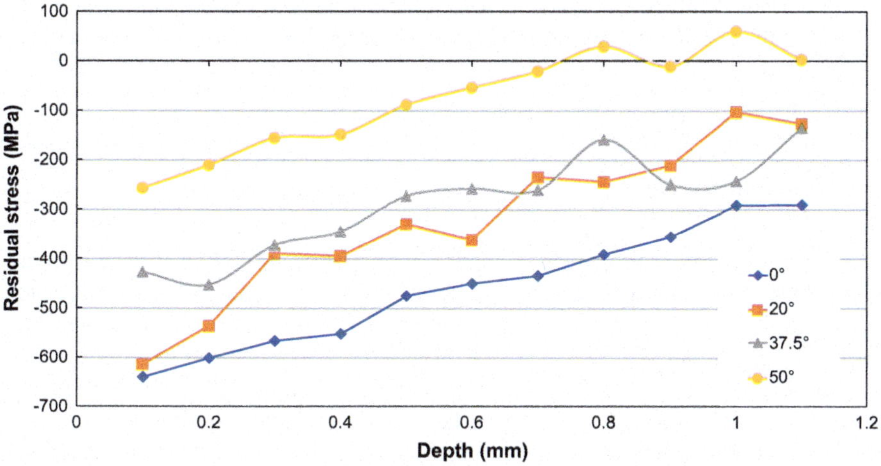

Fig. 4 Residual stress distribution after oblique laser peening at constant energy

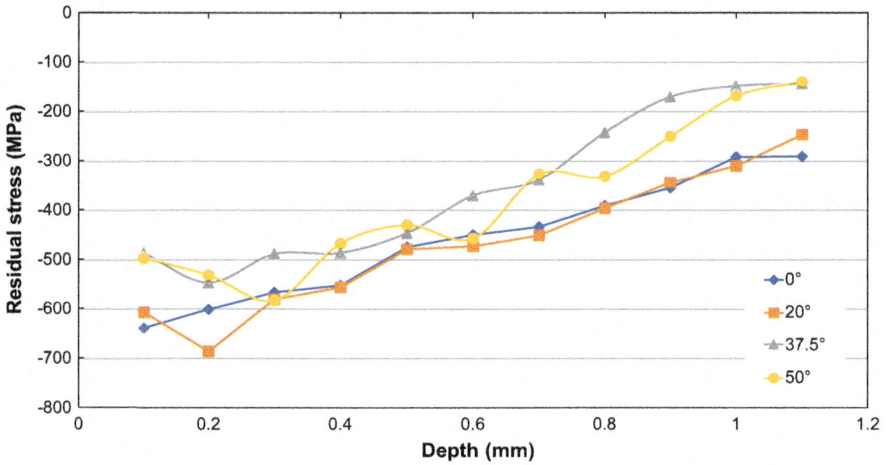

Fig. 5 Residual stress distribution after oblique laser peening at constant fluence

at 0° and 20° look similar, which suggest using same level of fluence can deliver similar effect. Similarly, the magnitude of residual stresses at 50° also increased significantly compared to constant energy case and showed similar values as with 37.5°. However, the stress magnitudes at higher angles (37.5° and 50°) were still lower than stresses at near normal incident angles by almost 100 MPa. It might be because of the different interaction mechanism between laser beam and material at larger incident angles [7, 8].

To compensate for energy loss at larger incident angles, further investigations were performed by peening samples with increasing laser fluence. In this case, the magnitudes were found to be similar to that in near normal incident angles as shown in Fig. 6. This exhibits that when peening at larger incident angles, fluence should be increased to ensure efficient laser energy transfer and similar level of peening effect is observed.

4 Discussion

Results from residual stress measurement demonstrate that peening effect is primarily dependent on laser fluence. Fluence measures the total amount of laser energy being delivered per unit area on the component and is calculated by dividing laser pulse energy by spot area.

For the sake of discussion, Fig. 7 compares the magnitude of residual stress at a depth of 1 mm from the peened surface under different conditions. This depth is chosen as laser peening is expected to have considerable effect up to at least 1 mm depth for industrial applications [9]. The residual stress under normal incident angle is represented as dotted line in the graph for comparison. As can be observed,

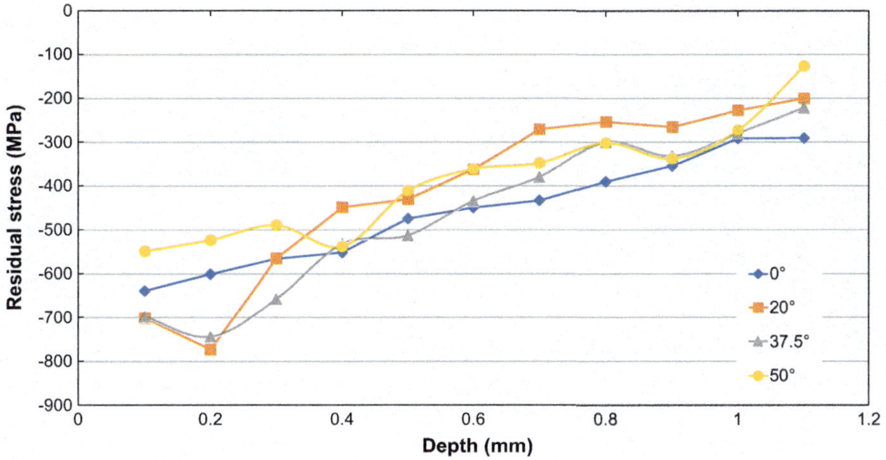

Fig. 6 Residual stress distribution after oblique laser peening at increasing fluence

under constant energy, the magnitude already reduced by more than half of the benchmark, even at a mere angle of 20°. At 50°, the residual stress magnitude at 1 mm depth is already tensile, suggesting shallow penetration depth at large incident angle. Upon delivering a constant fluence, residual stress magnitude at 20° matches the benchmark. However, the value quickly drops to around – 150 MPa at 37.5° and 50°. When peened under increasing fluence, the magnitude of residual stress become similar irrespective of incident angle.

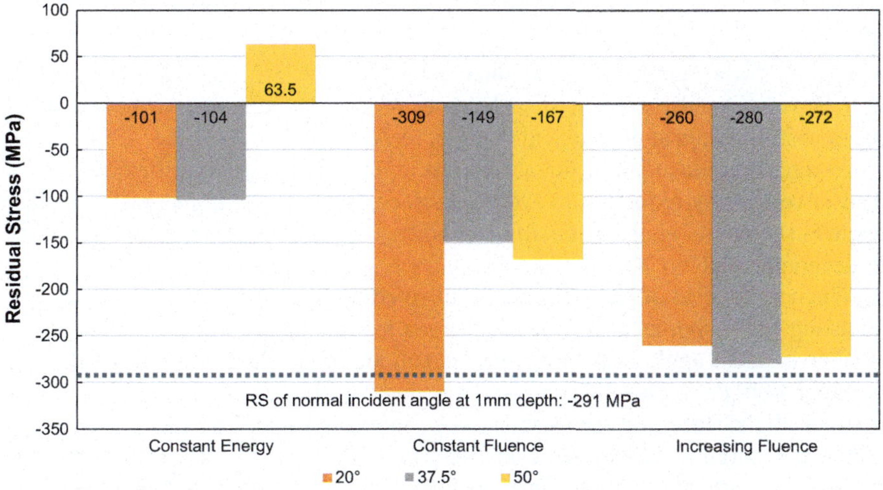

Fig. 7 Comparison of residual stress values at 1 mm depth

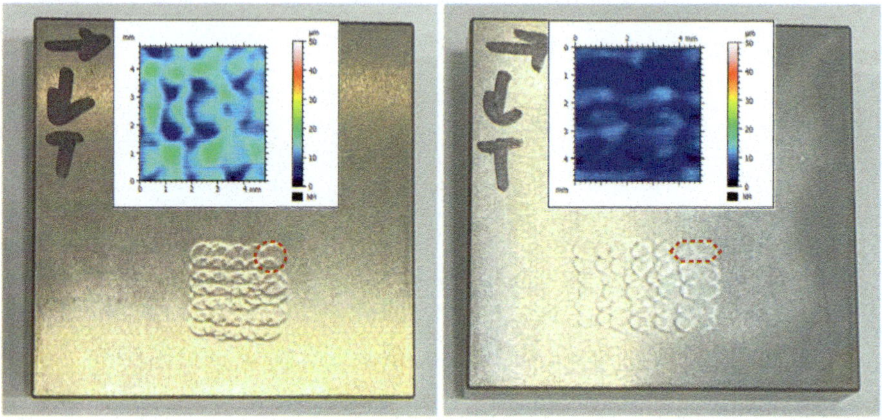

(a) (b)

Fig. 8 Surface topographical colour map of **a** 0° and **b** 50°

When laser peening is performed at angle of incidence other than normal, the beam footprint at the component's surface is not circular. As shown in Fig. 8, the change in beam footprint at the surface is easily observed by comparing the surface topography at 0° and 50° incident angle. The elongation of the beam results in elliptical spot footprint at oblique angles. Therefore, the total area receiving the laser pulse energy increases, which results in lowering of fluence. The reduction in fluence can be represented as

$$\frac{f_\theta}{f_0} = \frac{E/A_\theta}{E/A_0} == \frac{D^2}{D^2 \sec\theta} = \cos\theta \tag{1}$$

where, f_0 is laser fluence at normal incident angle, f_θ is the laser fluence at an angle of θ, E is laser pulse energy, D is diameter of laser beam spot and A_0 and A_θ are beam footprint area at normal and θ angle, respectively.

As a result, reduced laser fluence results in lower amount of plastic deformation. This is verified by reduction in average surface roughness with increasing incident angle as shown in Fig. 9. Eventually, it produces lower compressive residual stresses in the component.

When same level of laser fluence is delivered by increasing laser pulse energy or adjusting beam spot area, then it is observed that the stress profiles look similar at smaller incident angles. However, the magnitude of residual stresses is still lower at larger oblique angles. This is possibly due to different interaction mechanism between the surface and the beam at large incident angle. The beam spot is very elliptical at large incident angles and despite using same fluence, the geometry of the spot may influence shock dynamics [10]. In addition, the laser beam has to pass through a longer distance along the confinement medium at large oblique angles. This could also result in loss of energy along the beam path inside confinement medium.

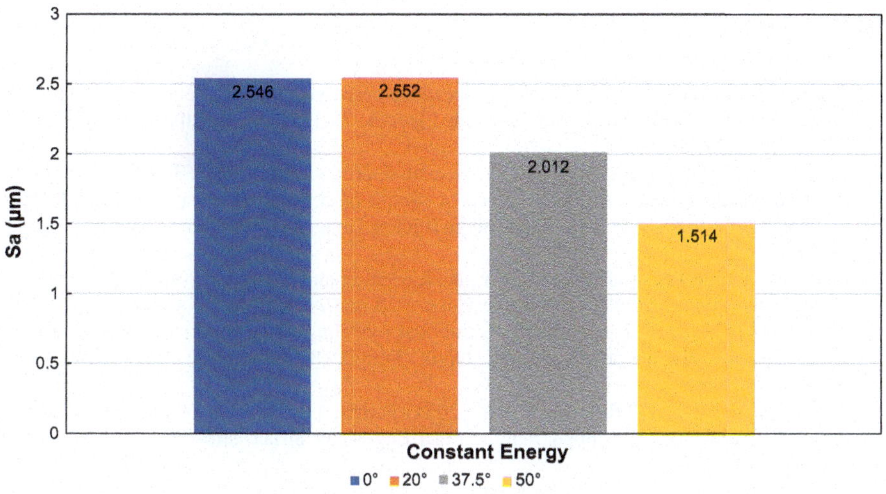

Fig. 9 Bar graph of surface roughness values at constant energy

All this necessitates the use of larger laser fluence to have similar peening effect at larger oblique angles.

5 Conclusions

Laser shock peening at different incident angles was investigated in this study. The results indicate that residual stress magnitude and penetration depth decreases with increase in incident angle due to reduction in laser fluence. On employing same level of laser fluence, the residual stresses showed similar distribution for near normal incident angles (i.e., up to 20°). At larger incident angles, however, the stresses were still lower than in normal peening by almost – 100 MPa, which is attributed to energy losses due to different laser-material interaction at large angles and long travel path along confinement medium. The losses were compensated by increase the laser fluence at larger incident angles, which produced similar stress profile distribution.

Acknowledgements The authors are grateful to Tyrida International and Advanced Remanufacturing and Technology Centre (ARTC) for providing the funding for this study under research project M20_01_TYR. The authors thank Jolynn Leong for her assistance in managing the project and Kyaw Thein Hla for his assistance in lab experimental trials.

Funding This research project was funded and supported by Tyrida International and Advanced Remanufacturing and Technology Centre (ARTC) under project number M20_01_TYR.

References

1. Montross CS, Wei T, Ye L et al (2002) Laser shock processing and its effects on microstructure and properties of metal alloys: a review. Int J Fatigue 24:1021–1036. https://doi.org/10.1016/S0142-1123(02)00022-1
2. Clauer AH (2019) Laser shock peening, the path to production. Metals (Basel) 9:626
3. Shen X, Wang C, An Z et al (2015) Effects of oblique laser shock processing on rotary bending fatigue of aero-engine fan shaft. Hongwai yu Jiguang Gongcheng/Infrared Laser Eng 44:3548–3553
4. R S, P G, Gupta RK et al (2019) Laser shock peening and its applications: a review. Lasers Manuf Mater Process 6:424–463. https://doi.org/10.1007/s40516-019-00098-8
5. Qiao H, Sun B, Zhao J et al (2019) Numerical modeling of residual stress field for linear polarized laser oblique shock peening. Optik (Stuttg) 186:52–62. https://doi.org/10.1016/j.ijleo.2019.04.083
6. Maharjan N, Chan SY, Ramesh T et al (2021) Fatigue performance of laser shock peened Ti6Al4V and Al6061-T6 alloys. Fatigue Fract Eng Mater Struct. https://doi.org/10.1111/ffe.13390
7. Allmen MV, Paul GL (1988) Laser beam interactions with materials physical principles and applications—springer series in materials science Volume 2 (1987), X, 232p., DM 86.00. Laser Part Beams 2. https://doi.org/10.1017/s0263034600005735
8. Kukko A, Kaasalainen S, Litkey P (2008) Effect of incidence angle on laser scanner intensity and surface data. Appl Opt. https://doi.org/10.1364/AO.47.000986
9. Zabeen S, Preuss M, Withers PJ (2015) Evolution of a laser shock peened residual stress field locally with foreign object damage and subsequent fatigue crack growth. Acta Mater 83:216–226. https://doi.org/10.1016/j.actamat.2014.09.032
10. Fabbro R, Peyre P, Berthe L, Scherpereel X (1998) Physics and applications of laser-shock processing. J Laser Appl 10:265–279

Surface Enhancements from Peening Effect on Inconel 718 Fabricated by Direct Energy Deposition

Nataniel Yong Syn Tham and Zhang Hao

Abstract Direct energy deposition Inconel 718 is increasingly explored and used for various industrial applications due to its near net shape and ability to build complex geometries. For post-processing, hammer peening process is a well-established surface finishing step to enhance the compressive stress of the metal, reducing and preventing cracks. Increasingly, laser has been explored for post-processing additive manufactured metal as an in-situ heat treatment process to relieve tensile stresses formed during the process. In this paper, an investigation into the effects of hammer peening in conjunction with laser is performed. Microstructural analysis, hardness values and tribological properties will be used to observe these effects. It was found that hammer peening increased the hardness of Inconel 718, and when used in tandem with laser, the hardness decreased, albeit higher than the non-post processed sample. Lastly, there is a disconnect between the correlation of higher hardness values with lower wear rate, as the lowest hardness value sample has the lowest wear rate in this study, and that hammer peening itself did not reduce the wear rate of direct energy deposited Inconel 718.

Keywords Direct energy deposition · Inconel 718 · Laser · Hammer peening

N. Y. S. Tham (✉)
Advanced Remanufacturing and Technology Center (ARTC), Agency for Science, Technology and Research (A*STAR), 3 Cleantech Loop, #01/01 CleanTech Two, Singapore 637143, Republic of Singapore
e-mail: nataniel_tham@artc.a-star.edu.sg

Z. Hao
School of Mechanical and Aerospace Engineering, Nanyang Technological University, 50 Nanyang Avenue, Singapore 639798, Singapore
e-mail: hzhang058@e.ntu.edu.sg

N. Maharjan and W. He (eds.), *Proceedings of the 3rd International Conference on Advanced Surface Enhancement (INCASE) 2023*, Lecture Notes in Mechanical Engineering, https://doi.org/10.1007/978-981-99-8643-9_5

1 Introduction

Directed energy deposition (DED) is a rapidly developing additive manufacturing technique that uses a high-energy laser to melt and fuse metallic powders into solid parts. Commonly, one of the materials fabricated with DED is Inconel 718 (IN 718), a nickel-based superalloy commonly used in high-temperature applications in the aerospace and power generation industries. However, the as-built surfaces of DED IN 718 parts often exhibit high residual stresses and have large laves phase compared to traditional fabrication process, which can negatively affect their fatigue life and mechanical properties. Consequently, hammer peening (HP) is a surface treatment technique that can improve the fatigue life of metallic components by inducing compressive residual stresses in the surface layer.

In recent years, laser technology has been used in combination with HP to increase the peening effects by softening the surface. Additionally, laser has been used as an in-situ heat treatment process to regulate the hard and brittle Laves phase, which are known to be sites of crack initiations [1]. At present, there is a lack of research of using HP in conjunction with laser on DED IN 718. As such, this paper presents an investigation into the effects of HP with and without laser on the surface characteristics and tribological properties of DED IN 718.

2 Methods

2.1 *Laser with Hammer Peening*

A laser system was attached at an angle besides the pneumatic hammer head, both of which were mounted onto an ABB industrial robotic platform (IRB 6660). The spherical tungsten carbide hammer head had a tool radius of 5 mm and a wobble angle of 8o, providing an impact energy of 90 mJ. A laser power of 1300 W was chosen as it provides the minimum temperature required to melt the surface. The overlap was fixed at 1200% with a frequency of 225 Hz and a longitudinal feeding rate of 50 mm/s.

2.2 *Materials Characterisation*

Four samples were prepared namely DED Inconel 718 (As-built), As-built with hammer peening (HP), As-built with laser followed by hammer peening (LAHP) and As-built with hammer peening followed by laser (HPLA). They were then hot mounted and then ground and polished for further analyses. The cross-sections were observed under the scanning electron microscope (Zeiss, Evo HD25). A hardness tester (InnovaTest, Falcon 500) was used to create ten indentations with a load of

500g. Lastly, coefficient of friction and wear rates were obtained using a tribometer (Anton Paar, TRB3).

3 Results and Discussion

Figure 1 shows the optical micrographs for LAHP, where a layer of columnar dendritic layer can be observed on the surface of the DED IN 718, which has a different structure from the equiaxed dendritic grains below. The cross sectioned area shows that the laser parameters were sufficient to remelt the surface before being peened.

Vickers hardness of As-Built, HP, LAHP and HPLA were obtained, as seen in Fig. 2. As-built has the lowest hardness among the four at 284 HV, whilst after undergoing HP the hardness significantly increased to 452 HV. It has been well-established that peening techniques would increase the surface hardness. Amanov et al. utilised a similar robotic hammer peening technique and found that the process refines coarse grains to nano-grains, which are responsible for the increase in hardness [2]. LAHP has similar hardness to As-built at 294 HV, however HPLA increased the hardness to 324 HV. LAHP was selected to relax the surface which has accumulated tensile stresses from the DED fabrication before introducing compressive stresses. Conversely, performing peening first before laser treatment was to observe if the enhancement from HP would be negated. The use of laser is likely to have multiple effects on the sample that could negate the processing effects such as HP, as the surface is rapidly melted and solidified. For instance, a study conducted found that previous thermal history from DED could be altered, creating stress relaxation of the surface and a lower hardness value [3].

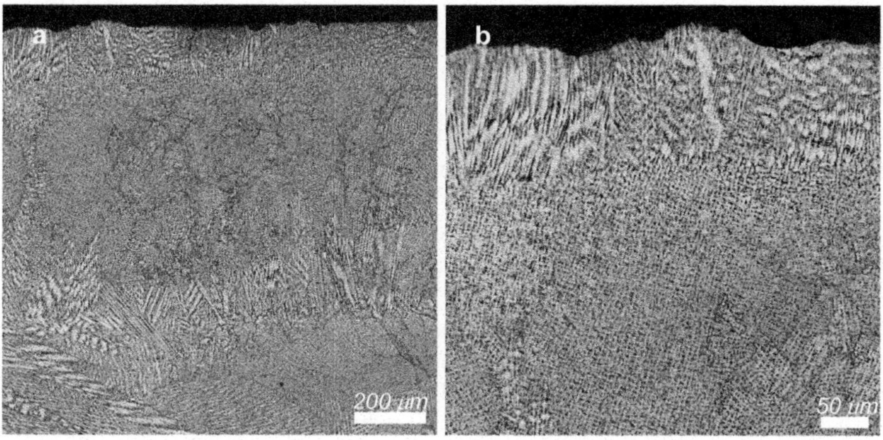

Fig. 1 Optical micrographs of LAHP at: **a** higher magnification; **b** lower magnification

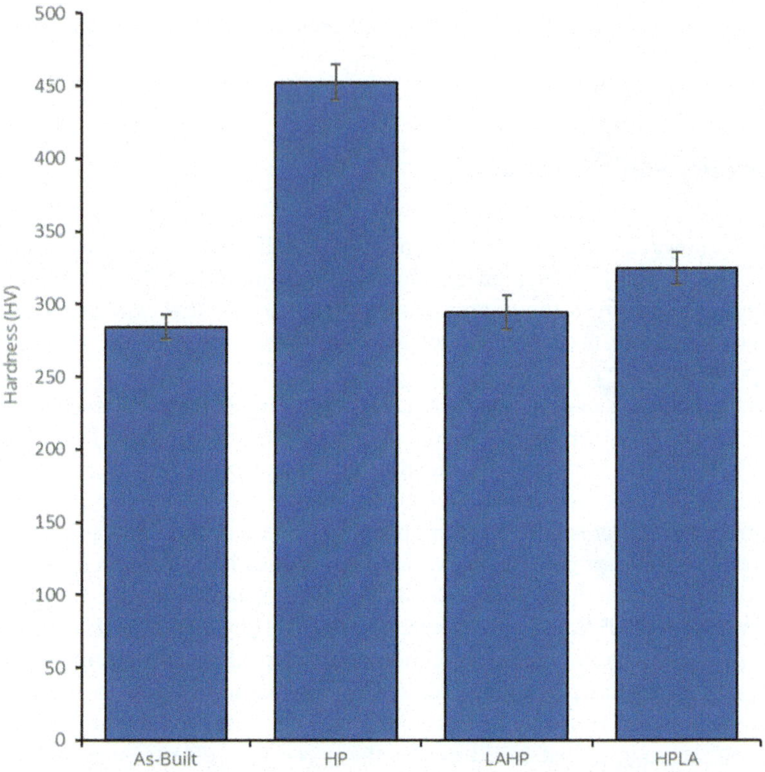

Fig. 2 Hardness values of As-built, HP, LAHP and HPLA specimens

Additionally in Fig. 3, it was found that the long Laves phase were broken into smaller granular ones after hammer peening process. However, after LAHP and HPLA, there were dissolution and decrease in size of the granular Laves phase. Sui et al. conducted an in-situ laser heat treatment process to reduce laves after laser cladding process, and found that it changed the morphology and size of the Laves phase from a long one to granular shape. After performing the laser treatment for 180 s, the author found that the volume fraction of the Laves phase decreased significantly from its original state [1]. This lack of laves phase could result in a reduction of hardness, although compensated by the hammer peening process, resulting in a similar hardness to As-Built.

Pin-on-disk wear tests were conducted with the corresponding coefficient of friction (COF) and wear rate shown in Figs. 4 and 5 respectively. There were not a significant difference in COF across the various samples seen in Fig. 4, with As-Built achieving a COF of 1.11 whilst HP, LAHP and HPLA obtained a COF of 1.16. However, the parameters used in this study were sufficient to observe trends in the wear rate for the four samples. Wear rates found in this study for As-Built, HP, LAHP and HPLA were $5.6 \times 10^{-4}\,\text{mm}^3\,\text{N}^{-1}\text{m}^{-1}$, $5.8 \times 10^{-4}\,\text{mm}^3\,\text{N}^{-1}\text{m}^{-1}$,

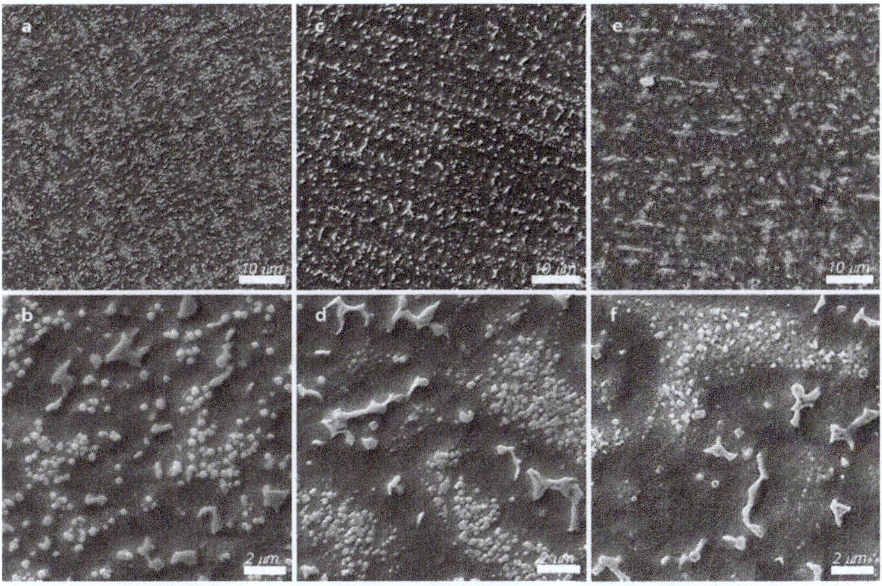

Fig. 3 SEM images of **a**, **b** HP; **c**, **d** LAHP; **e**, **f** HPLA at different magnification

6.15×10^{-4} $mm^3 N^{-1}m^{-1}$ and 6.75×10^{-4} $mm^3 N^{-1}m^{-1}$ respectively, as illustrated in Fig. 5.

It is worth noting that whilst As-Built has the lowest hardness value and HP has the highest hardness value, As-Built has the lowest wear rate among the four samples.

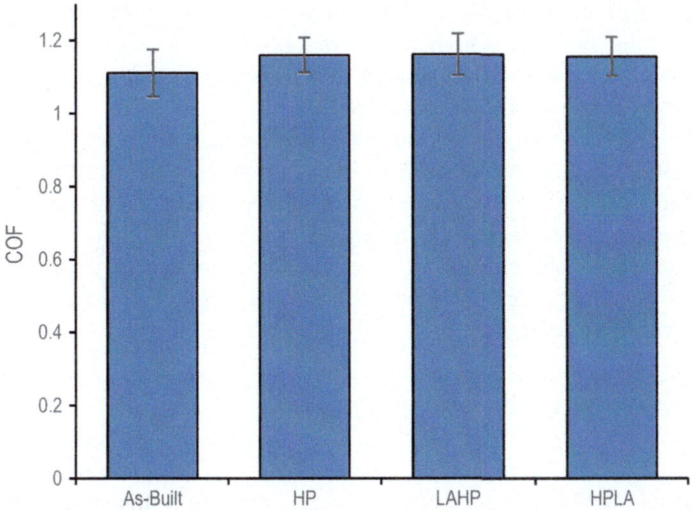

Fig. 4 Coefficient of friction of specimens

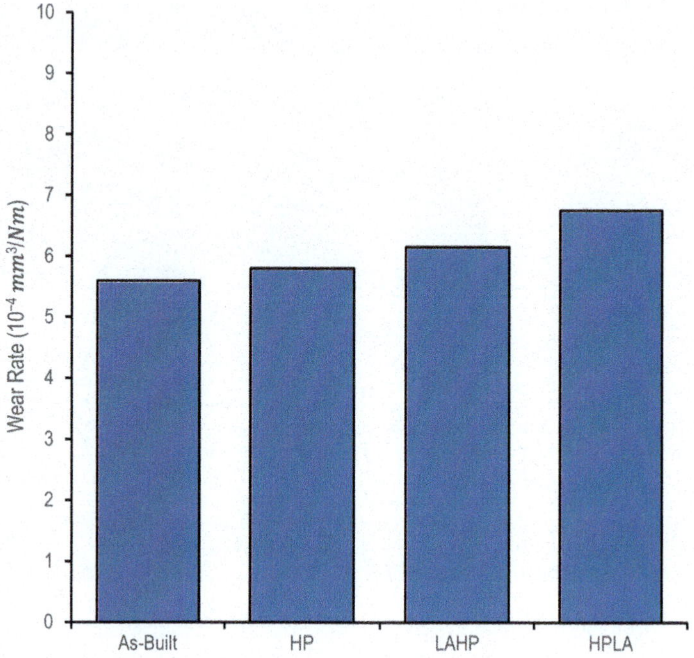

Fig. 5 Wear rate of specimens

Typically, a high hardness surface would obtain a low wear rate, however with DED IN 718 this may not be the case. Joshy et al. and Karabulut et al. conducted studies on IN 718 fabricated using selective laser melting (SLM), a powder bed additive manufacturing process. Joshy et al. found that IN 718 after heat treatment increased in hardness value but have a higher wear rate compared to As-Built [4]. Karabulut found that solutionising IN 718 caused a 15% reduction in hardness value compared to As-Built yet obtained similar wear rate [3].

One explanation could be that the brittle Laves phase helps to increase the wear resistance of IN 718, preventing direct contact between the tip of the wear tester and the matrix. After laser treatment, there is a reduction in the amount of Laves phase on the surface, thereby allowing greater contact between the two surfaces and reducing the wear resistance despite the inherent compressive stresses created by the hammer peening process. Further studies observing the wear track surfaces and residual stresses are needed.

4 Conclusion

In this study microstructure, hardness values and tribological properties of DED IN 718 were processed using various combinations of HP and laser. The findings were summarised below:

- HP process breaks up the long Laves phase into smaller granular phase and with the introduction of laser, dissolution of the Laves phase occurred.
- Although HP garnered the highest hardness value, it does not significantly increase the wear resistance of IN 718. Similarly although LAHP and HPLA has a higher hardness value compared to As-Built, the wear rate are found to be higher.
- More studies are needed to observe the residual stress and wear tracks, however the dissolution of the Laves phase could have increased the wear rate due to the increased surface contact of the wear tester and matrix.

References

1. Sui S, Li H, Li Z et al (2022) Introduction of a new method for regulating laves phases in Inconel 718 Superalloy during a laser-repairing process. Engineering 16:239–246. https://doi.org/10.1016/j.eng.2021.08.030
2. Amanov A, Karimbaev R, Li C, Wahab MA (2023) Effect of surface modification technology on mechanical properties and dry fretting wear behavior of Inconel 718 alloy fabricated by laser powder-based direct energy deposition. Surf Coat Technol 454. https://doi.org/10.1016/j.surfcoat.2022.129175
3. Karabulut Y, Tascioglu E, Kaynak Y (2021) Heat treatment temperature-induced microstructure, microhardness and wear resistance of Inconel 718 produced by selective laser melting additive manufacturing. Optik (Stuttg) 227. https://doi.org/10.1016/j.ijleo.2019.163907
4. Joshy J, Kuriachen B (2023) Effect of heat treatment and cryo-treatment on dry tribological behavior of Inconel 718 fabricated using laser powder bed fusion. Wear. https://doi.org/10.1016/j.wear.2023.204819

Picosecond Laser Surface Texturing of Al2024-T3 Substrate for Super-Hydrophobicity

Xincai Wang, Kaidong Ye, Min Qian, Raahgini Chandrasegaran, and Hong Xie

Abstract In the paper, 1.06 μm picosecond (ps) laser is employed to fabricate micro/nano surface textures on Al2024-T3 substrate. The effects of different laser parameters including laser fluence, repetition rate, scanning speed, pulse number, focal position, etc. on the topography, feature profile, feature dimensions of the induced surface textures have been studied. The optimum laser parameters have been identified and the uniform-distributed 2 scale micro-nano hierarchical bump array, sub-micron-ripple and micro-dimple array surface patterns are fabricated on the Al2024-T3 substrates. The laser textured Al2024-T3 sample shows stable super-hydrophobic property with a water contact angle (WCA) of more than 150 degree and contact angle hysteresis (CAH) of less than 10 degree. The mechanism involved in the process is discussed in terms of surface morphology and surface chemistry. The developed process has the potential to be applied in improving the performance of the Al-based components and/or products in terms of superhydrophobicity and easy cleaning.

Keywords Picosecond laser · Surface texturing · Al2024-T3 · Super-hydrophobicity

1 Introduction

Controlled modification of material's surface wettability has been the subject of significant scientific research due to its importance for various applications, which include easy-cleaning, adhesion mitigation and microfluidic and biomedical application. The wettability property in terms of water contact angle and contact angle hysteresis are affected by both material surface chemistry and topology [1]. Based on the Wenzel [2] and Cassie-Baxter [3] models, the roughness of the sample surface is

X. Wang (✉) · K. Ye · M. Qian · R. Chandrasegaran · H. Xie
Singapore Institute of Manufacturing Technology Singapore (SIMTech), A*STAR, 2 Fusionopolis Way, #08-04, Innovis 138634, Singapore
e-mail: xcwang@SIMTech.a-star.edu.sg

N. Maharjan and W. He (eds.), *Proceedings of the 3rd International Conference on Advanced Surface Enhancement (INCASE) 2023*, Lecture Notes in Mechanical Engineering, https://doi.org/10.1007/978-981-99-8643-9_6

critical to increase its hydrophobicity or hydrophilicity. Surface morphology is one of the key factors determining the wettability of a solid surface. Various techniques have been used for surface texturing to create the desired surface morphologies, which can be classified into 2 major categories, namely top-down processes such as lithographic process, template-based method [4], plasma treatment [5] and laser surface treatment process [6], and bottom-up processes mainly consisting of self-assembly and self-organization processes [7]. As a top down approach, laser micro- and nano-processing has been extensively investigated and is developing rapidly. The pico- or femto-second ultrafast lasers have been the most employed tools for surface texturing. The pico- or femto-second laser pulse induced ablation is able to produce well controlled 2 scale micro- and nano-scaled surface textures [8] so as to fabricate some types of surfaces with the improved surface functional properties. So far, a significant amount of extensive research work has been carried out on surface treatment and modification on different types of materials such as metallic [8, 9], semiconductor material [10], and polymer material [11] to modify their wetting properties with different lasers. The mechanisms involved in the wettability conversion of structured metallic surfaces from super-hydrophilicity to super-hydrophobicity have also been well studied. [8, 12]. The mechanism for the wettability change is found to be dependent on the specific material to be laser-beam-treated. Regarding the laser surface treated stainless steel, Kietzig et al. [8] proposed that the conversion in wetting property to superhydrophobicity from original hydrophilicity after laser treatment was due to the accumulation of certain amount of carbon layer on the laser treated surface through the dissociation of carbon dioxide into carbon with the active magnetite as catalyst.

In the paper, 1.06 μm ps laser is employed to fabricate micro/nano surface textures on Al2024 substrate. The effects of different laser parameters including laser fluence, rep rate, scanning speed, pass number, focal position, etc. on the topography, feature profile, feature dimensions of the induced surface textures have been studied. The optimum laser parameters have been identified that the uniform-distributed 2 scale micro-nano hierarchical surface patterns are fabricated on the Al2024 substrates. The fabricated 2-scale micro/nano hierarchical textured Al2024 sample shows stable superhydrophobic property with a water contact angle (WCA) of more than 150 degree and contact angle hysteresis (CAH) of less than 5 deg. The mechanism involved in the process is discussed in terms of surface morphology and surface chemistry.

2 Experimental

Al2024-T3 plates with dimensions of 25 mm × 25 mm × 5 mm with good finishing on one side was used for the laser surface texturing. Before laser texturing, the Al2024 plates underwent sonication cleaning with DI water for 10 min and dried with compressor air. After laser texturing process, the samples were also going through the same cleaning process with ultrasonic bath in the DI water for 10 min, then dried with compressor air. A TIME BANDWIDTH; DUETTO ps laser system was employed to do the laser surface treatment. The laser has Gaussian profiled spatial mode (TEM00) with a beam quality of M2 < 1.3 and the beam divergence of less than 0.3 mrad. The laser wavelength is near-infrared 1.06 μm, the output power is up to 10W, and the laser pulse width is 10.3 ps. The pulse repetition frequency can be set at a minimum value of 50 kHz and continuously tuned up to 8200 kHz. The incident laser beam was focused and manipulated by a galvo-scanner, which was equipped with a telecentric f-theta lens. The focus length of the f-theta focus lens is 100 mm. The linear polarized raw laser beam has a diameter of 7 mm (@ 1/e2), which was focused and directed on to the sample surface with a beam diameter of 25 μm. The laser machining control software can produce various different machining patterns. The patterns can be lines, circular and rectangular shapes, and also hatched patterns, etc. The maximum scanning speed is 2000 mm per second. For the surface treatment process, the focused laser beam was raster scanned a designed pattern as shown in Fig. 1. The effects of the following were investigated—the rate of the repetition, velocity of the scanning, the density of laser power, the pass number, the pulse number on each spot as well as the pitch of the hatched pattern on properties of the produced surface textures such as the morphology, the feature size, and the uniformity of the produced pattern. The sample surfaces were characterized by using the Scanning Electron Microscopy (SEM). The surface wettability was characterized via measuring the water contact angle (WCA) and contact angle hysteresis (CAH). With the sessile drop method, the WCA was obtained with VCA Optima (VCA-2500XE AST products, Inc.). During the WCA measurement, a specific volume of water droplet was created with a syringe connected to a capillary tip. The droplet was then detached gently onto the substrate. The image of the droplet was recorded with a camera and analyzed with the software to determine the tangent line for the contact angle. Contact angle hysteresis (CAH) was tested with KR"USS SDA100. The sessile drop method was applied; where water is pumped into the syringe and a droplet is dispensed out of the needle to achieve the advancing angle, and then the receding angles with a water volume loop of 1μL – 4μ – 2μL at a constant dosing/ withdraw rate at 8μL/min. The program recorded the measured angles with time till the end of the program. The difference of the advancing angle and the receding angle is defined as CAH and is automatically calculated.

3 Results and Discussion

Figure 1 schematically shows the designed pattern to be employed to produce the desired micro-bumps based texturing. The process is conducted through scanning the focused laser beam along the regular hatched raster pattern so as to produce the desired surface textures such as micro-bumps array pattern.

Figure 2 showed some surface patterns produced on the Al2024 alloy substrate with different laser parameters such as different laser energy fluence, the repetition frequency, the scanning velocity, repetition pass number, and the hatched densities.

Fig. 1 Schematic of the designed surface patterns

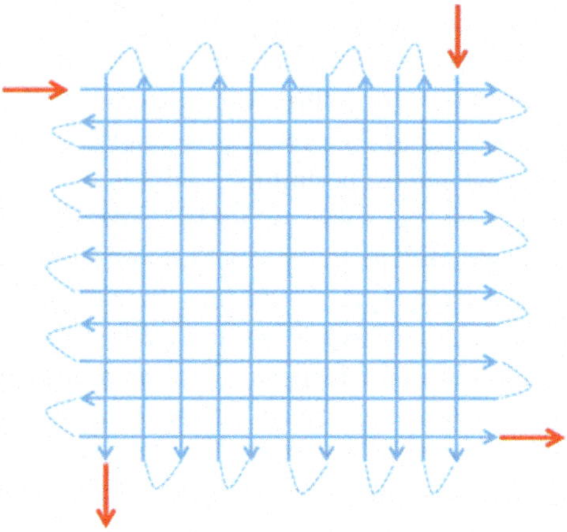

It can be seen that the surface morphology and feature dimensions of the produced surface textures depend on the laser surface parameters.

With the laser parameters of 1 MHz/0.9, 25/25 μm, and 500 mm/s × 75, it was observed that a uniform and consistent 2D micro-bump array surface pattern was produced, where the micro-bump size was measured to have a diameter of 18.45 ± 1.38 μm, and the height of the micro-bump pattern was 13.68 ± 0.74 μm; the period of the pattern was 25 μm. From the magnified SEM image of the micro-bump as shown in the inset of Fig. 2a, some type of sup-micro features were observed to be superimposed on the surface of the micro-bumps. With the laser parameters of 500 kHz/0.5; 20/20 μm; 1000 mm/s × 75, a uniform and consistent 2D micro-bump

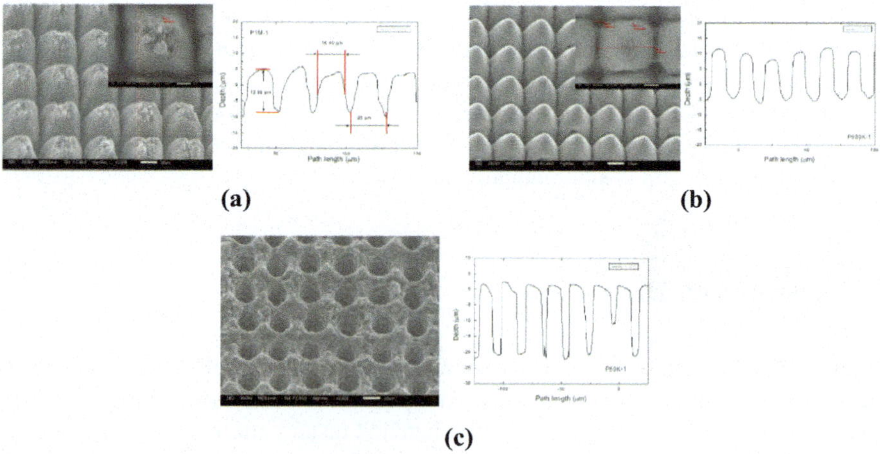

Fig. 2 The SEM images and feature profiles of the laser produced surface textures with different laser parameters; **a** 1 MHz/0.9; 25/25 μm; 500 mm/s × 75; **b** 500 kHz/0.5; 20/20 μm; 1000 mm/s × 75; **c** 50 kHz/0.7; 20/20 μm; 1000 mm/s × 75

array surface pattern was produced with the micro-bump size of 9.99 ± 0.52 μm, and the height of the micro-bump pattern was 22.19 ± 1.69 μm; the period of the pattern was 20 μm. From the magnified SEM image of the micro-bump as shown in the inset of Fig. 2b, sub-micron periodic ripples with a period of around 600 nm were produced to be superimposed on the surface of the micro-bumps. So, a 2-scale micro/nano hierarchical surface structure was achieved. When the laser pulse repetition rate was set at 50 kHz with other parameters to be kept same, it was found that a uniform and consistent 2D micro-dimple array surface pattern instead of micro-bumps was produced with dimple size of 6.52 ± 1.33 μm, and the depth of 21.73 ± 0.67 μm, and the period of 20 μm as shown in Fig. 2c.

With the optimum laser surface texturing process parameters of laser power of 0.5P, the repetition frequency at 500 kHz, the scanning velocity at 1000 mm/s, and pass number of 75, a micro/nano 2 scale surface texture was fabricated on the Al2024 substrate with a textured area of 25 mm × 25 mm as shown in Fig. 3. It can be seen that the induced surface textures are quite uniform and consistent. From the magnified SEM images, the 2-scale micro/nano surface structures have been achieved with nano-ripples to be formed on the surface of the bumps. It was also observed that some tiny holes with the dimeter of around 1.5–5 μm were formed in-between the micro-bumps.

After laser surface texturing, the WCA on the textured surface was measured as a function of the timing as shown in Fig. 4. It can be seen that immediately after laser texturing process, the textured Al2024-T3 surface is superhydrophilic with a WCA of almost zero due to the laser treatment induced fresh high surface energy of the textured surface. However, with an increase in the timing, the WCA increase gradually and reached up to 150 deg on the 25 days after laser texturing process. Then, the WCA on the textured Al2024-T3 surface was found to be saturated and stabilized to be more than 150 deg with further increasing the timing. Figure 5 showed the measured WCAs on the 3 different locations of the textured sample surface indicating the uniformity of the laser textured pattern in terms of superhydrophobicity.

Figure 6 showed the measured contact angle hysteresis (CAH) on laser textured Al2024-T3 substrate surface. As shown in Fig. 6, the CAH on the laser textured

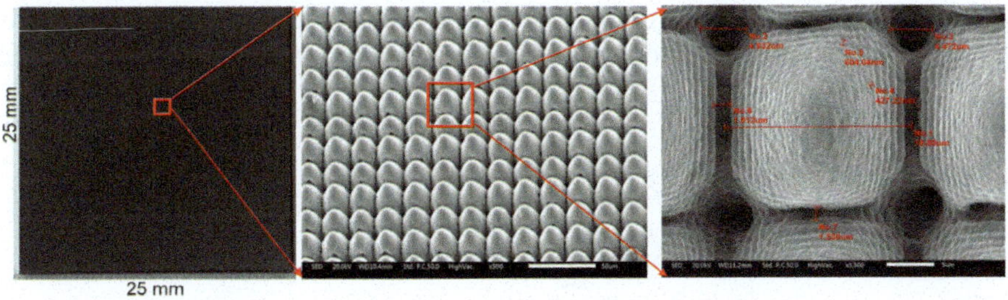

(a) (b)

Fig. 3 Laser surface texturing on Al2024-T3 with a textured area of 25 mm × 25 mm with laser parameters of 500 k/0.5; 20/20 μm; 1000 mm/s × 75; **a** photo of the textured sample, **b** SEM images

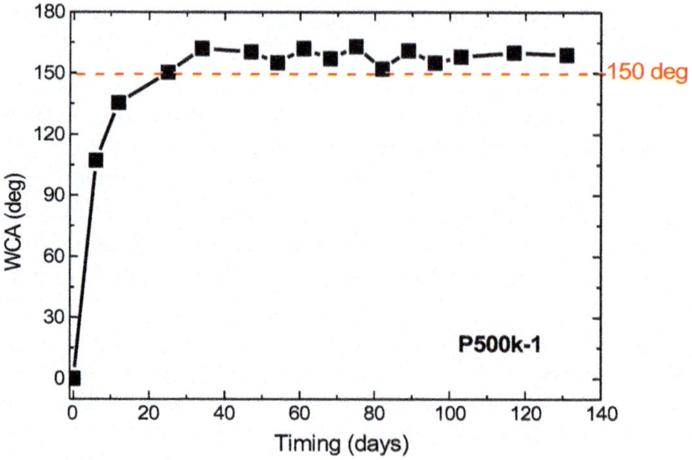

Fig. 4 Measured WCA on the laser textured surface as a function of the timing

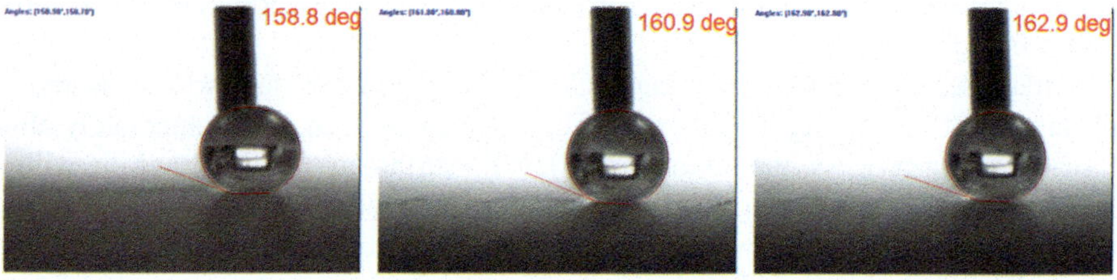

Fig. 5 Measured WCA on the 3 different locations of the laser textured substrate surface

Al2024-T3 surface was measured to be less than 5 deg. So, a superhydrophobic textured Al2024-T3 surface has been achieved with a WCA of more than 150 deg and CAH of less than 5 deg and a Cassie Baxter partial wetting superhydrophobic state is achieved.

The water droplet repellence test was also conducted on laser textured substrate surface. The sample substrate was held with a tilt angle of less than 5 deg. It was found that when the water droplets were falling and hit the textured sample surface,

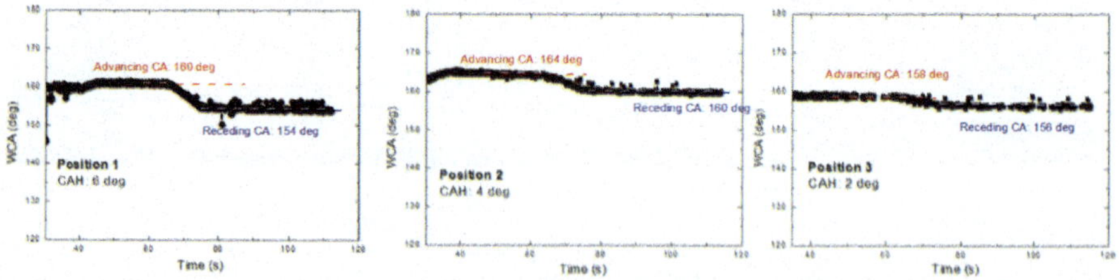

Fig. 6 Measured CAH on the 3 different locations of the laser textured P500k-1 substrate surface

the water droplet was rolling off the textured surface. So the roll off angle (ROA) is less than 5 degree.

With laser surface texturing, the WCA on Al2024 substrate surface was increased up to more than 150 degree after some timing of 25 days, indicating that laser surface treatment was able to convert the Al2024 surface into superhydrophobic surface. The surface wetting property on a solid substrate is known to be determined by its surface morphological structure and the surface chemistry property. The change of the WCA as a function of time (aging effect) is supposed to be due to the variation in the chemical composition of the surface, either oxidation on the textured surface or carbon-related content formation on the textured surface because the surface topography is not changing with the timing. Aluminum oxide (Al_2O_3) have a large number of polar sites that are composed of unsaturated aluminum and oxygen atoms and lead to a hydrophilic condition. And the possible formation of carbon content increases the number of non-polar sites and thus the surface energy is reduced so as to lead to hydrophobic property with an increased WCA. So, the WCA increment on the textured Al2024-T3 surface with the timing are most probably due to the formation of non-polar C-related layer on the textured surface. So, we proposed that the WCA on the textured Al2024-T3 surface to increase with the timing is probably due to the formation of non-polar C-related layer on the textured surface. The carbon may come from the decomposition of CO_2 into C with ultrashort laser irradiation induced catalytic effect.

4 Conclusion

In conclusion, a systematic investigation on ps laser surface texturing of Al2024 substrate had been conducted. The 2 scale micro-nano hierarchical surface patterns was fabricated on the Al2024 substrate surface. It was found that the WCA on the laser textured Al2024 surface was increased with the timing and reached up to more than 150 deg after 25 days. Also the CAH on the laser textured Al2024 surface was measured be less than 5 deg. So, a superhydrophobic Al2024 substrate surface has been obtained with a WCA of more than 150 degree and CAH of less than 5 deg. It was proposed that the WCA increment on the textured Al2024-T3 surface with the timing are most probably due to the formation of non-polar C-related layer on the textured surface in combination with the produced 2 scale micro-nano hierarchical surface textures.

The developed laser textured superhydrophobic Al alloy surface has potential to be applied in aerospace industry for its icephobic performance. Further investigation work will be conducted on its icephobic properties such as icing accretion delay, icing adhesion strength reduction test and its durability test including sand erosion and rain erosion testing.

References

1. Johnson RE, Dettre RH, Brandreth DA (1977) Dynamic contact angles and contact-angle hysteresis. J Colloid Interface Sci 62:205–212
2. Wenzel RN (1949) Surface roughness and contact angle. J Phys Colloid Chem 53:1466–1467
3. Cassie BD, Baxter S (1944) Wettability of porous surfaces. Trans Faraday Soc 40:546–550
4. Li J, Fu J, Cong Y, Wu Y, Xue LJ, Han YC (2006) Macroporous fluoropolymeric films templated by silica colloidal assembly: a possible route to super-hydrophobic surfaces. Appl Surf Sci 252:2229–2234
5. Kim SH, Kim JH, Kang BK, Uhm HS (2005) Superhydrophobic CFx coating via in-line atmospheric RF plasma of He−CF4−H2. Langmuir 21:12213–7
6. Wagterveld RM, Berendsen CWJ, Bouaidat S, Jonsmann J (2006) Ultralow hysteresis superhydrophobic surfaces by excimer laser modification of SU-8. Langmuir 22:10904–10908
7. Kulinich SA, Farzaneh M (2004) Hydrophobic properties of surfaces coated with fluoroalkylsiloxane and alkylsiloxane monolayers. Surf Sci 573:379–390
8. Kietzig AM, Hatzikiriakos SG, Englezos P (2009) Patterned superhydrophobic metallic surfaces. Langmuir 25:4821–4827
9. Kietzig AM, Mirvakili MN, Kamal S, Englezos P, Hatzikiriakos SG (2011) Laser-patterned super-hydrophobic pure metallic substrates: Cassie to Wenzelwetting transitions. J Adhesion Sci Tech 25:2789–2809
10. Her TH, Finlay RJ, Wu C, Deliwala S, Mazur E (1998) Microstructuring of silicon with femtosecond laser pulses. Appl Phys Lett 73:1673–1675
11. Wang B, Wang XC, Zheng HY, Lam YC (2015) Surface wettability modification of cyclic olefin polymer by direct femtosecond laser irradiation. Nanomaterials 5:1442–1453
12. Wang XC, Wang B, Xie H, Zheng HY, Lam YC (2018) Picosecond laser micro/nano surface texturing of nickel for superhydrophobicity. J Phys D: Appl Phys 51:115305, 8

Hybrid Ultrasonic Cavitation Abrasive Peening and Electrochemical Polishing on Additively Manufactured AlSi10Mg Components

K. W. S. Goh, K. L. Tan, and S. H. Yeo

Abstract Additive manufacturing processes face challenges when producing complex metallic components. Surface defects, such as incomplete melting, balling, and geometric inaccuracies, are common in printed metallic components with complex structures, such as lattice structures. These defects make it unfeasible to use these components commercially without further processing. Limitations such as inaccessibility and limited improvement in surface quality in existing post-processing techniques make the motivation for this study. This paper is largely based on the partial results of an ongoing thesis in the development of a novel hybrid process. The hybrid process combines electrochemical polishing (ECP) and ultrasonic cavitation abrasive peening (UCAP). It takes advantage of ECP's ability to brighten and smoothen surfaces and UCAP's ability to generate compressive residual stresses, enhance surface hardness, and improve the efficiency of ECP synergistically. The first part of this paper gives an overview of the need for an alternative post-processing technique and the benefits of a hybrid process. Further, results and discussion of investigative experimentation will be highlighted describing the effects and mechanisms of the hybrid process.

Keywords AM post-processing · Surface modification · Electrochemical polishing · Ultrasonic cavitation abrasive peening · Hybrid processes

K. W. S. Goh (✉) · S. H. Yeo
School of Mechanical and Aerospace Engineering, Nanyang Technological University, 50 Nanyang Avenue, Singapore 639798, Singapore
e-mail: M210038@e.ntu.edu.sg

K. W. S. Goh · K. L. Tan
Advanced Remanufacturing and Technology Centre, 3 Cleantech Loop, #01/01 CleanTech Two, Singapore 637143, Singapore
e-mail: tan_kai_liang@artc.a-star.edu.sg

N. Maharjan and W. He (eds.), *Proceedings of the 3rd International Conference on Advanced Surface Enhancement (INCASE) 2023*, Lecture Notes in Mechanical Engineering, https://doi.org/10.1007/978-981-99-8643-9_7

1 Introduction

Additive manufacturing (AM) is well known for its ability to produce components of high complexity. Among metallic AM processes, powder-bed fusion (PBF) is favored in the industry to produce lightweight porous structures. PBF processes include Direct Metal Laser Sintering (DMLS), Selective Laser Melting (SLM), and Electron Beam Melting (EBM). The fine resolution of PBF makes it possible to produce highly complex components unattainable by other manufacturing processes. Despite its capabilities, components produced by PBF have been reported with several surface defects. Due to thermal gradients formed during PBF processes, defects such as residual stress, balling, and spatter can occur. Residual stresses within components are known to cause deformation, delamination, and cracking [1]. Spatters are splashes from the laser melt pool creating unwanted structure of odd shapes while balling is a phenomenon of partially melted powders adhered to the surface [2]. These defects, along with the inherent staircase effect of AM [3], cause the surface to require post-processing before industrial application.

Post-processing of highly complex PBF components such as porous lattice structures can be challenging due to inaccessibility. In recent years, abrasive flow machining (AFM) and its many variations such as magnetic-assisted AFM [4], ultrasonic-assisted AFM [5], and chemical-assisted AFM [6] have been increasingly popular for post-processing PBF components. However, as the AFM process uses a hydraulic press to push an abrasive slurry through the components, it is mostly limited to continuous internal and external profiles. Laser polishing (LP) has great potential as an in-situ post-processing technique for PBF components due to a common source of power to DMLS and SLM. However, further residual stress defects caused by interlayer laser remelting can be difficult to address [7]. LP as a post-processing technique, outside of the PBF machine, is limited mostly to freeform external surfaces due to its line-of-sight nature [8]. Chemical etching (CE), electrochemical polishing (ECP), and hot isostatic pressing (HIP) are suitable post-processing techniques for lattice structures owing to their non-contact, non-line-of-sight nature. However, these processes have disadvantages. CE removes material indiscriminately, with a lack of dimensional control [9], ECP is affected by the edge effect which requires pre-processing to reduce surface undulations [10], and HIP has limited effect on surface roughness improvement, with no material removal capability [11]. Existing techniques for post-processing lattice structures are generally unable to improve multiple features in a single step. Hence, hybridisation becomes a viable option to combine benefits of processes, reduce number of steps, while potentially counteracting each of their limitations.

Cavitation has traditionally been studied for its damaging erosion effects, leading to premature component degradation. In recent years, researchers have found ways to utilise cavitation for advantageous material removal. Pecnik et al. [12] made use of ultrasonic cavitation for scale deposit removal, while Tan et al. [13] made use of ultrasonic cavitation with abrasive media to remove partially melted powders of Inconel 625 PBF components with surface hardening effects. Ultrasonic cavitation has been useful in other domains. Ultrasonic cavitation has been reported to have peening effects, capable of inducing compressive residual stress (CRS) into metallic materials [14]. Ultrasonic cavitation has been found useful in sonochemical reactors where chemical reactions have found new pathways and increased efficiency [15]. Due to the chemical benefits of ultrasound, ultrasound in electroplating (EP) has been studied. It was found that besides the increased rate of plating, plated coating layers were found to be harder, thicker, and brighter [16]. Despite the benefits of ultrasonic cavitation, and ultrasound on EP, there is limited research on the use of ultrasound and abrasives on ECP which is the focus of this study.

In this study, the effects of ultrasound and abrasives on ECP on DMLS-built AlSi10Mg specimens will be demonstrated using the developed process. The developed process will be a hybrid process between ECP and ultrasonic cavitation abrasive peening (UCAP). According to side surface photographic images, the effect of ultrasound and abrasives on ECP will be discussed.

2 Methods

2.1 *Work Material*

Regular cubic AlSi10Mg specimens of dimensions 15 × 15 × 15 cm produced by Direct Metal Laser Sintering (DMLS) using EOS M290, EOS GmbH were used in this study. Powders proprietary to EOS with size ranging from 25 to 70 μm were used in the DMLS printing process. DMLS specimens were printed in an argon inert environment to prevent oxidation during the printing process. Scanning orientation of the printing process alternated at 67° for subsequent layers, and print parameters were according to the default recommended settings. Specimens were wire-cut from the build plate upon completion of printing process. Specimens were ultrasonically cleaned for 15 min each and air-dried. Only the side surfaces were selected for this preliminary study due to the dominance of partially fused melted powders.

2.2 *Hybrid UCAP-ECP Setup*

An ultrasonic wave of 20 kHz frequency was transmitted using a transducer, generator, and horn assembly from Fisherbrand Model 505 Sonic Dismembrator, Thermo Fisher Scientific Inc. that can deliver a peak-to-peak vibration amplitude of range 24 to 120 μm at a horn tip of diameter 12.7 mm. The process medium was a mixed concentration of aqueous 1 M sodium hydroxide (NaOH) and glycerol ($C_3H_8O_3$). Silicon carbide (SiC) micro-particles of diameter 50 μm were added into the media to form an abrasive slurry with a 5 wt % concentration for the hybrid UCAP-ECP process and were distributed homogeneously using a magnetic stirrer maintained

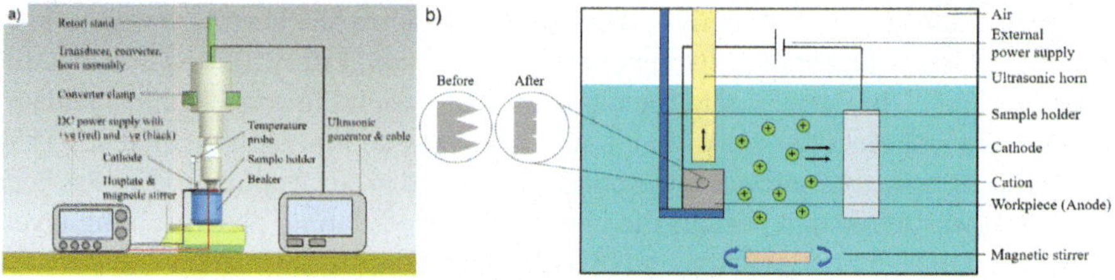

Fig. 1 Aa Schematic of hybrid experimental setup. b) blown-up schematic of hybrid experimental setup

at 300 rpm. All reagents were analytical grade and used directly without further purification. ECP was carried out by continuous electrochemical polishing (CECP). Vinyl tape was used to mask surfaces that were not targeted for polishing. DC power supply used was Kikusui PWR801MH, Kikusui Electronics Co. Ltd. A constant voltage of 5 V was used. The schematic of the hybrid UCAP-ECP experimental setup is illustrated in Fig. 1.

2.3 Material Characterisation

After individual processing, components were soaked in acetone for 10 min to remove masking residue before the specimens were ultrasonically cleaned for 15 min to remove residual abrasive particles and masking adhesive on the surface.

3 Results

3.1 Surface Quality

A typical as-printed PBF AlSi10Mg side surface, built at 90° sloping angle is shown in Fig. 2. The surface asperities can be classified into three categories. First, approximately spherical particles around 25 μm were found to be spread out across the surface. These were believed to be partially fused melted powders as they correspond to the EOS AlSi10Mg powder size distribution. These partially melted powders attach to the sideskin surfaces due to poor heat dissipation creating a thermal gradient [2]. Second, overlap discontinuities could be observed between the other irregularities. Overlap discontinuities can be difficult to spot on the as-printed surface due to overcrowding of other surface defects. These discontinuities are a possible result of interlayer cooling creating thermal gradients or a lack of laser input energy [2]. Third, larger irregular structures known as balling were found spread across the surface. Balling is an agglomeration of numerous partially melted powders. These typical

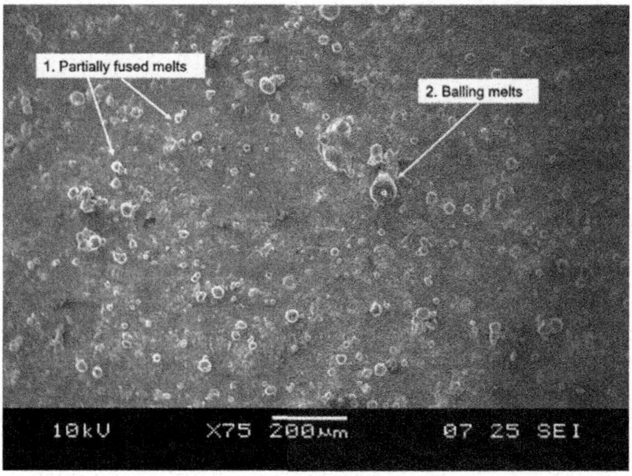

Fig. 2 Typical side surface characteristics of as-printed DMLS AlSi10Mg component

surface asperities on an as-printed side surface contribute to a high initial surface roughness which forms the need for post-processing.

The experimental conditions are shown in Table 1. Across specimens 1–2, and 3–4, the parameters were varied as illustrated in Fig. 3. The following parameters were varied: (i) temperature, (ii) electrolyte composition, (iii) standoff distance, and (iv) vibration amplitude. Side surface photographic images at the approximate central location of the specimens were captured in Fig. 3. At the initial parameters, with pure 1 M NaOH electrolyte at 70 °C, specimen 1 processed by ECP for 5 min was observed having a bright and shiny surface. However, specimen 2 processed by hybrid UCAP-ECP at 1 mm standoff distance, had the side surface observed with severe pitting marks, with the most severe ones being at the peripherals of the ultrasonic horn's circumference. At the varied parameters, with an electrolyte mix of 75/25 vol. % of 1 M NaOH/Glycerol at 50 °C, specimen 3 processed by ECP for 5 min was observed with a dark and dull surface, a stark contrast from specimen 1. On the contrary, specimen 4 processed by hybrid UCAP-ECP at 6 mm standoff distance, had the side surface observed with high shine with no signs of pitting to the naked eye.

4 Discussion

In this section, the postulated material removal and pitting mechanism will be discussed. Under the initial parametric conditions, with a 1 mm standoff distance, 70 °C temperature, and 100 vol. % of 1 M NaOH electrolyte, surfaces processed by hybrid UCAP-ECP sustained severe pits. From the severe pitting sustained in specimen 2, the anodic viscous layer may have been broken down which had caused the accelerated evolution of pitting. The viscous layer is a passive layer which protects the anodic metal surface during ECP, concurrently maintaining an ionic gradient for

Table 1 Experimental conditions

Specimen	Process type	Media	Amplitude (μm)	Standoff distance (mm)	Abrasives	Temperature (°C)	Duration (min)
1	ECP	NaOH/ Glycerol (100/0 vol%)	N/A	N/A	No	70	5
2	Hybrid	NaOH/ Glycerol (100/0 vol%)	60	1	Yes	70	5
3	ECP	NaOH/ Glycerol (75/25 vol%)	N/A	N/A	No	50	5
4	Hybrid	NaOH/ Glycerol (75/25 vol%)	60	6	Yes	50	5

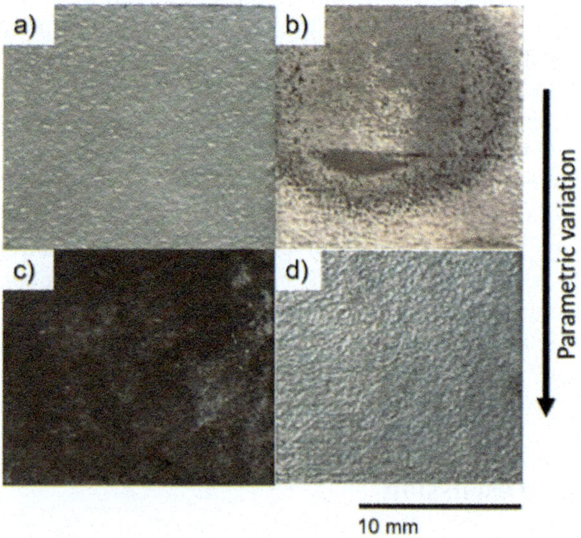

Fig. 3 **a** Side surface photograph of specimen 1. **b** side surface photograph of specimen 2. **c** side surface photograph of specimen 3. **d** side surface photograph of specimen 4

diffusion to occur [17]. There are a few possible reasons as to why the viscous layer may have been broken down. During ultrasonic cavitation, heat was generated which decreased the viscosity of the electrolyte. In addition, the ultrasonic horn may have been positioned too close to the anode, where the cavitation impact disturbed the viscous layer excessively.

To prevent pitting, the viscous layer must be maintained. From the analysis of the severe pitting sustained in specimen 2, the following parameters were varied for various reasons. Temperature was reduced as cavitation generates and accumulates heat, viscosity was increased to maintain the viscous layer, and the standoff distance was increased to reduce the anode-horn proximity. With the varied parameters, the surface of specimen 4 showed no visible pitting, with a bright shine. This shows evidence of the existence of an inertial cavitation threshold where the viscous layer could be optimally disturbed, increasing the rate of ECP while preventing pitting. As the diffusion layer thickness decreases, the limited current density of ECP will increase, thereby increasing the rate of ECP [18]. At the varied parametric conditions, specimen 3 after 5 min of ECP had incomplete polishing. This was likely due to the reduced temperature which caused the current density to drop significantly [17], slowing the ECP process. This showed that the presence of cavitation enabled ECP to polish the surface at a lower processing temperature due to the heat generation and cavitation-viscosity interaction.

5 Conclusion

ECP is a process popular for post-processing metal components to a bright, shiny, and smooth surface. However, the edge effect of ECP is a common problem that requires a certain level of pre-processing. UCAP has been reported to be efficient at targeted removal of partial melts and loose balling structures [19], which could potentially reduce the edge effects of ECP through hybridisation. Ultrasound and abrasive micro-particles appear to have the capability to disturb the anodic viscous layer which should improve the ECP rate. Furthermore, UCAP has been reported to have surface strengthening benefits with capability to induce CRS to extend the fatigue life cycle of a component [13]. In addition, both UCAP and ECP use non-contact material removal and modification approach which does not have problems with tool accessibility. Overall, hybridisation of UCAP with ECP presents good synergistic potential for the surface finishing of complex component geometry. Controlling the parameters is of paramount importance to prevent pitting on the surface, achieve a good surface finish, and improve mechanical properties. The use of ultrasound and abrasives on ECP, along with its material removal mechanism, and synergistic benefits are not well explored which makes the focus of the future direction of this study.

Acknowledgements *Declaration of Competing Interests:* The author declares that he has no known personal or financial interests that would influence or appear to influence the work presented in this paper.

The funding support from Advanced Remanufacturing Technology Centre (ARTC), Singapore and Nanyang Technological University, Singapore is gratefully acknowledged. The financial support from A*STAR Graduate Academy (AGA), Singapore in the form of a PhD scholarship is acknowledged.

References

1. Zhang B, Li Y, Bai Q (2017) Defect formation mechanisms in selective laser melting: a review. Chin J Mech Eng 30(3):515–527
2. Echeta I et al (2019) Review of defects in lattice structures manufactured by powder bed fusion. Int J Adv Manuf Technol 106(5–6):2649–2668
3. Grasso M, Colosimo BM (2017) Process defects and in situ monitoring methods in metal powder bed fusion: a review. Measure Sci Technol 28(4)
4. Sehijpal S, Shan H S, Kumar P (2002) Wear behavior of materials in magnetically assisted abrasive flow machining. J Mates Processg Technol 128:155–161
5. Sharma AK et al (2015) Experimental investigations into ultrasonic-assisted abrasive flow machining (UAAFM) process. Int J Adv Manuf Technol 80(1–4):477–493
6. Mohammadian N, Turenne S, Brailovski V (2018) Surface finish control of additively-manufactured Inconel 625 components using combined chemical-abrasive flow polishing. J Mater Process Technol 252:728–738
7. Vaithilingam J et al (2016) The effect of laser remelting on the surface chemistry of Ti6al4V components fabricated by selective laser melting. J Mater Process Technol 232:1–8
8. Yung KC et al (2018) Laser polishing of additive manufactured CoCr alloy components with complex surface geometry. J Mater Process Technol 262:53–64
9. Pyka G et al (2012) Surface modification of Ti6Al4V open porous structures produced by additive manufacturing. Adv Eng Mater 14(6):363–370
10. Dong G, Marleau-Finley J, Zhao YF (2019) Investigation of electrochemical post-processing procedure for Ti-6Al-4V lattice structure manufactured by direct metal laser sintering (DMLS). Int J Adv Manuf Technol 104(9–12):3401–3417
11. Agarwala MK, Bourell DL, Beaman JJ (1994) Densification of selective laser sintered metal parts by hot isostatic pressing. Int Solid Freeform Fabrication Symp
12. Pečnik B et al (2016) Scale deposit removal by means of ultrasonic cavitation. Wear 356–357:45–52
13. Tan KL, Yeo SH (2020) Surface finishing on IN625 additively manufactured surfaces by combined ultrasonic cavitation and abrasion. Additive Manuf 31
14. Gao Y et al (2014) Ultrasonic cavitation peening of stainless steel and nickel alloy. J Manuf Sci Eng 136(1)
15. Lukevics E, Gevorgyan VN, Goldberg YS (1984) Ultrasound-induced heterogeneous reduction of halo, alkoxy and amino derivatives of group ivb elements with lithium aluminium hydride. Tetrahedron Lett 25(13):1415–1416
16. Walker R, Walker CT (1975) New explanation for the brightness of electrodeposits produced by ultrasound. Ultrasonics 79
17. Yang G et al (2016) Electropolishing of surfaces: theory and applications. Surf Eng 33(2):149–166
18. Mason TJ, Lorimer JP (2002) Sonoelectrochemistry. Appl Sonochem 225–266
19. Tan KL, Yeo SH (2017) Surface modification of additive manufactured components by ultrasonic cavitation abrasive finishing. Wear 378–379:90–95

New Electronic Peening Intensity Sensor: Effect of Impingement Angle

Sylvain A. Forgues, Brigitte Labelle, Adel Alouani, Ramzi Ben Moussa, and Sébastien Legrand

Abstract In this paper, the previously presented theoretical equations governing the utilization of a new sensor to measures peening intensity will be expanded to include the effect of impingement angle. The equations are derived from the Impulse–Momentum principle and use the input velocity, the coefficient of restitution, the media particle mass and the impact impulse time to determine the maximum impact force at any angle. The equations are verified with results from experimental testing at impingement angles of 85°, 65°, 55°, 45° and 35° at five different input velocities using AWCR28 media propelled from a Shockform designed single particle cannon. Results show that for all angles, the sensor output offers a good linear correlation with the input velocity. The sensor can be used to assess the quality of the particle stream when peening at different angles and with different media flow rates. Examples of sensor output graphs will be presented showing how collisions between incoming and outgoing media particles can affect the distribution of impacts on a part.

Keywords Electronic · Peening · Intensity · Sensor · Impingement · Angle

1 Introduction

The method to evaluate peening intensity has not evolved much in the last 80 years. The bending of a small strip of metal is still used today to determine and verify the peening intensity. Using these strips of metal is very labor intensive as they must be pre-measured with a gage, installed on a special holder, peened for a period of time and measured once more with that gage. Intensity determination requires peening a minimum of four metal strips at different times. The bending of a strip is measured for each peening time and the data is curve fitted to an exponential equation. The «10% rule» is then used to calculate the peening intensity from the equation. Many people find this procedure long and confusing and it is estimated that 20–30% of

S. A. Forgues (✉) · B. Labelle · A. Alouani · R. B. Moussa · S. Legrand
Shockform Aeronautique Inc., 4482 boul. de la Grande-Allee, Boisbriand, QC J7H 1R9, Canada
e-mail: sylvain.forgues@shockform.com

N. Maharjan and W. He (eds.), *Proceedings of the 3rd International Conference on Advanced Surface Enhancement (INCASE) 2023*, Lecture Notes in Mechanical Engineering, https://doi.org/10.1007/978-981-99-8643-9_8

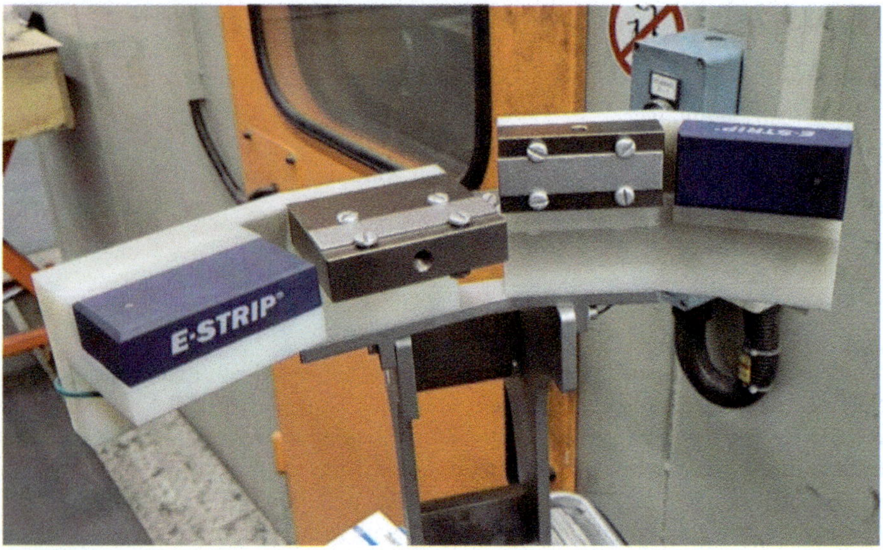

Fig. 1 Sensor on a test fixture with metal strips

peening machine time is spent on intensity determination and verification. In the Aerospace industry, most peening machines are highly automated but still require an operator to manually insert and remove the metal strips before a part can be peened. What if a peening intensity sensor could speed up and simplify this whole process?

A small sensor was recently introduced to do just that. The sensor, the size of a standard metal strip holder [1], illustrated in Fig. 1, measures the impact force of individual peening particles using a custom designed load cell. The magnitude of the impacts are recorded and sent wirelessly to a ruggedized computer. The computer can display the output of several sensors in the form of voltage or equivalent intensity. An impact distribution can also be displayed to show the quality of the blast stream. This very interesting feature will be discussed in Sect. 4.

2 Theoretical Equations for Impingement Angle

The theoretical equations governing the utilisation of the sensor are based on the Impulse–Momentum principle explained in most basic physics handbooks [2]. Equations and experimental data for a perpendicular impact to the sensor have been previously presented [3–5]. The equations are expanded in this Section to include impacts at any impingement angle.

Consider a peening particle of mass m moving in a xy plane at speed $\vec{v}$ and acted on by a force $\vec{F}$ during a period of time t. From Newton's second law, the force acting on the particle can be defined as follows:

$$\vec{F} = m\frac{d\vec{v}}{dt} \quad (1)$$

By rearranging, we get

$$\int_{t1}^{t2} \vec{F}\,dt = \int_{v1}^{v2} m\,d\vec{v} \quad (2)$$

The Impulse–Momentum principle is derived from Eq. (2) with impulse on the left and change of momentum on the right of the equation. It states that the Impulse defined as the force acting on a particle over a period of time is equal to the change in momentum of the particle defined as the mass multiplied by the change in velocity of the particle.

In our application, the peening particle has a velocity v_i just before the impact with the load cell pin at an angle θ_i as illustrated in Fig. 2. Just after impact, the particle has a velocity v_f and is moving away from the pin at an angle θ_f. The load cell of the sensor can only measure the normal component F_N of the applied force on the pin's surface. The normal components of the velocity before and after impact will therefore be used for the change of momentum.

Since the mass of the particle is assumed constant during the impact and the normal components of the initial and final velocity are in opposite directions, Eq. (2) becomes:

$$\int_{t1}^{t2} F_N dt = m(v_{fN} - (-v_{iN})) \quad (3)$$

or

$$\int_{t1}^{t2} F_N dt = m(v_{fN} + v_{iN}) \quad (4)$$

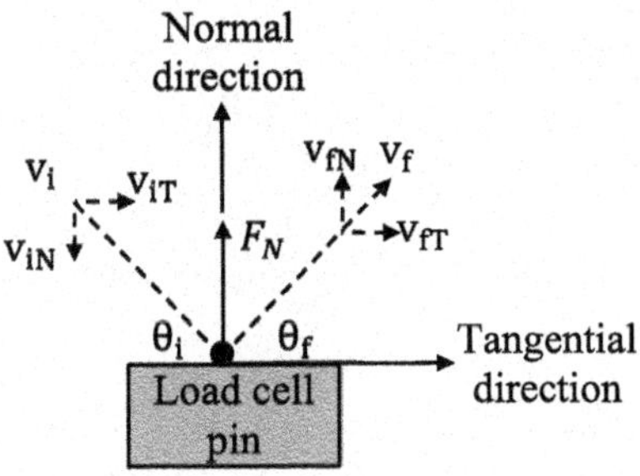

Fig. 2 Particle impact at an angle on load cell pin

A relation exists between v_i and v_f known as the coefficient of restitution. The coefficient of restitution is defined as the ratio of the velocity after impact to the velocity before impact. Coefficients of restitution depend on factors including material properties, body geometry, and impact velocity. During peening some kinetic energy is dissipated during the impact and $0 < C_R < 1$. For the normal components of velocity, the coefficient of restitution is defined as:

$$C_{RN} = \frac{v_{fN}}{v_{iN}} \tag{5}$$

Equation (4) can be re-written as

$$\int_{t1}^{t2} F_N dt = m(C_{RN} + 1)v_{iN} \tag{6}$$

From testing, and from the literature [6, 7], we also know that there is a direct relationship between the normal component of shot velocity before impact and peening intensity.

$$v_{iN} = K_B \; I_A \tag{7}$$

where I_A is the Almen intensity and K_B is a factor associated with particle properties.

Replacing Eq. (7) in Eq. (6) gives us a direct relation between the impulse and the peening intensity:

$$\int_{t1}^{t2} F_N dt = m(C_{RN} + 1)K_B I_A \tag{8}$$

where the mass m, the coefficient of restitution C_{RN} and the factor K_B are properties of the peening particle involved in the impact of the peening process.

The integral on the left side of Eq. (8) can be evaluated when the force applied on the particle is known as a function of time. During impact, this interaction can be simplified as illustrated in Fig. 3. As the particle hits the sensor, the force gradually increases as the velocity decreases until the force reaches a maximum value F_{max}. The force then gradually decreases as the particle departs in the other direction until the force reaches zero.

The area under the curve in Fig. 3 is equal to the impulse of the impact. In this case, it can be calculated as the area of a triangle with:

$$F_N \Delta t = A = \frac{1}{2}(\text{base})(\text{height}) \tag{9}$$

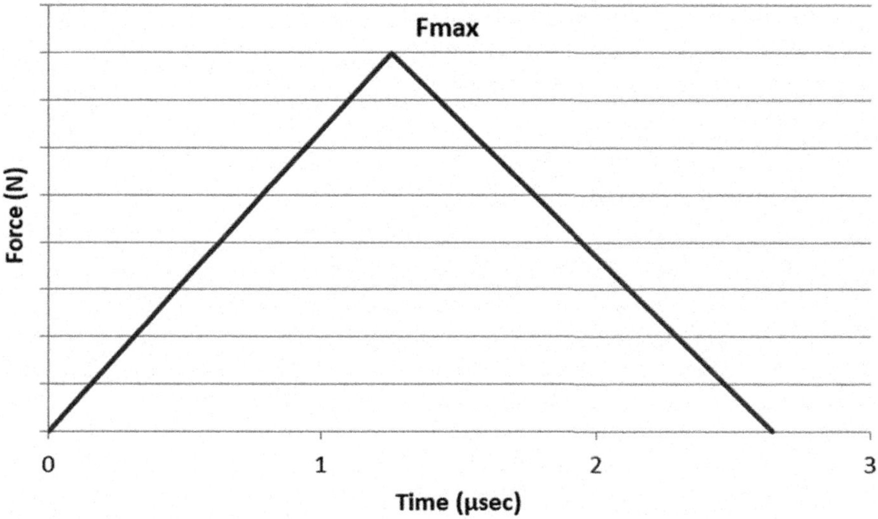

Fig. 3 Graphical illustration of particle force of impact over time

Where the base of triangle is the duration of the impact and the height of the triangle is the maximum force of impact F_{Nmax}. Equation (9) then becomes:

$$F_N \Delta t = A = \frac{1}{2}\left(t_{impact}\right)\left(F_{Nmax}\right) \tag{10}$$

Merging Eqs. (10) and (6) we get a relationship between the maximum impact force and the velocity before impact.

$$\frac{1}{2}\left(t_{impact}\right)\left(F_{Nmax}\right) = m(C_{RN} + 1)v_{iN} \tag{11}$$

Rearranging, the maximum impact force can be calculated as follows:

$$F_{Nmax} = \frac{2\,m\,(C_{RN} + 1)\,v_{iN}}{t_{impact}} \tag{12}$$

The max force F_{Nmax} can be calculated since the mass m, the coefficient of restitution C_{RN}, the initial velocity v_{iN} and the time of impact t_{impact} can be measured. Merging Eqs. (12) and (7), we see that F_{Nmax} has a direct relationship with Almen intensity I_A.

$$F_{Nmax} = \frac{2\,m\,(C_{RN} + 1)\,K_B\,I_A}{t_{impact}} \tag{13}$$

For our sensor, the normal maximum force F_{Nmax} is measured directly by the load cell. This will be confirmed if a linear relationship exists between the sensor output and the normal input velocity v_{iN} as described by Eq. (12).

3 Experiment Setup and Results

A special cannon was designed to propel individual peening particles onto the sensor (Fig. 4). This cannon uses a camera, LED strobe lighting and an image analysis software. The software measures the input velocity just before impact, the output velocity after impact, the incoming angle of impingement and the outgoing angle of impingement. The sensor provides the time of impact and the maximum sensor output.

Different types and size of media have been tested with the cannon. In this paper, data is presented for AWCR28 peening media as per AMS2431\3 [8]. Testing was performed at five different incoming angles of impingement θ_i of 85°, 65°, 55°, 45°, 35° and Fig. 5 illustrates typical images captured at 85°, 65°, 55°and 45°. Each image shows a particle as it is approaching the sensor's pin in white at the right side of the image. The velocity of the particle is evaluated by the image analysis software at several instances as it is travelling towards the pin. The software then uses the same image to evaluate the particle velocity away from the pin. For clarity, this is not

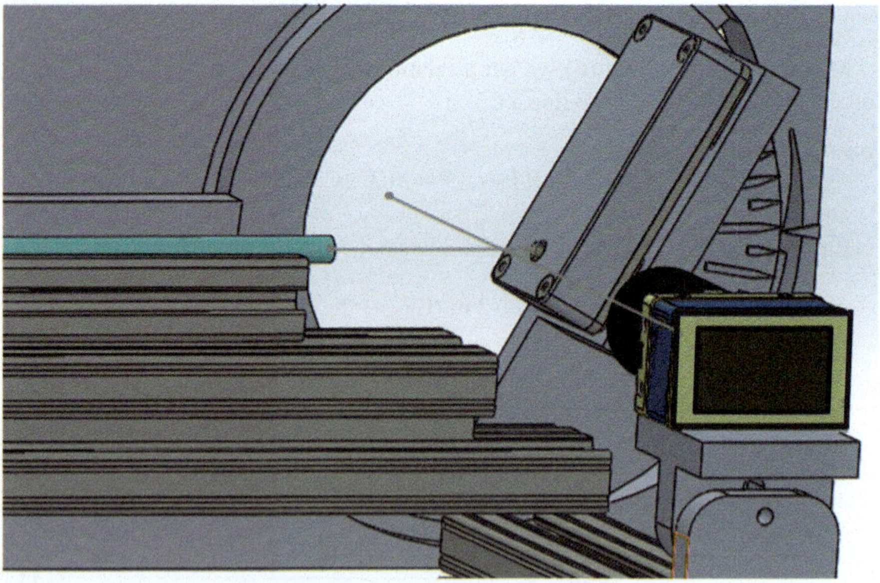

Fig. 4 Experimental setup for media impact velocity and angle measurement

Fig. 5 Particle Image with Input Velocity Measurement at 85°, 65°, 55°, 45° impingement angles

shown in the images. In the experimental setup, the incoming angle is pre-selected and the output angle is measured from the image.

For each angle, five different input velocities were tested to see the evolution of the coefficient of restitution at different speeds. Testing was performed at 2, 20, 31, 40 and 51 m/s. Ten sample points were used to calculate the average at each condition for a total of 250 measurements.

Table 1 shows data measured at 40 m/s input velocity. The first column is the input angle of impingement. The input angle decreases from nearly perpendicular to the pin surface at 85° all the way down to 35°. The second column shows the average input velocity followed by the calculated normal and tangential component of the input velocity in column 3 and 4. Column 5 and 6 show the output angle of the particle and its output velocity. Output angle decreases from 85° to 20° and follows a similar trend to the input angle. Columns 7 and 8 show the normal and tangential component of the output velocity. Columns 9, 10 and 11 show the calculated coefficient of restitution and its normal and tangential components respectively. Finally, column 12 shows the impact intensity measured by the sensor.

Figure 6 illustrates how the normal component of the coefficient of restitution varies with the normal component of the initial velocity at different angles of impingement and different velocities. The normal component of the coefficient of restitution decreases slightly as the normal input velocity increases. This same relationship was observed for 90° impingement angles [3] and illustrated as red dots in Fig. 6.

Plotting the sensor output in terms of the normal component of the initial velocity shows a linear relationship (Fig. 7). The sensor responds well to the normal component of impact force at different velocities and angles as predicted by Eq. 12. Since Almen intensity also has a direct relation with the normal component of initial

Table 1 Results for input velocity Vi = 40 m/s (each data point from 10 measurements)

Avg. input angle θi (deg.)	Avg. input Vel. Vi (m/s)	Norm. input Vel. ViN (m/s)	Tang. input Vel. ViT (m/s)	Avg. Out. angle θf (deg.)	Avg. Out.Vel.Vf (m/s)	Norm.Out. Vel. VfN (m/s)	Tang. Out. Vel. VfT (m/s)	Avg. Coeff. of Rest. Cr	Avg. Norm. Coeff. of Rest. CrN	Avg. Tang. Coeff. of Rest. CrT	Out. volt
85	40.1	39.9	3.5	85	18.0	17.8	1.7	0.44	0.44	0.24	15.5
65	40.7	36.8	17.2	56	18.3	15.1	10.1	0.45	0.41	0.59	14.7
55	40.7	33.4	23.4	40	26.4	17.0	20.1	0.65	0.51	0.86	13.5
45	39.5	28.0	28.0	27	28.6	12.9	25.4	0.72	0.46	0.91	10.5
35	40.1	23.0	32.8	20	31.7	10.6	29.8	0.79	0.46	0.91	8.4

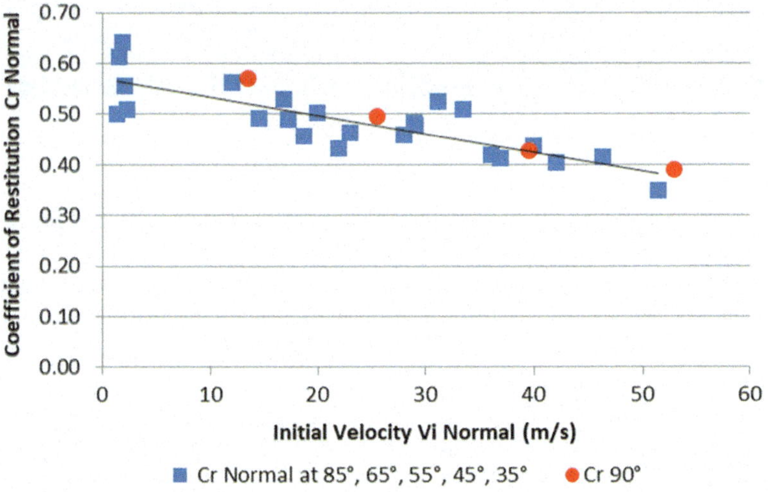

Fig. 6 Normal coefficient of restitution C_{RN} versus normal initial velocity v_{iN} at impigement angle 85°, 65°, 55°, 45°, 35° and at different velocities compared with coefficient of restitution at 90°

velocity, as described in Eqs. 7 and 13, the sensor output also provides a direct measure of Almen intensity.

4 Media Collisions at Different Angles and Media Flow Rates

With the new sensor, it is now possible to assess the quantity and velocity of the impacts in a particle stream. This was previously impossible with metal strips since they produce a single bending value from the sum of all the impacts on their surface. Peening particles impacting a surface have a wide range of velocities. The particles

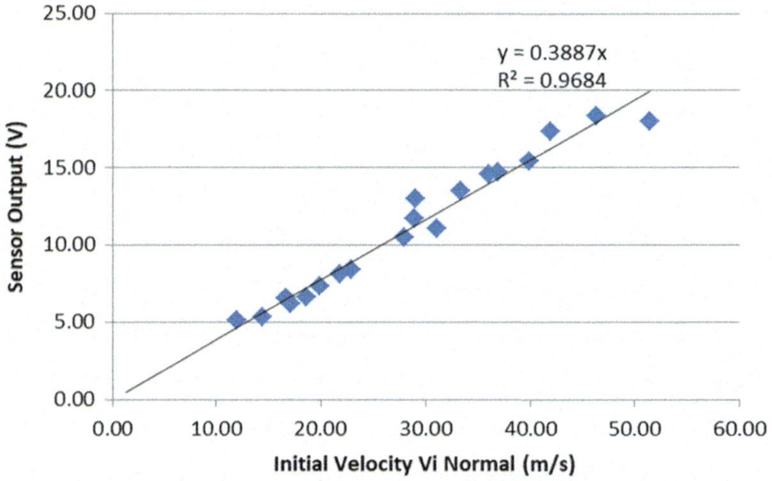

Fig. 7 Sensor output versus normal initial velocity v_{iN} and impingement angle 85°, 65°, 55°, 45°, 35°

at the center of the stream are moving faster than the particles at the edge of the stream [9]. Furthermore, several particles collide with rebounding particles before hitting the sensor and therefore reach the sensor with reduced velocity.

This phenomenon was predicted by simulation models [10–14]. Figure 8, taken from [14], shows the interaction of peening particles in the particle stream as the nozzle is moving over a surface. The curve on the right of Fig. 8 represents a distribution of velocities in a peening stream calculated by simulation. The simulation shows that only 30% of impacts reach the surface at the full initial velocity standardized to a value of 1. The rest of the impacts collide to some level with other particles before hitting the surface.

Figure 9 shows the distributions of particle velocities generated from the sensor output. The curves represent the impacts from particle streams at different angles and media flow rates but all for the same Almen intensity of 0.203mmA. The X axis

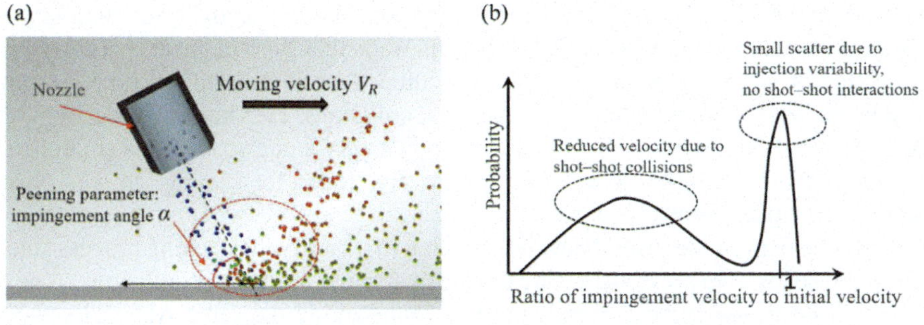

Fig. 8 Computer simulations to determine the shot-shot collisions during peening [14]

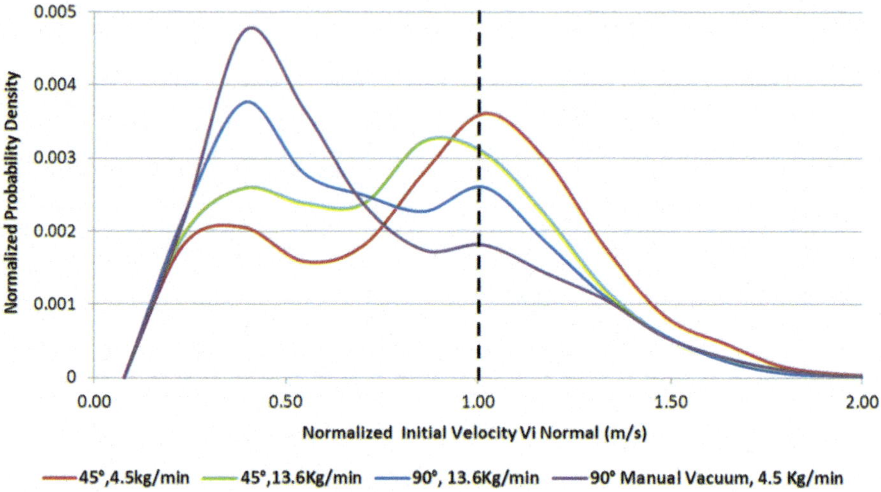

Fig. 9 Impacts distribution on sensor at different angles and media flow rates for same Almen Intensity of 0.203 mmA

represents the normalized impact velocity and the Y axis represents the normalized probability of impact for a specific velocity.

The four distributions are characterized by two separate peaks. The peak on the right is assumed to represent the most common direct impacts on the sensor from particles that have had no collision. These impacts have the most influence on the Almen intensity. The peak on the left is assumed to represent the particles that reach the sensor after collision with another particle. The number of collisions is influenced by the impingement angle and the particle flow rate. At 45° and 4.5 kg/min flow rate, a small number of collisions occur as illustrated by small left peak of the red curve. When the flow rate at 45° increases to 13.6 kg/min, as illustrated by the green curve, the number of collisions increases because more particles are in the peening stream and collisions are more likely. Hitting the sensor at a 90° impingement angle instead of 45° but with the same 13.6 kg/min flow rate (blue curve) increases the number of collision since at 90° impingement, many particles rebound directly into the incoming particle stream. Finally, the purple curve represents the impact distribution obtained using a manual vacuum peening head [15]. With this type of head, particles are propelled at 90° impingement and then vacuumed back into the machine directly inside the head. This setup produces the most collisions, even at flow rate of 4.5 kg/min, since all the particles are contained inside a very small space. This small head increases the probability of collision but is very useful for keeping an area clean when performing manual peening.

The sensor produces useful new information to quantify the peening process since the metal strips used previously were unable to account for particle collisions. A next step will be to see if these impact profiles representing the same Almen intensity produce the same residual stress distribution.

5 Conclusion

A new sensor was presented that quickly and easily measures the impact intensity of a peening particle stream. The sensor is the size of an Almen strip holder for easy integration into existing peening fixtures. The theoretical equations governing the sensor were developed to include the influence of the peening angle of impingement. A special cannon used to validate the equations was presented along with images and experimental data from the large number of tests performed at different input velocities and angles of impingement.

Experimental data shows that at angle, the normal component of the coefficient of restitution behaves just like the coefficient of restitution at 90°. This means that when the input velocity is known, the coefficient of restitution at 90° can be used for the calculations at all angles of impingement. Data was also presented showing that the sensor output is directly proportional to the normal component of initial velocity. This was predicted by the theoretical equations and confirms that the sensor can measure peening intensity.

Output distributions from the sensor were presented showing how the angle of impingement and the media flow rate influence the probability of impacts at the same Almen intensity. The distributions are characterized by two distinct peaks. The lower intensity peak assumed to represent the particle impacts that have collided with another particle before reaching the sensor and the higher intensity peak for the particle impacts without interference. These distributions were favorably compared with published distributions obtained by computer simulation.

The sensor was shown to produce useful new information that will simplify the control and improve understanding of the peening process. Future work will study how the impact profile will influence the peening compressive residual stress distribution.

References

1. SAE J442, Test strip, holder and gage for shot peening
2. Sears, Zemansky, University Physics, 1970, Addison-Wesley Publishing Company
3. Forgues SA et al (2022) New electronic peening intensity sensor: theory and experimental validation. In: Conference proceedings, international conference on shot peening ICSP14, Milan
4. Forgues S. et al (2022) New electronic peening intensity sensor: comparison with Almen arc height verification. In: Conference proceedings, international conference on shot peening ICSP14, Milan
5. Forgues SA et al (2021) New electronic peening intensity sensor: Almen intensity correlation and time variation study. In: Conference proceedings, INCASE2021, Singapore
6. Barker, Young, Pouliot (2005) Particle velocity sensor for improving shot peening process control. Conf Proc: ICSP-9:385–391
7. Könitzer, Polanetzki (2011) Implementation of velocity measurement as intensity verification in Production, 2011. Conf Proc ICSP-11:19–24

8. SAE AMS 2431/3 Peening Media—Conditioned Carbon Steel Cut Wire Shot, Regular Hardness (45 to 52 HRC) (AWCR)
9. David Kirk (2011) Variability of a shot stream's measured peening intensity. The Shot Peener Magazine 25(3, Summer)
10. Hong T et al (2008) A numerical simulation to relate the shot peening parameters to the induced residual stresses. Eng Fail Anal 15(8):1097–1110
11. Gariepy A et al (2017) Simulation of the shot peening process with variable shot diameters and impacting velocities. Adv Eng Softw 114:121–133
12. Fubin T et al (2017) A sequential DEM-FEM coupling method for shot peening simulation. Surf Coat Technol 319:200–212
13. Wang C et al (2021) A new methodology to establish the relationship between equivalent shot velocity and air pressure by surface roughness for shot peening. Int J Adv Manuf Technol 112:2233–2247
14. Wang Y et al (2022) Study on the measurement method of shot velocity and mechanical properties based on conventional shot peening. ACS Omega 7:33233–33247
15. Shockform web site, SmartPeen® controlled shot peening machine with a media recollection vacuum system. https://shockform.com/app/uploads/2021/06/Shockform-smartpeen-pdf.pdf

Hydrophobic Surface of HVOF Sprayed Tungsten Carbide Based Coating

Aw Poh Koon, Aaron Teo Zhi Hao, and Raahgini Chandrasegaran

Abstract Tungsten carbide coating deposited by High Velocity Oxy-Fuel (HVOF) spray process possesses high microhardness and dense coating structure. It has very good wear and corrosion resistant properties which are used on components in various industries such as aerospace, marine, offshore, oil and gas. Beside these known properties, there is one interesting surface property—the coating surface has potential to reach near superhydrophobic state which enable a durable antistick surface. This experimental study revealed that the HVOF sprayed WC-CoCr coating transiting from hydrophilicity during as-sprayed to hydrophobicity after 15 days of exposure to atmospheric condition.

Keywords Hydrophobic · High Velocity Oxy-Fuel (HVOF) · Tungsten Carbide Coating · Thermal Spray

1 Introduction

HVOF-sprayed tungsten carbide-based coatings are commonly used for wear and corrosion protection in industries such as aerospace, oil and gas, and marine [1]. While their effectiveness in these applications is well-known, some of their surface properties, such as hydrophobicity, have not been fully explored. Some published works have reported on related surfaces, such as plasma-sprayed ceramic coatings

A. P. Koon (✉)
Singapore Institute of Manufacturing Technology, Singapore, Singapore
e-mail: pkaw@simtech.a-star.edu.sg

A. T. Z. Hao
GE Vernova, Singapore, Singapore
e-mail: Zhihaoaaron.teo@ge.com

R. Chandrasegaran
Makino Asia Pte Ltd, Singapore, Singapore
e-mail: raahgini.c@makino.com.sg

N. Maharjan and W. He (eds.), *Proceedings of the 3rd International Conference on Advanced Surface Enhancement (INCASE) 2023*, Lecture Notes in Mechanical Engineering, https://doi.org/10.1007/978-981-99-8643-9_9

[2] and HVOF-sprayed cermet coatings [3, 4], suggesting that thermal spray coatings may have intrinsic hydrophobic surfaces due to their surface chemistry and morphology. However, the formation mechanism of these hydrophobic surfaces has not been fully understood.

It may be assumed that such a surface property can be obtained as soon as the coating is formed, but recent laboratory studies have revealed a different observation. As-sprayed cermet coating surfaces were found to transition from hydrophilic to hydrophobic after being exposed to air for an extended period of time. This paper aims to share such observation made and further work may be required to study the underlying mechanisms responsible for this transition.

The water contact angle is commonly used as a measure of the hydrophilicity or hydrophobicity of a surface. A water contact angle of 90° is often used as a cut-off point to define a surface as either hydrophilic or hydrophobic [5]. A surface is considered hydrophilic if the water contact angle is less than 90°, indicating that the surface is wetted easily by water. Conversely, a surface is considered hydrophobic if the water contact angle is greater than 90°, indicating that the surface repels water and is difficult to wet. When the water contact angle exceeds 150°, the surface is referred to as superhydrophobic, indicating an extremely strong hydrophobic effect.

2 Methods

2.1 Powder Material

The powder material, WC-CoCr, used in this study was obtained from Oerlikon Metco. Table 1 tabulates the WC-CoCr powders information including its chemistry, manufacture route, morphology and particle size range. The as-received powder material was prepared metallographically for analysis using scanning electron microscope (SEM).

Table 1 Powder information

Parameters	Characteristics
Chemistry	86WC 10Co 4 Cr
Manufacture	Agglomerated and sintered
Morphology	Spheroidal
Nominal powder size	15–45 μm

2.2 HVOF Spray Process

The WC-CoCr coating was applied to an aluminum substrate using an Oerlikon Metco Diamond Jet series spray torch. This process involved using oxygen and propane as fuel gas to generate a supersonic spray jet, which uniformly heats the powder spray material introduced axially. The gas stream then propels the powder particles toward the substrate, where they undergo plastic deformation upon impact and create a strong bond between the coating and the substrate. The substrate was cleaned in IPA and the surface was sand blasted with sand grit mesh size of 60 prior to coating deposition.

2.3 Coating Characterisation

SEM analysis was performed on the as-sprayed coating surface for analysis of the surface morphology. To study the microstructure and porosity of the coating, the samples were sectioned using an abrasive cutter and prepared metallographically according to ASTM E1920-03. Optical microscopy was then used to analyze the coating, and the porosity of the coating was measured using imaging processing software in accordance with ASTM E2109. Measurements were taken on the cross-sectioned coating, and the average value of at least three measurements per coating was taken at three random locations. Surface roughness was measured using surface roughness profilometer from Taylor Hobson with the settings below:

- Filter: Gaussian
- Measurement speed: 0.5 mm/s
- Data length: 20 mm
- Cut offs: 2.5 mm

Static water contact angle was measured using a VCA-optima system. 1 μl of DI water droplet was dispensed on the sample surface through the motorized syringe. At least 3 measurements were taken on the sample surface randomly.

3 Results and Discussion

3.1 Powder Analysis

Figure 1 shows the SEM images of top view and cross-sectional view of the as-received WC-CoCr powder. Figure 1a shows that the morphology of powders is spheroidal which has good flowability and enable repeatable coating property. Figure 1b shows the cross-sectional view of the as-received WC-CoCr powder. Voids

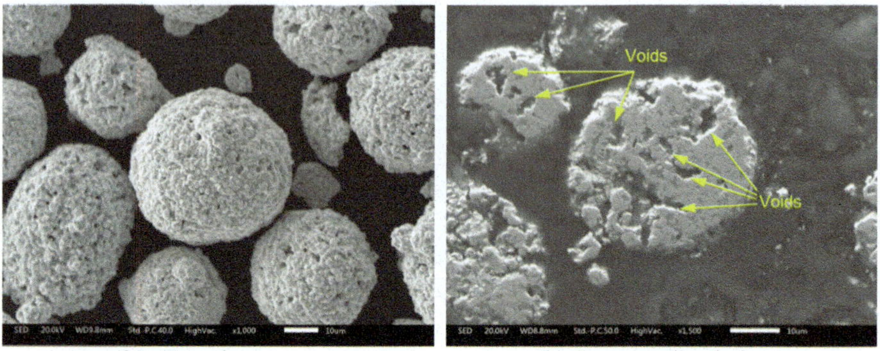

(a) Top view (b) Cross sectional view

Fig. 1 SEM analysis of as-received powder

can be observed within each powder particles, which revealed that the powders were agglomerated and sintered [1].

3.2 As-Sprayed Coating Analysis

Figure 2 shows the surface morphology of the as-sprayed WC-Co-Cr coating, as analyzed by SEM. At a high magnification of 1500 times, the irregular surface roughness caused by the formation of molten and semi-molten particles can be observed. The surface roughness range was measured to be between 6.3 and 6.5 μm Ra. Figure 3 shows the optical image of the cross sectional view of the metallographically prepared coating with a measured porosity of $0.8\% \pm 0.17$.

Fig. 2 SEM images of surface morphology of as-sprayed coating

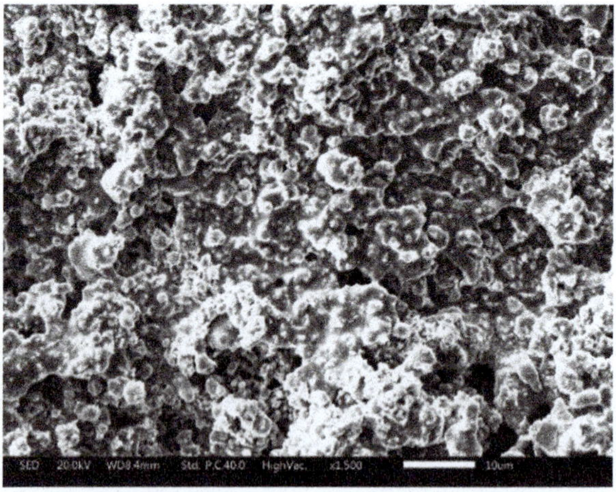

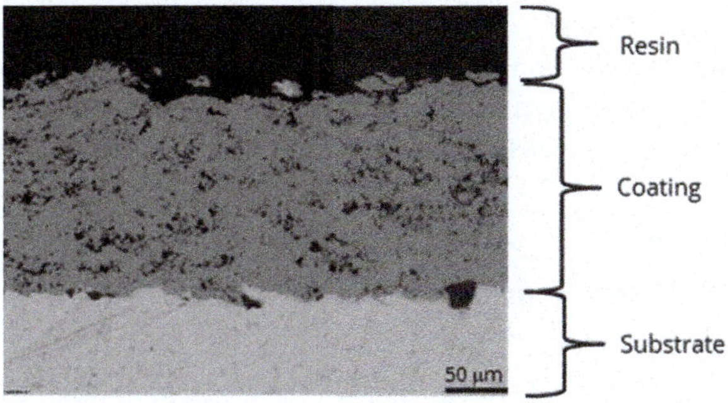

Fig. 3 Optical image of as-sprayed coating

3.3 *Water Contact Angle Analysis*

Figure 4 shows the Water Contact Angle (WCA) of the coating as a function of days after the deposition. The plot shows that the hydrophobic surface did not form immediately after the coating was deposited, and it took around 23 days to reach saturation at approximately 120° WCA. Hydrophobicity of the surface was occurred 15 days after coating deposition. It is noted that the variation of the WCA readings reduced over time. This behavior was not observed in other studies where ceramic coatings, such as Y_2O_3, ZrO_2-Y_2O_3, and Al_2O_3, were deposited by plasma spraying [2]; or when cermet coatings 25NiCr-75Cr_2C_3 [3] and WC-Co were deposited by HVOF spraying [4]. It is also noted that the measurement for the cermet 25NiCr-75Cr_2C_3 was performed using salt water [3].

Further investigation is needed to fully understand the mechanism behind the observed hydrophobicity of the WC-CoCr coating over time. While surface roughness has been suggested as a possible explanation in other studies [2–4], the lack

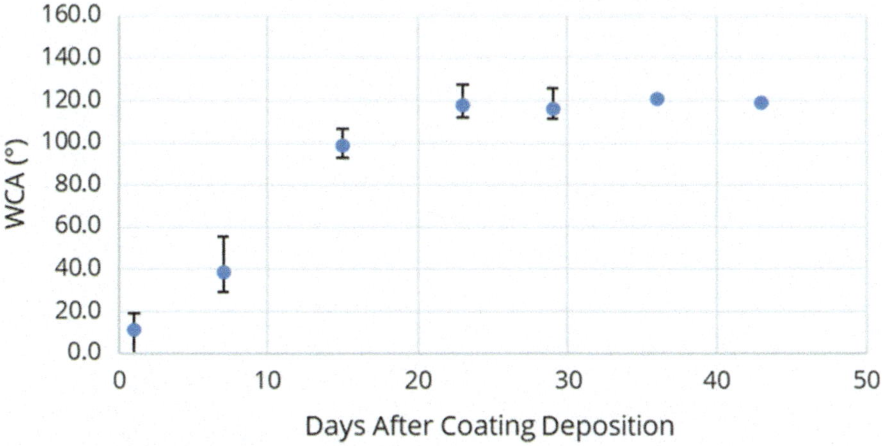

Fig. 4 Water contact angle (WCA) as a function of days after coating deposition

of detailed research on the surface chemistry evolution of thermal sprayed coatings after deposition leaves room for exploration in this area. It is possible that changes in surface chemistry due to exposure to air and other environmental factors may play a role in the observed transition from hydrophilic to hydrophobic behavior.

4 Conclusion

The HVOF-sprayed WC-Co-Cr coating surface was found to transition from hydrophilic to hydrophobic after being exposed to air for over 15 days, as evidenced by the measurement of water contact angle using DI-water. Further studies are required to understand the underlying mechanisms responsible for this transition.

References

1. Pawlowski L (2008) The science and engineering of thermal spray coatings, 2nd Edn. John Wiley & Son Ltd
2. Mazumber A, Alangi N, Sethi A, Prabhu KN, Mukherjee J (2009) Study on wettability of plasma spray coated oxide ceramic for hydrophobicity. Surf Interf 20:100591
3. Amudha A, Nagaraja HS, Shashihala HD (2021) Mechanical and wetting properties of 25%NiCr-75%Cr_2C_3 cermet coated on low carbon steel using HVOF thermal spray technique. Physica B: Phys Condens Matter 602:412409
4. Ghosh G, Sidpara A, Bandyopadhyay PP (2020) Fabrication of mechanically durable slippery surface on HVOF sprayed WC-Co coating. Surf Coat Technol 394:125886
5. Law KY (2014) Definitions for hydrophilicity, hydrophobicity and superhydrophobicity: getting the basic right. J Phys Chemistr Lett 20:686–688

Effect of Laser Peening on Surface Morphology and Deformation Level of Additively Manufactured 316L Stainless Steel

Abeer Mithal, Niroj Maharjan, and Sridhar Idapalapati

Abstract Laser peening (LSP) is a surface enhancement technology that utilizes a short-pulsed laser to improve the fatigue life of components. With the advent of additive manufacturing (AM), there has been an interest in studying LSP of AM parts. In this experimental work, we examine the effect of LSP on the microstructure and surface morphology of AM 316L stainless steel. LSP at moderate and extreme peak power densities of 7.86 and 17.68 GW/cm^2 respectively, was performed on a 316L steel sample fabricated by directed energy deposition (DED). The samples were characterized for their surface morphology and near-surface microstructure using a range of analytical techniques. The results indicate that all LSP conditions had no significant effect on the surface topography or oxide level. When no ablative coating was used, the surface residual stress was tensile whilst with coating the surface residual stress state was compressive. The plastic strain (as measured by EBSD) was not significantly different for all LSP conditions. The use of extremely high peak power density (17.68 GW/cm^2) showed no significant increment in plastic strain, hardness or surface residual stress compared to moderate peak power density (7.86 GW/cm^2) possibly due to the dielectric breakdown of water. The findings indicate that a very high peak power density does not necessarily translate to larger peening effects and may not be required for material processing.

Keywords Laser shock peening · Directed energy deposition · Surface enhancement · Materials characterization · EBSD

A. Mithal (✉) · S. Idapalapati
School of Mechanical and Aerospace Engineering, Nanyang Technological University, 50 Nanyang Avenue, Singapore 639798, Singapore
e-mail: MITH0003@e.ntu.edu.sg

A. Mithal · N. Maharjan
Advanced Remanufacturing and Technology Centre, Agency for Science Technology and Research, 3 Cleantech Loop, Singapore 637143, Singapore

N. Maharjan and W. He (eds.), *Proceedings of the 3rd International Conference on Advanced Surface Enhancement (INCASE) 2023*, Lecture Notes in Mechanical Engineering, https://doi.org/10.1007/978-981-99-8643-9_10

1 Introduction

Plastic deformation-based surface enhancement techniques have long been employed in industry to improve the mechanical performance of engineering components. These technologies rely on near surface plastic deformation to induce work hardening and compressive residual stresses which improve the fatigue strength by delaying crack initiation and retarding crack propagation. With the recent interest in industrialization of additive manufacturing (AM), these techniques have also been applied to AM parts. Several studies have shown the beneficial effects of surface enhancement techniques on the fatigue strength of AM components [1–6].

One such technique is laser shock peening (LSP). Laser shock peening uses a high power, short-pulsed laser to induce a pressure wave in the target material. If the magnitude of the pressure wave is sufficiently high, it induces plastic deformation, leading to the development of residual stresses. LSP is considered an ultra-high strain rate process owing to the extremely small pulse time of the laser (generally of the order ~10^{-9} s). This means that less cold work occurs, and induced stresses are more stable at higher temperatures. The downside of this is that it is difficult to induce sufficient strain energy into the material to achieve recrystallization during thermomechanical processes.

In recent years, a new class of processes has developed where plastic deformation processes are applied to intermediate built layers in AM parts [7–16]. The intrinsic heating from the fusion-based AM processes is sufficient to trigger recrystallization in the deformed volumes and an equiaxed grain structure is achieved. These processes are commonly referred to as hybrid interlayer or coupled processes. To date, most of these processes have used rolling or hammering/forging type tool to induce deformation. The use of a laser would be highly beneficial as it is non-contact and easy to integrate with AM machines as laser beams can be delivered via thin flexible fiber optic cables.

In this work, we investigate the effect that LSP has on the stored strain energy of an AM component. LSP was performed both with and without ablative coating as well as at an extremely high energy density to study how sufficient strain energy can be induced into the target AM material to trigger recrystallization if used as a hybrid interlayer process. We also evaluate the effect that the various LSP conditions had on the surface morphology as past work has indicated this can influence the microstructural evolution in hybrid interlayer processes [15]. The results from this work aim to explain the microstructural and topographical changes that occur as a result of LSP on AM parts as well as guide future development of hybrid interlayer processes using laser shock peening instead of rolling/hammering.

2 Materials and Methods

In this experimental work, a thin block (thickness ~4 mm) of 316L stainless steel was extracted from a larger block fabricated using wire electrical discharge machining (WEDM) as shown in Fig. 1. 316L stainless steel was chosen due to its versatility and applicability to several industries including automotive, marine and biomedical amongst others. The block was built using laser based directed energy deposition (DED-LB) using the parameters listed in Table 1. DED-LB was done in a Lasertec 65 3D machine from DMG Mori. The thin block was then subjected to small area LSP as shown in Fig. 1b. LSP was done at three different process parameters as summarized in Table 2, using a pulsed Nd:YAG laser with 1064 nm wavelength, and 18 ns pulse width. All LSP operations were performed at an overlap of 50% of laser spot diameter.

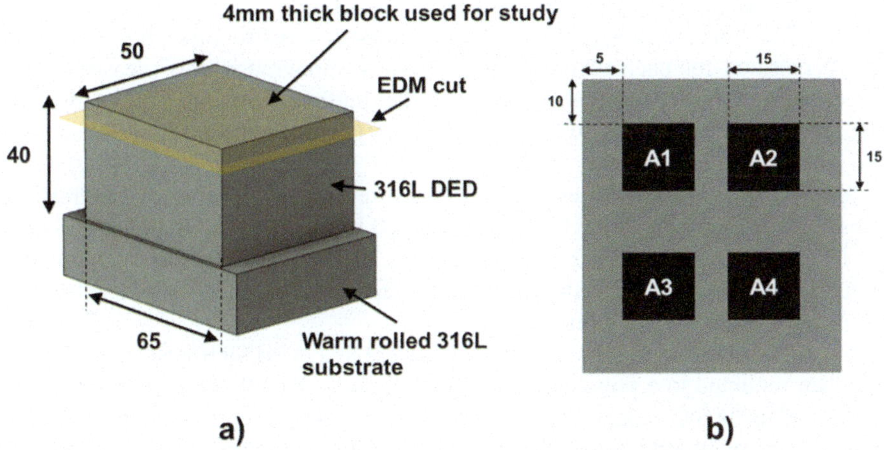

Fig. 1 **a** Schematic of sample build and block extracted by WEDM, **b** details of regions subjected to various LSP parameters. LSP parameters for A1-A4 are summarized in Table 2. All dimensions are in mm

Table 1 Parameters used for DED-LB

Parameter	Value
Laser power	1500 W
Laser spot size	3 mm
Scanning speed	1100 mm/min
Powder flow rate	15.6 g/min
Nominal layer height	1 mm
Hatch distance	1.5 mm
Scanning pattern	Bidirectional raster with 90^0 rotation between layers

Table 2 Parameters used for LSP

Area	Pulse energy (J)	Spot size (mm)	Ablative coating used?	Power density (GW/cm^2)
A1	10	3	N	7.86
A2	10	3	Y	7.86
A3	10	2	N	17.68
A4	Control-no LSP			

It should be noted that for the A3 condition, the peak power density used was higher than the dielectric breakdown limit of water (~10 GW/cm^2). Thus, at peak power densities beyond 10 GW/cm^2 a shielding effect would occur due to loss of confining medium transparency [17, 18]. Unfortunately, due to limitations of the LSP setup used, a power density of 10 GW/cm^2 could not be achieved to verify the dielectric breakdown limit independently. The ablative coating used for the A2 condition was 130 μm thick 3M® 471 black vinyl tape.

After LSP surface topographical and sub-surface microstructural characterization was performed for each of the 4 areas. Surface topography was measured by a Keyence confocal microscope (VK X1000). The surface elemental composition was also studied using a scanning electron microscopy (JEOL 5600) and energy dispersive spectroscopy (EDS).

X-ray diffraction (XRD) was used to measure the macroscopic surface residual stresses of the samples through the $cos\alpha$ method [19] using Pulstec μ-360S system. To increase the number of diffracting grains and thereby improve the counting statistics of the XRD data, the in-plane averaging method [20] was used with 3 measurement areas per condition.

For sub-surface microstructure, the areas were first cross sectioned using WEDM and then mounted in a polymer resin. Metallographic sample preparation was then performed using SiC papers of increasingly finer grit. The samples were then polished with a 3 μm diamond suspension and finally a 0.04 μm colloidal silica solution to obtain a mirror surface finish and finally etched with Kalling's waterless reagent. After sample preparation, the samples were analyzed using optical microscopy (OM), scanning electron microscopy (SEM), electron backscatter diffraction (EBSD) and Vickers hardness indenter to fully characterize the material condition.

EBSD was carried out using a Nordlys detector attached to the SEM mentioned previously, using a step size of 1.5 μm, and the data was analyzed using the ATEX software package. Vickers microhardness was measured using an Innovatest Falcon 500 machine using a load of 200 gf and a dwell time of 10 s.

3 Results and Discussion

3.1 Surface Morphology

Surface topography maps obtained from confocal microscopy are presented in Fig. 2. For A1, A2 and A3, the dimples due to the LSP process are clearly visible (in dark blue) with an associated material pile up around the dimple crater. The dimples for A3 look smaller due to the use of a smaller spot size. No significant differences are observed in surface topography between the three LSPed regions. On the other hand, the control (unpeened) area shows a different morphology with the absence of clear peening dimples. Instead, several microcraters are visible which were formed by the WEDM process (see Fig. 2). 2D roughness parameters (Ra, Rz, Rku, and Rdq) extracted from the confocal microscopy scans are presented in Fig. 3.

Within the experimental scatter, none of the LSP parameters used had any significant influence on the surface topography. Previous work showed that LSP can reduce the slope and kurtosis of an as-deposited DED surface [15]. The results from this study indicate that LSP on a WEDM DED surface (as in this work) does not affect roughness parameters significantly.

The surface morphology was also studied in terms of elemental composition. LSP, DED and WEDM all have a large thermal component in the process which can lead to significant oxidation of the surface. The objective was to understand whether LSP influences the inherent level of oxidation present on a DED surface. Fig. 4a–c shows the EDS spectra from an as-built DED, WEDM (A4) and LSPed (A1) surface respectively. No significant compositional difference was observed between the LSPed and DED surfaces, however the WEDM surface showed Cu and Zn peaks.

The Cu and Zn observed on the WEDM surface is likely contamination from the WEDM process (brass wire with Zn coating was used) as reported in previous work [21–23]. Uncoated LSP i.e LSP without any ablative coating (areas A1 and A3), was able to remove this contamination as can be seen by the significantly

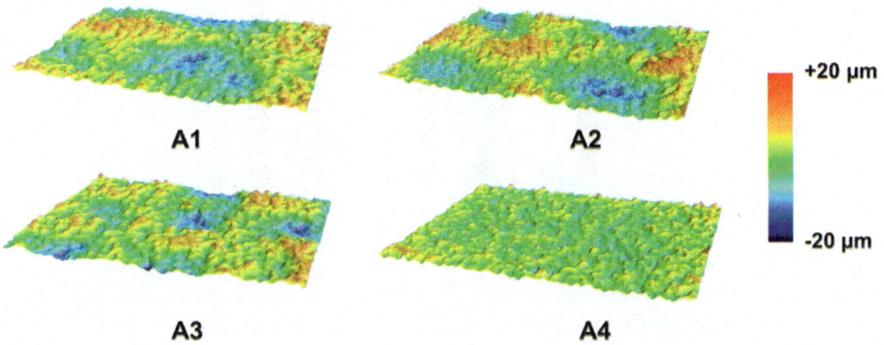

Fig. 2 Surface topography maps for all 3 LSPed areas (A1, A2 and A3) and the control area (A4). Field of view is 3 by 2 mm

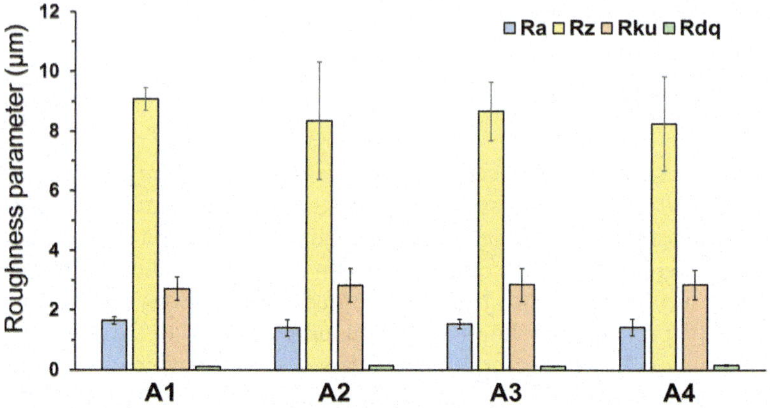

Fig. 3 2D roughness parameters extracted from results surface topography maps. Each reported value is the average of 6 readings taken over two different scan locations

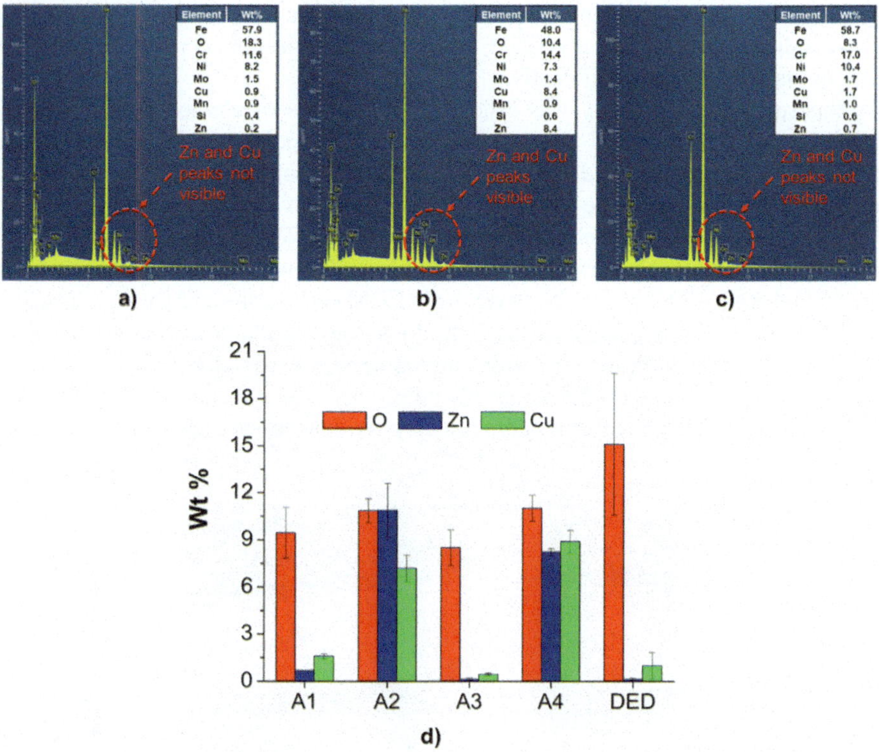

Fig. 4 EDS spectra for **a** DED surface, **b** WEDM surface and **c** LSP (A1) surface. **d** Elemental composition of O, Zn and Cu for various samples. Measurement for each sample taken twice in two different locations

lower concentration of Zn and Cu in areas A1 and A3 in Fig. 4d. The LSP with coating (A2) was not effective in removing the contaminants. This difference can be attributed to the nature of uncoated LSP process. When an ablative coating is used, the coating gets ablated during LSP. The tamping medium then confines the generated plasma, preventing it from escaping which leads to a high-pressure shockwave being generated. On the other hand, when no coating is used, the surface layer of the sample itself gets ablated. In the context of this work, LSP with coating, 'protected' the surface Zn and Cu contaminants from being ablated and led to them being present in the surface even after LSP. These results indicate that uncoated LSP could have interesting applications in surface cleaning whilst simultaneously inducing beneficial LSP effects such as plastic deformation and compressive residual stresses.

No significant difference was observed in the oxidation level of all LSPed areas (A1, A2 and A3). The WEDM area (A4) also showed a similar oxidation level indicating that LSP did not increase or decrease the oxidation level present on the WEDM surface.

It was observed that the as deposited DED surface (prior to WEDM) had a higher oxide level. EDS mapping revealed that the as built DED surface consisted of a thin oxide crust in certain regions whilst other regions were devoid of this crust as shown in Fig. 5. This explains the higher variation in measured oxygen level in Fig. 4d for the DED surface as the oxygen concentration would be strongly dependent on the proportion of oxide crust present in the scanned area. Although not studied in this work, application of LSP directly to a DED surface would likely break away and peel off this oxide crust leading to a reduced surface oxide level. A similar effect has been reported previously for shot peening [24].

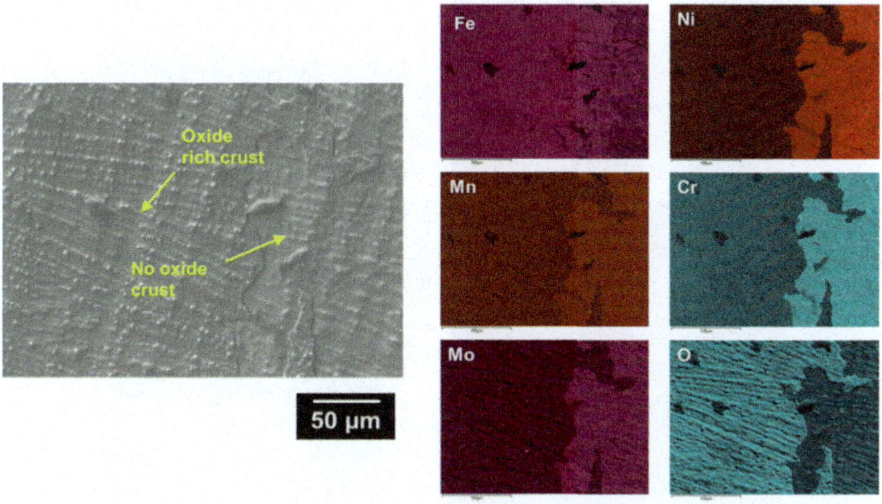

Fig. 5 SEM image and EDS maps of as built DED surface

3.2 Microstructure

The microstructures of the various conditions were analyzed in cross section. Results from optical microscopy are shown in Fig. 6. All observed areas showed sub-grain, dendritic features typical of the rapid solidification conditions in DED-LB. The control area A4 and area A2 showed highly directional dendrites parallel to the build direction. On the other hand, both A1 and A3 areas appear to have reduced directionality of the dendrites which could be indicative of grain refinement and rotation due to plastic deformation.

EBSD analysis was carried out to further understand the grain structure as it was not clearly visible from the OM images. The results are shown in Fig. 7. The top row of Fig. 7 shows IPF maps of all 4 analyzed areas with respect to the build direction while the bottom row shows the kernel average misorientation (KAM) maps. A <100> texture can be seen in the build direction for all four areas which is due to the thermal gradient present during solidification. The texture seems to be reduced and the grains are refined for A1 and A3 areas which agrees well with the OM results. There was no clear difference in the microstructure between the A2 and the control sample (A4). The differences in uncoated and coated LSP could either be due to effect of the mechanical loading only in coated LSP [25] or due to the inherent variation

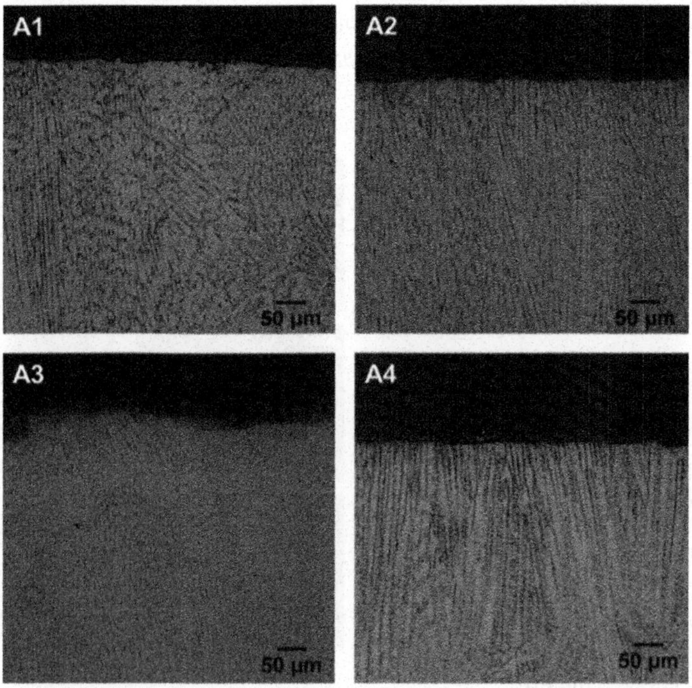

Fig. 6 Optical microscopy images of four different sample sections

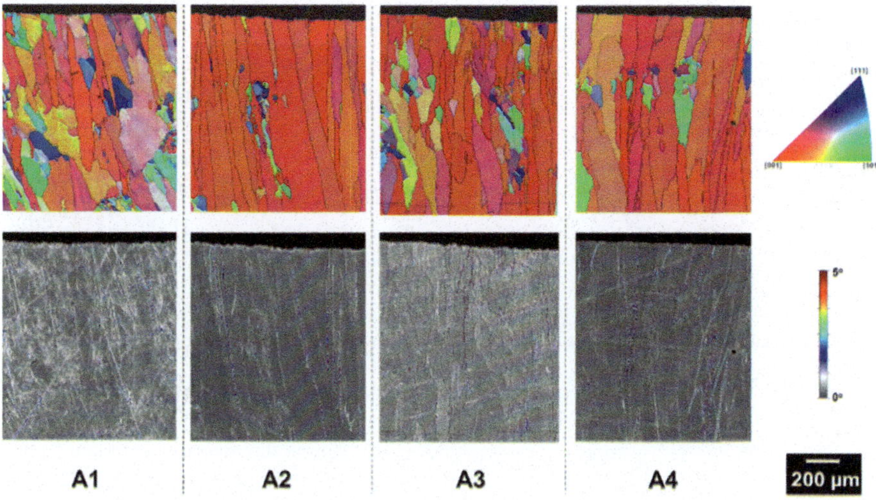

Fig. 7 EBSD results for the various LSP conditions. Top row IPF Y maps, bottom row KAM maps

in the original microstructure produced by DED process itself or a combination of both.

The KAM maps also show varying degrees of misorientation induced into the sub-surface. Interestingly, there was no significant difference in the EBSD results of A1 and A3 areas indicating that use of extremely high energy density had no significant benefit. This is likely due to the dielectric breakdown of water which can significantly shield the target material from the laser radiation. It is possible that between the energy densities used for A1 and A3 (7.86 and 17.68 GW/cm^2 respectively), there lies an optimum power density where deformation level would increase. However, this was not verified in the present study.

The KAM was also analyzed as a function of depth to understand the variation of plastic deformation as shown in Fig. 8a. KAM was increased in the sub-surface regions for both areas subjected to uncoated LSP (A1 and A3). LSP with coating showed almost no difference in KAM as compared to the control.

It should be noted that the KAM gives an approximation of geometrically necessary dislocations (GNDs) and is insensitive to statistically stored dislocations (SSDs). Thus, the KAM results may only reveal part of the overall deformation profile. It is possible that coated LSP (A2) induced a higher number of SSDs than A1 and A3 leading to equal or even greater amounts of plastic strain than that revealed from the KAM maps. This is reinforced by the fact that a similar increment in near surface hardness was recorded for A2 as A1 and A3 (Fig. 8b). This suggests that the hardness increment in both cases could be attributed to the increased plastic deformation and the introduction of compressive residual stresses beneath the surface (Fig. 2). Precise quantification of the effects of both are outside the scope of this work.

Fig. 8c shows the surface residual stresses (RS). The control sample (A4) shows a high surface tensile RS of 500 MPa due to rapid resolidification from the WEDM

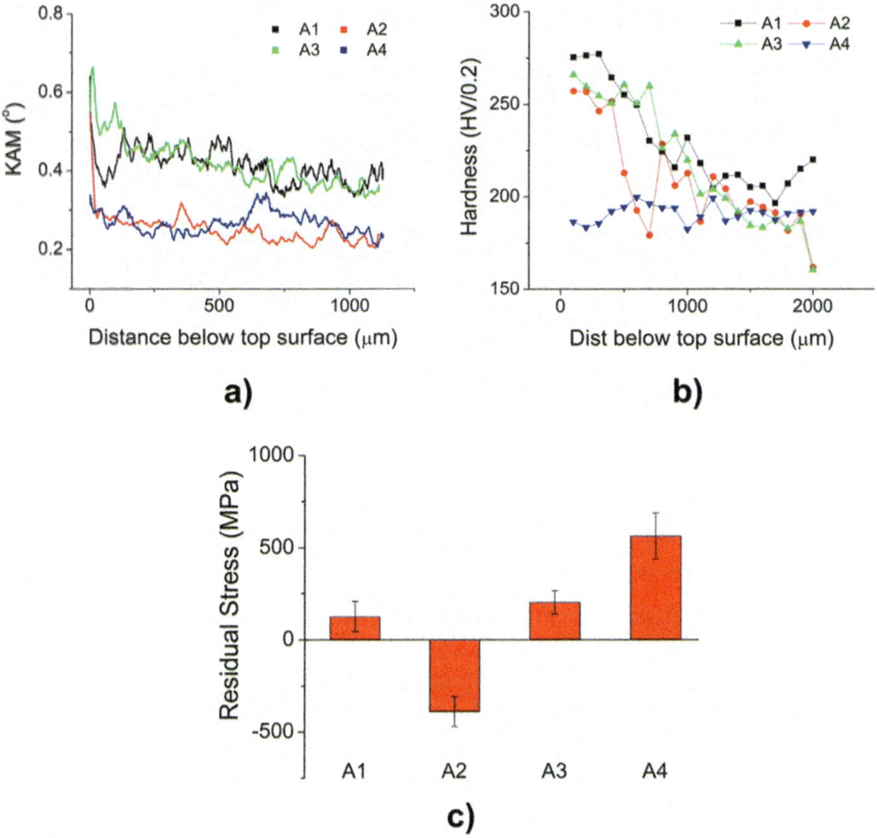

Fig. 8 Variation of **a** KAM, **b** micro-hardness as a function of distance from the top surface, and **c** surface residual stresses measured by XRD

process [26]. LSP with coating (A2) resulted in compressive RS of 480 MPa in the surface whilst uncoated LSP produced tensile RS with magnitude almost 1/4th that of the control sample. This is line with previous work which showed absence of an ablative layer tends to create a thin layer of tensile residual stress due to thermal effects [27].

It is noteworthy that despite a 2.24 times higher peak power density area A3 showed no significant difference in plastic deformation, microhardness, or surface RS from area A1. This can be explained by the dielectric breakdown limit of water which is generally considered to be around 10 GW/cm^2. It is unclear whether use of incident power density of 17.68 GW/cm^2 resulted in an effective power density of 10 GW/cm^2 (the saturation limit) or was even lower due to increased shielding effects. It is possible that the similar deformation levels between A1 and A3 are on either side of a local maxima, and use of an intermediate power density (such as 10 GW/cm^2) would result in a higher deformation level. Unfortunately, this was not verified in the present work due to setup limitations and will be considered in subsequent studies.

4 Conclusion

The objective of this work was to analyze the effect of LSP on the microstructure and surface topography of DED-LB 316L stainless steel. The following conclusions were drawn:

(i) The process parameters for LSP used in this study had no significant effect on surface topography or surface oxidation level. This highlights the potential of LSP to induce near surface compressive stresses without damaging surface quality.
(ii) Uncoated LSP (LSP without ablative coating) was able to clean the surface and remove impurities induced during the WEDM process, while LSP with coating did not remove the impurities. The removal of impurities during uncoated LSP was ascribed to ablation.
(iii) LSP increased microhardness in the near surface region due to a combination of increased dislocation density (strain hardening) and residual stresses. Surface residual stress was compressive only when ablative coating was used, however hardness measurements were quite similar for all LSP conditions investigated.
(iv) When laser power density was increased from 7.86 to 17.68 GW/cm^2, no significant difference was observed in terms of hardness, GND density or residual stresses. Due to the dielectric breakdown of water, use of extremely high power density did not yield any benefit as compared to a more moderate power density.

Acknowledgements This work was supported by an A*STAR RIE2020 advanced manufacturing and engineering (AME) programmatic grant through the structural metal alloys program (SMAP, Grant No. A18B1b0061). AM acknowledges the support from A*STAR in the form of a PhD scholarship.

References

1. Carlos Navarro P et al (2020) Effect of surface treatment on the fatigue strength of additive manufactured Ti6Al4V alloy. Frattura ed Integrità Strutturale 14(53):337–344
2. Jin X et al (2020) Effects of laser shock peening on microstructure and fatigue behavior of Ti–6Al–4V alloy fabricated via electron beam melting. Mater Sci Eng A 780
3. Soyama H, Takeo F (2020) Effect of various peening methods on the fatigue properties of titanium alloy Ti6Al4V manufactured by direct metal laser sintering and electron beam melting. *Materials (Basel)* 13(10)
4. Aguado-Montero S et al (2022) Fatigue behaviour of PBF additive manufactured TI6AL4V alloy after shot and laser peening. Int J Fat 154

5. DiCecco LA, Mehdi M, Edrisy A (2021) Fatigue improvement of additive manufactured Ti–TiB material through shot peening. Metals 11(9)
6. Maleki E et al (2022) Fatigue behaviour of notched laser powder bed fusion AlSi10Mg after thermal and mechanical surface post-processing. Mater Sci Eng A 829
7. Gao S et al (2022) A hybrid directed energy deposition process to manipulate microstructure and properties of austenitic stainless steel. Mater Design 213
8. Martina F et al (2015) Microstructure of interpass rolled wire + arc additive manufacturing Ti-6Al-4V components. Metall Mater Trans A 46(12):6103–6118
9. Li C et al (2021) Hierarchical layered and refined grain structure of Inconel 718 superalloy produced by rolling-assisted directed energy deposition. Addit Manufact Lett 1
10. Kan WH et al Effect of in-situ layer-by-layer rolling on the microstructure, mechanical properties, and corrosion resistance of a directed energy deposited 316L stainless steel. Addit Manufact 55
11. Wang Y, Shi J (2019) Microstructure and properties of Inconel 718 fabricated by directed energy deposition with in-situ ultrasonic impact peening. Metall and Mater Trans B 50(6):2815–2827
12. Hönnige JR et al (2020) The effectiveness of grain refinement by machine hammer peening in high deposition rate wire-arc AM Ti-6Al-4V. Metall Mater Trans A 51(7):3692–3703
13. Farias FWC et al (2023) In situ interlayer hot forging arc-based directed energy deposition of Inconel® 625: process development and microstructure effects. Addit Manufact 66
14. Wang Y, Shi J (2020) Recrystallization behavior and tensile properties of laser metal deposited Inconel 718 upon in-situ ultrasonic impact peening and heat treatment. Mater Sci Eng A 786
15. Mithal A, Maharjan N, Idapalapati S (2023) Enhanced cooling rates in laser directed energy deposition with interlayer peening. Rapid Prototyp J 29(6):1289–1298
16. Duarte VR et al (2020) Hot forging wire and arc additive manufacturing (HF-WAAM). Addit Manufact 35
17. Peyre P et al (1998) Experimental study of laser-driven shock waves in stainless steels. J Appl Phys 84(11):5985–5992
18. Berthe L et al (1996) Laser shock processing of materials: study of laser-induced breakdown in water confinement regime. Lasers Opt Vis Product Manufact I 2789:SPIE
19. Tanaka K (2019) The cosα method for X-ray residual stress measurement using two-dimensional detector. Mech Eng Rev 6(1):18–00378–18–00378
20. Miyazaki T, Fujimoto Y, Sasaki T (2016) Improvement in X-ray stress measurement using Debye-Scherrer rings by in-plane averaging. J Appl Crystallogr 49(1):241–249
21. Bleys P et al (2006) Surface and sub-surface quality of steel after EDM. Adv Eng Mater 8(1–2):15–25
22. Manupati VK et al (2019) Investigation of copper and zinc contamination on the work piece surface with WEDM. In: Innovation, Engineering and Entrepreneurship. Springer International Publishing, Cham
23. Fan H et al (2023) Effects of substrate surface treatments on hybrid manufacturing of AlSi7Mg using die casting and selective laser melting. J Mater Sci Technol 156:142–156
24. Zhang X, Li D, Geng J (2021) An approach to reduce stress and defects: a hybrid process of laser cladding deposition and shot peening. Rapid Prototyp J 27(9):1609–1616
25. Ramesh T, Maharjan N (2022) Effect of ablative layer thickness on laser shock peening process intensity. In: Proceedings of the 2nd international conference on advanced surface enhancement (INCASE 2021). Springer Singapore, Singapore
26. Sidhom H et al (2013) Effect of electro discharge machining (EDM) on the AISI316L SS white layer microstructure and corrosion resistance. Int J Adv Manufact Technol 65(1):141–153
27. Gujba AK, Medraj M (2014) Laser peening process and its impact on materials properties in comparison with shot peening and ultrasonic impact peening. Materials 7(12):7925–7974

Complementary Effect of Metal Shot Peening Over Deep Cold Rolled Ti-6Al-4 V Surface as a Two-Step Mechanical Surface Treatment

Henry Kuo Feng Cheng, Wai Luen Chan, Nazihah Shariff, and Augustine Teo

Abstract A two-step process of the mechanical surface treatments (MST) on Ti-6Al-4 V using a deep cold rolling (DCR) followed by a metal shot peening (MSP) was studied to understand the complementary effect of MSP over DCR. DCR was used for the introduction of higher and deeper compressive residual stress on the Ti-6Al-4 V coupons while MSP was used for the further improvement of the compressive residual stress near the surface of the coupons. The study shows that the two-step process reduces the anisotropy of Ti-6Al-4 V in both surface roughness and compressive residual stress compared the standard single step DCR process. Furthermore, the two-step process improves the near-surface compressive residual stress owing to the secondary MSP process without compromising the depth of influence by the first DCR treatment.

Keywords Ti-6Al-4 V · Deep cold rolling · Metal shot peening · Anisotropy · Compressive residual stress · Surface roughness

1 Introduction

Deep cold rolling (DCR) process is very similar to the roller burnishing process, in which a hydrostatically controlled ball is employed to induce surface and subsurface property enhancement. The key process parameters are rolling pressure, ball diameter, number of passes and the overlap between the two adjacent rolling tracks. DCR can produce smoother surfaces, deeper compressive stresses (in some cases up to

H. K. Feng Cheng (✉) · W. Luen Chan · A. Teo
Advanced Remanufacturing & Technology Centre (ARTC), Agency for Science, Technology and Research (A*STAR), 3 Cleantech Loop, #01-01 CleanTech Two, Singapore 637143, Singapore
e-mail: henry_cheng@artc.a-star.edu.sg

N. Shariff
School of Mechanical and Aerospace Engineering, Nanyang Technological University, Singapore, Singapore

N. Maharjan and W. He (eds.), *Proceedings of the 3rd International Conference on Advanced Surface Enhancement (INCASE) 2023*, Lecture Notes in Mechanical Engineering, https://doi.org/10.1007/978-981-99-8643-9_11

2 mm below the surface) and a moderate cold work than shot peening process, while restricting both the crack nucleation and growth [1–3]. DCR typically produces subsurface residual stresses, in accordance with Hertzian theory. Increasing the rolling force enhances the compressive residual stresses until they reach a saturation point, determined by the material's yield strength or work hardening state. However, excessive rolling force causes the compression zone to shift deeper, resulting in subsurface compressive residual stresses and tensile residual stresses on the surface [1].

On the other hand, metal shot peening (MSP) is a mechanical surface treatment process used to improve the mechanical performance of metal components. It is commonly used in industries such as aerospace, automotive, and manufacturing to increase the lifespan and reliability of metal components subjected to cyclic loading and harsh environments. MSP involves bombarding the surface of a metal component with small, spherical pellets called shots. The shot particles create controlled plastic deformation on the surface, inducing compressive residual stresses and improving fatigue strength, stress corrosion resistance, and wear resistance. The compressive residual stress is controlled by the process parameters, including shot size, shot velocity and peening coverage. Compared to DCR, the depth of compressive residual stress induced by MSP is significantly shallower, typically less than 1 mm below the surface [4].

Therefore, in this work, DCR was applied for the introduction of higher and deeper compressive residual stress on the subsurface while MSP was used for a further improvement of the compressive residual stress near the surface of the coupons. The effects of the individual processes on Ti-6Al-4V coupons were first evaluated. Then, the complementary effects of MSP on DCR treated coupons were reported based on the surface roughness and hardness together with their related compressive residual stress.

2 Methodology

The material used for the study was commercially available Ti-6Al-4V (ASTM B265-95 Grade-5) coupons with the dimensions of 56 × 18 × 8mm (L × W × H). The coupons were machined, polished and heat treated prior to the experiments. The purpose of machining and polishing was to make the surface roughness of the coupons below 5 μm. The coupons were heat-treated at 550 °C for 2 h to release their pre-existing residual stresses. Then, they were cooled to room temperature in nitrogen [5].

DCR process was conducted on HAAS VF-2 CNC machine. A 6 mm diameter hydrostatic ball (ECOROLL HG6) was used. The rolling parameters were selected as 50% overlap and two rolling pressures of 250 and 400 bars, respectively. MSP treatments were conducted after DCR treatments using a robotic shot peening machine manufactured by Abrasive Engineering Pte. Ltd. The robotic arm was equipped with a straight nozzle with a diameter of 14 mm at an impingement angle of 70° and a standoff distance of 120 mm. The intensity of 0.176 mmA and 98% coverage were selected. An AMS 2431/1 certified ASR110 steel shots, were used. The shots were made of cast steel with hardness between 45 and 52 HRC.

Surface hardness was measured in Vickers' hardness using an Innovatest's Falcon 500. Surface roughness measurements were conducted using a Taylor Hobson's Form Talysurf. Both surface roughness and hardness were measured at five different locations on each specimen and the average values are reported. The surface hardness measurements were used a Vickers micro-hardness tester with 0.5 N load. Residual stresses were evaluated by a Stresstech's Prism® system, an advanced hole drilling technique.

3 Results and Discussion

3.1 Surface Hardness

Figure 1 shows surface hardness of the coupons before and after different treatments. Mechanical surface treatment processes, such as DCR and MSP, result in work hardening of the surface layer of the material after the treatments. This work hardened layer increases the material's hardness. Furthermore, both DCR and MSP can also cause grain refinement together with generation of compressive residual stress due to severe plastic deformation at the surface [3, 6].

After MSP, the surface hardness of the coupons increases from 415 to 440 Hv. For DCR, the surface hardness of the coupons treated at rolling pressures of 250 bars and 400 bars are 420 and 480 Hv, respectively. Rolling with a same ball diameter, the higher the rolling pressure, the higher the work hardening. At a higher rolling pressure, the higher amount of plastic deformation is generated if the ball diameter is constant. Therefore, the coupon rolled at 400 bars shows a higher surface hardness.

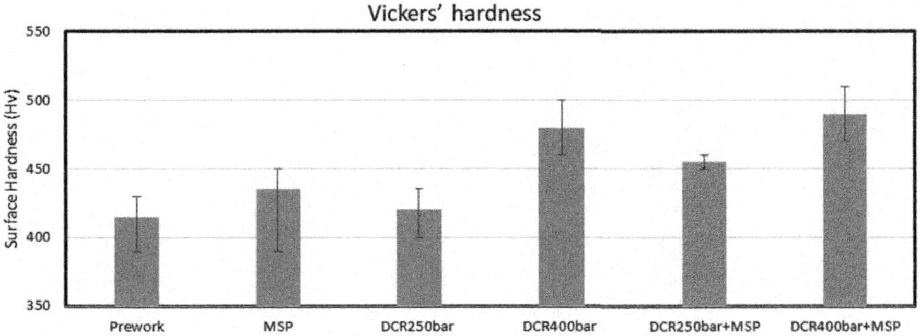

Fig. 1 Comparison of surface hardness for prework and different surface enhancement processes

MSP treatments were conducted on the coupons after treated with DCR to study the complementary effects of MSP process on DCR treated coupons. For DCR250bar+MSP, there is a significant increase in the hardness after MSP treatment. However, for DCR400bar+MSP, a little increase in hardness is observed. For DCR, increasing the rolling pressure would increase the surface hardness until they reach a saturation point, determined by the material's yield strength or work hardening state [1]. For DCR400bar+MSP, the surface work hardening state might be very close to its saturation point and therefore, the addition treatment with MSP may not be able to increase its hardness value further.

3.2 *Surface Roughness*

Figure 2 summarises the surface roughness of the coupons before and after different treatment. The surface roughness values were measured along the length (longitudinal) and across the length (transverse) of the coupons. Due to the machining and polishing processes prior to the experiments, the surface roughness values for prework coupons were as low as 0.41 μm for longitudinal direction and 0.2 μm for transverse direction, respectively.

For MSP process, due to bombarding the surface of a metal part with small spherical pellets, known as shot, under high velocity, the shot impacts the surface, creating numerous small indentations or dimples. As a result, surface roughness increases significantly compared to the polished surface of the prework coupons. Furthermore, after metal shot peening, the surface roughness for both directions show similar since, under the same shot size and peening intensity, surface roughness would be comparable [7].

On the other hand, deep cold rolling tends to reduce the surface roughness of the coupons especially if the initial surface roughness of the coupons is relatively high. DCR typically gives an anisotropic roughness characteristic in which the transverse roughness increases, and the longitudinal roughness decreases significantly [8].

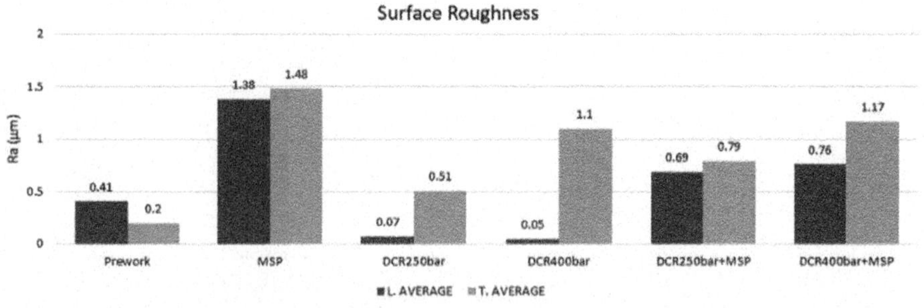

Fig. 2 Comparison of surface roughness for prework and different surface enhancement processes—where L represents the longitudinal direction and T for the transverse direction

Furthermore, typical surface topography by DCR is characterized by a plasticized zone. At the same ball size and overlap, increasing rolling pressure would increase the plasticized zone at the surface [9]. Therefore, the anisotropic roughness characteristic of DCR was more dominant at the rolling pressure of 400 bars than at 250 bars. However, the final surface roughness of DCR treated coupons were lower than those treated by MSP.

Finally, the second process of MSP on the DCR treated coupons would reduce the anisotropic roughness characteristic of DCR. The complementary effect of MSP was more significant on the coupons treated at 250 bars of rolling pressure than those at 400 bars. However, both coupons show higher roughness than their respective DCR treated coupons but lower than MSP treated coupons.

3.3 Compressive Residual Stress

Residual stresses of the test coupons are shown in Figure 3. Due to the heat treatment, the prework coupons shows very small magnitude of residual stresses. A significant amount of compressive residual stress was induced after MSP. The depth of compressive residual stress of MSP treated coupon is up to 0.2 mm.

However, for DCR, the amount of compressive residual stresses and their depth of influence are significantly higher and deeper than those of MSP coupons. This is because DCR can induce deeper compressive residual stresses compared with MSP. Furthermore, the depth of influence for DCR coupons treated at 250 and 400 bars are 0.5 and 0.7 mm, respectively. Both coupons show a similar range of maximum compressive stresses around −1000 MPa. However, the coupons rolled at 400 bars show lower near-surface compressive residual stresses compared to those rolled at 250 bars of rolling pressure.

According to Altenberger [1], for DCR, once a certain threshold of compressive residual stress is reached, increasing the rolling force no longer enhances the

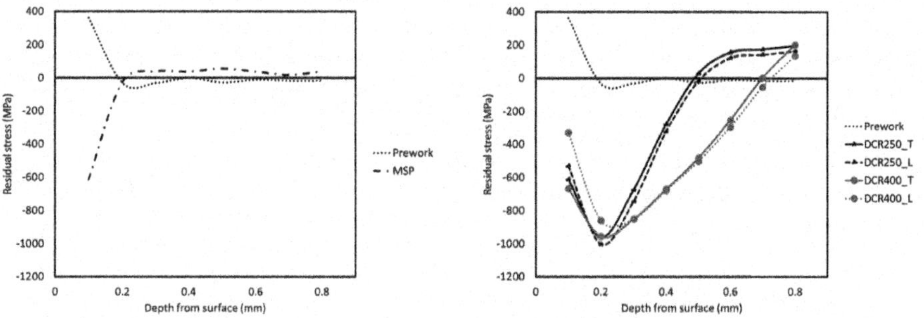

(a) MSP (b) DCR (T-transverse and L-longitudinal)

Fig. 3 Compressive residual stress profiles for prework and different surface enhancement processes

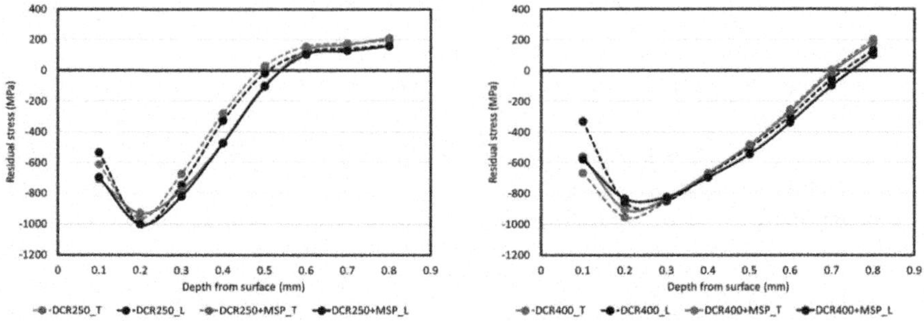

(a) DCR250bar and DCR250bar+MSP (b) DCR400bar and DCR400bar+MSP

Fig. 4 Compressive residual stress profiles for single DCR (dotted) and two-step (solid) processes

maximum (saturation) compressive residual stress. Instead, it pushes the formation of residual stresses deeper into the material. Increasing the rolling force initially increases the compressive residual stresses until a saturated level is reached, which is determined by the material's yield strength or work hardening state. Further increasing the rolling force shifts the compression area to greater depths, and very high forces can lead to subsurface compressive residual stresses but result in tensile residual stresses at the surface.

Figure 4 shows the complementary effects of MSP process on the compressive residual stresses of DCR treated coupons. At 250 bars, MSP can improve the compressive residual stresses of DCR treated coupons. Using MSP as a secondary process, the near-surface the compressive residual stresses as well as their depth were significantly improved although there was no improvement in the maximum compressive residual stresses in both their magnitude and location. This is because once the compressive residual stresses reach to its saturated level, the effect of surface enhancement shifts the compression area into greater depths, in which the saturation is determined by the material's yield strength [1].

However, at 400 bars of rolling pressure, the effects of the secondary MSP process would be more dominant at the near-surface compressive residual stresses. The more tensile residual stresses at the near surface induced by DCR at the longitudinal direction were improved after MSP. Besides, the anisotropy between DCR induced residual stresses was significantly reduced after MSP without compromising much in their magnitude. Therefore, MSP could complement DCR as a secondary process to further improve the compressive residual stresses of the component which would leads to more uniform residual stress profiles in both near-surface and sub-surface properties.

4 Conclusion

The complementary effect of metal shot peening over deep cold rolled Ti-6Al-4 V surface was investigated as a two-step mechanical surface treatment. The two-step process showed a significant improvement the near-surface compressive residual stress owing to the secondary MSP process without compromising the depth of influence by the first DCR treatment. MSP process could complement DCR as a secondary process to further improve the compressive residual stresses of the component in which the DCR-induced anisotropy between compressive residual stresses was significantly reduced without compromising on the state of work hardening at the surface of Ti-6Al-4 V coupons. For a future work, the microstructure comparison between the traditional single process and the two-step process should be studied in detail to further understand the complementary effect of metal shot peening over deep cold rolling.

References

1. Altenberger I (2005) Deep rolling—the past, the present and the future. In: Proceedings of 9th international conference on shot peening, pp 6–9
2. Mader S, Klocke F (2005) Fundamentals of the deep rolling of compressor blades for turbo aircraft enginees, modelling and simulation. In: Proceedings of 9th international conference on shot peening, pp 125–130
3. Nalla RK, Altenberger I, Noster U, Liu GY, Scholtes B, Ritchie RO (2003) On the influence of mechanical surface treatments—deep rolling and laser shock peening—on the fatigue behavior of Ti-6Al-4V at ambient and elevated temperatures. Mater Sci Eng, A 355(1–2):216–230
4. Ezeilo A, Webster G, Webster PJ, Webster PS (1993) Comparison of shot peening residual stress distributions in a selection of materials. Shot Peener 079:275–282. Retrieved from https://www.shotpeener.com/library/pdf/1993079.pdf. Accessed on May 2023
5. Tsuji N, Tanaka S, Takasugi T (2007) Characterization of surface modified Ti-6Al-4V alloy by combination of plasma-carburizing and cold-rolling. In: Ninomi M, Akiyama S, Ikeda M, Hagiwara M, Maruyama K (eds)Ti-2007 science and technology, The Japan Institute of Metals (2007), pp 1663–1666. Retrieved from https://cdn.ymaws.com/titanium.org/resource/resmgr/ZZ-WCTP2007-VOL2/2007_Vol_2_Pres_190.pdf. Accessed on May 2023
6. Liu H, Wei Y, Tan CKI, Ardi DT, Tan DCC, Lee CJJ (2020) XRD and EBSD studies of severe shot peening induced martensite transformation and grain refinements in austenitic stainless steel. Mater Charact 168:110574
7. Yao CF, Dou XT, Wu D, Zhou Z, Zhang JY (2016) Surface integrity and fatigue analysis of shot-peening for 7055 aluminum alloy under different high-speed milling conditions. Adv Mech Eng 8(10):1–10. Retrieved from https://www.researchgate.net/publication/309323079_Surface_integrity_and_fatigue_analysis_of_shot-peening_for_7055_aluminum_alloy_under_different_high-speed_milling_conditions. Accessed on May 2023
8. Cheng HKF, Chan R, Hu J, Chan WL (2022) Deep cold rolling (DCR) as potential surface enhancement process for mold inserts and cores. In: Wei Y, Chng S (eds) Proceedings of the 2nd international conference on advanced surface enhancement (INCASE 2021), lecture notes in mechanical engineering. Springer, Singapore. https://doi.org/10.1007/978-981-16-5763-4_8

9. Abrao AM, Denkena B, Kohler J, Breidenstein B, Morke T (2023) The influence of deep rolling on the surface integrity of AISI 1060 high carbon steel. Procedia CIRP 13:31–36. Retrieved from https://www.researchgate.net/publication/262491558_The_Influence_of_Deep_Rolling_on_the_Surface_Integrity_of_AISI_1060_High_Carbon_Steel. Accessed on May 2023

Femtosecond Laser Interference Micro-patterning on Mold Surface to Improve Hydrophobicity for Microfluid Applications

Kaidong Ye, Xincai Wang, Jiangfeng Hu, and Zhenfeng Wang

Abstract Fluidic flow in micro-channels is driven by capillary filling actuation mechanism. Therefore, surface hydrophobicity plays a significant role in controlling the fluidic flow. Surface hydrophobicity can be tuned by depositing hydrophobic material onto the surface, or by creating surface micro-/nano-textures that follow Cassie-Baxter model (hydrophobic). To facilitate fluidic flow in a microchannel, it is critical to design and create superhydrophobic surfaces with high contact angle and low contact angle hysteresis, which require accurate control of two-scale micro-/nanostructures. In this study, we propose to use laser interference to create deterministic micro- and or nano-level patterns on a metal mold surface. Femtosecond laser is chosen due to its high machining precision and minimum heat-affected zone. The surface structures are characterized by optical and scanning electron microscopy. The surface wettability is analyzed by measuring surface water contact angles. Through hot embossing, the surface structures are replicated onto surfaces of polymer microfluidic channels. The proposed method is shown to be effective and easy to implement for microfluidic applications.

Keywords Laser interference patterning · Femtosecond interference · Microfluidics

K. Ye (✉) · X. Wang · J. Hu
Advanced Remanufacturing and Technology Centre (ARTC), A*STAR, 3 CleanTech Loop, CleanTech Two, #01/01, Singapore 637143, Singapore
e-mail: yek@artc.a-star.edu.sg

Z. Wang
Singapore Institute of Manufacturing Technology Singapore (SIMTech), A*STAR, 2 Fusionopolis Way, #08-04, Innovis, Singapore 138634, Singapore

N. Maharjan and W. He (eds.), *Proceedings of the 3rd International Conference on Advanced Surface Enhancement (INCASE) 2023*, Lecture Notes in Mechanical Engineering, https://doi.org/10.1007/978-981-99-8643-9_12

1 Introduction

The fabrication of periodic structures at micron and submicron length scales provides surface-enhanced properties. For surface process that is down to micron level and below that, we need to rely on laser interference method.

Laser-induced periodic surface structures (LIPSS) is a nanofabrication technique that has received much attention. The principle of operation is simple: when two laser beams overlap in time and space, an interference pattern is generated that can be used to create surface gratings. Its setup is simple, only split the laser beam into 2 or more and let them to re-join again on the material surface to affect the process. The optical parts used are common and low cost. Depending on the pulse energy, multiple or even single laser pulses can create patterns on material surface, therefore it is a high-efficient process. It can fabricate periodically patterned micro/nanometer-scale structures in large-area and in air conditions.

The investigation and research of laser interference process have two major categories. One is using nano-second pulse laser to do the interference process. German scientists had used laser with high pulse energy (several hundred milli-joules to several joules per pulse) to achieve fast and large area processing [1–3].

The other category of laser interference process is to use ultra-fast (femto-second) laser. Femto-second lasers are powerful tools for these processes, owing to their considerable advantage over long-pulse duration lasers. Due to their extremely short pulse duration, a very high peak power is achieved leading to intense non-linear effects. As a result, almost any type of material can be processed, including polymers and metals, without undesirable collateral thermal effects [4, 5].

An important application of surface modification is on microfluidic applications. Microfluidic devices, like debubbler, capillary valve, need to use hydrophobic surface. Traditional way of fabrication polymer devices is by means of mechanical ways like milling, drilling etc. or standard semiconductor manufacturing methods, like lithography and ion etching. Popular processing are direct printing (hydrophobic material on device surface) and micro-patterning (expensive, multi-step and multi-component assembly). They are all time consuming and costly.

Currently industry use imprint technology to process metals for hot embossing, roll-2-roll printing and inject molding applications.

In this paper, we have implemented the laser interference patterning with femto-second lasers. Interference patterns have been created on the surfaces of metals like Stainless Steel and Nickel, to investigate its application on microfluidics. This method has not been carried out by others. A brand-new way of adjusting two laser pulses in the time domain is also reported.

2 Experimental Setup

In our experiments, a commercial femtosecond laser (Quantronix IntegraC-1.5W) was used. The system generates pulses at a wavelength of 795 nm with pulse duration of 130 fs and a repetition rate of 1 kHz. The maximum output power of the laser is around 1 W.

The setup of the experiment is shown in Fig. 1. Femto-second laser beam was split into 2 by the beam splitter. Beam 1 was reflected by mirror 1 and irradiated on to the sample after being focused by a lens. Beam 2 was reflected by mirror 2, 3 and 4 and irradiated at the same position at the sample surface. The two beams will create interference patterns on the sample. Mirrors 2 and 3 were fixed on top of a manual stage. This is to enable us to change the optical path of beam 2 to implement time coincident of the two laser beams.

Femto-second laser interference process has a big challenge on the interference application. The 130-fs laser pulse has a length of 39 μm in the space domain. Therefore, it needs almost identical length for both beam paths 1 and 2 to let the two pulses to meet on the sample surface at the same spot and at the same time. If they cannot meet both in time domain and space domain, an interference pattern will not be formed.

Though other researchers had found out some ways to do the alignment [4–6], they are either need expensive equipment or not implementable in our lab. We need to find our own way to solve the alignment difficulty.

We decided to use a photonic crystal, Cadmium Sulfide (CdS) to help with the alignment. CdS is a photonic material which is widely used in sensor applications.

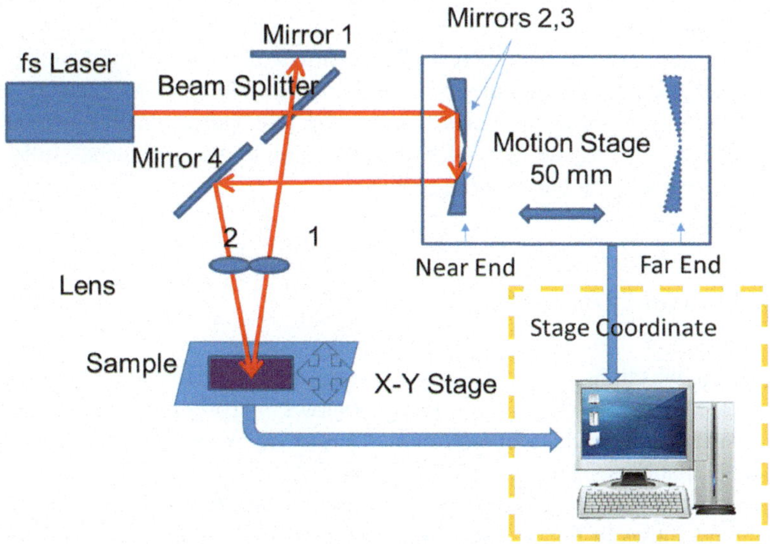

Fig. 1 Experimental setup of the laser interference patterning process

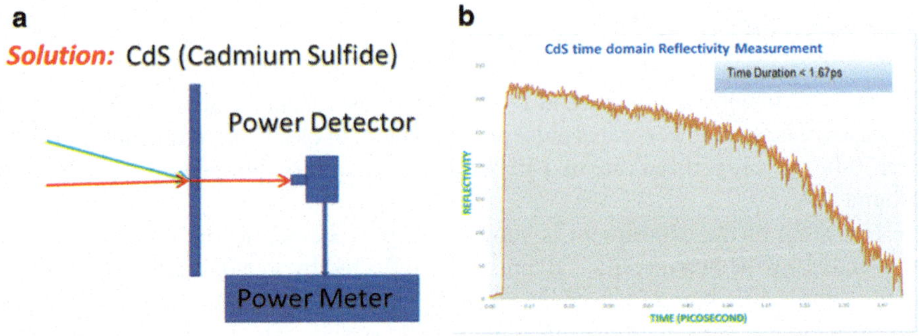

Fig. 2 **a** Experimental setup to verify the time coincidence of two laser pulses, **b** CdS reflectivity change with time

It has almost immediate time response to ultrafast laser pulse. When a laser pulse arrived at the CdS surface, its reflectivity will reach its maximum value instantly, and decrease sharply within 2 picoseconds as we measured and showed in Fig. 2b. If at this moment a second laser pulse arrives, its energy which can penetrate through the CdS crystal reduces to a minimum value due to the reflectivity change by the first pulse. Therefore, we can use this phenomenon to detect the time coincident of the two femto-second laser pulses.

Figure 2a shows the method we used to verify the time coincidence of two femto-second laser pulses. A slice of photonic material CdS was fixed on the position of interest. When the two femto-second laser pulses arrive at the same time and at the same position on the CdS, its reflectivity will increase drastically, causing the transmitting light power to drop accordingly. Therefore, the reading on the power meter will also drop.

When we adjust the manual stage back and forth to change the length of the optical path 2, we also keep an eye on the readings from the power meter. When the reading reaches a minimum value, we know that the two laser pulses reach space and time coincidence. Replacing the CdS slice with the sample, we can successfully create interference patterns on the material surface.

3 Results and Discussion

After solving the alignment problem, we have successfully created patterns on Stainless Steel surface and Nickel surface. Figure 3 shows microscopic image of the pattern created on stainless steel surface. The laser power is 1W and it only needs 5 pulses to create the patterns. The feature size is around 2 μm.

To create large area pattern using laser interference method, we can use the scanning method shows in Fig. 4. After one laser exposure, we move the sample to the next position to expose the adjacent position and continue the process.

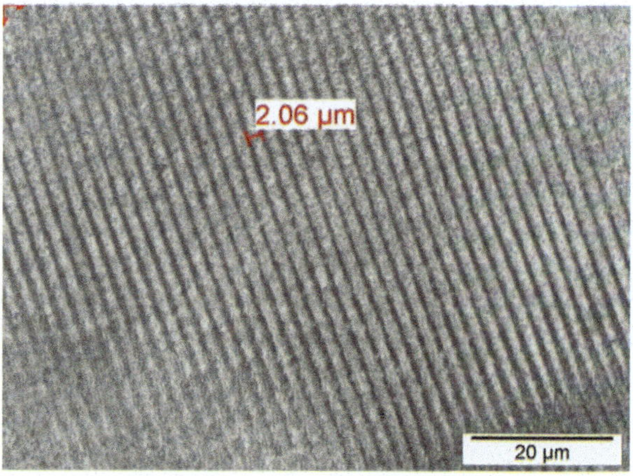

Fig. 3 Microscopic image of the interference pattern created on stainless steel surface

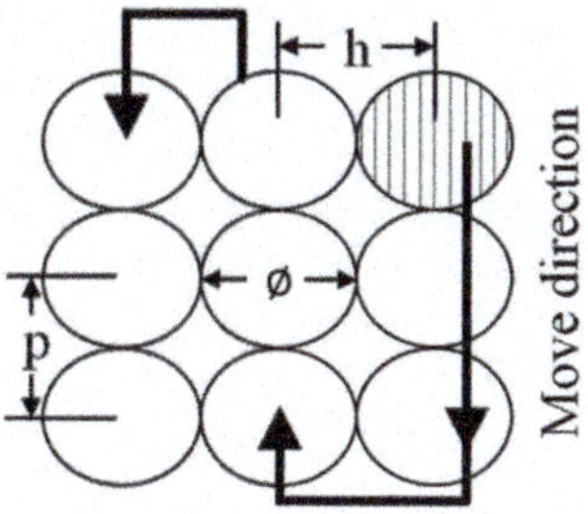

Fig. 4 Laser scanning method to process material in large area

Using this method, we have successfully created patterns in larger areas. Figure 5 shows a 25 × 25 mm patterns created on stainless steel and 5 × 5 mm on nickel surface.

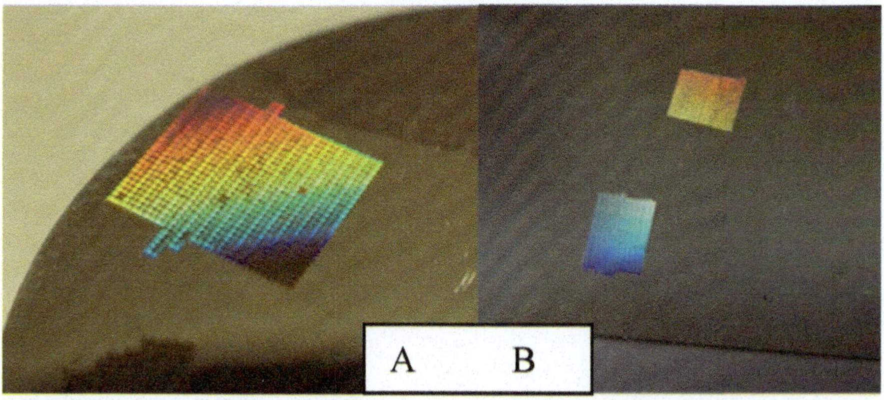

Fig. 5 Image of 5 × 5 mm patterns created on **a** stainless steel and **b** nickel surface

We've also created two-dimensional dot array structures using the laser interference patterning process. First, we expose the laser on the sample surface to create line patterns; then the sample is rotated 90 degrees to have the second laser exposure, resulting the dot array patterns. Figure 6 shows the dot array created on the stainless-steel surface after two laser exposures. The feature size is around 1.7 μm.

To see the effect of surface patterning on material's wetting property, we've measured the water contact angle of the processed materials and observed great improvement on hydrophobicity for both materials as shown in Fig. 7.

The aim of creating patterns on the metal surface is to use it as a molding material to hot-emboss the patterns on to the polymer surface for microfluidic applications. To verify the feasibility of such application, we've used the processed

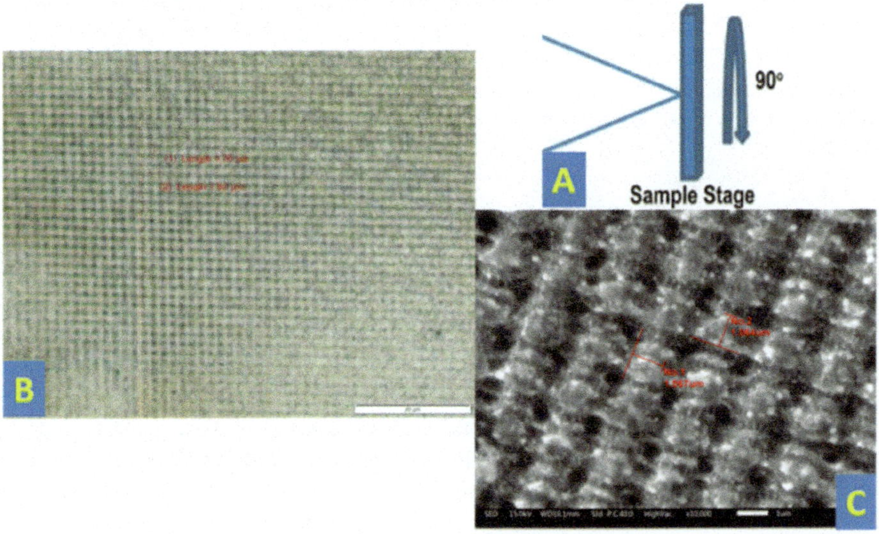

Fig. 6 **a** Method to create dot array pattern. **b** Microscopic image. **c** SEM image of dot array pattern

Fig. 7 Water contact angle measurement of stainless steel and nickel after laser processing

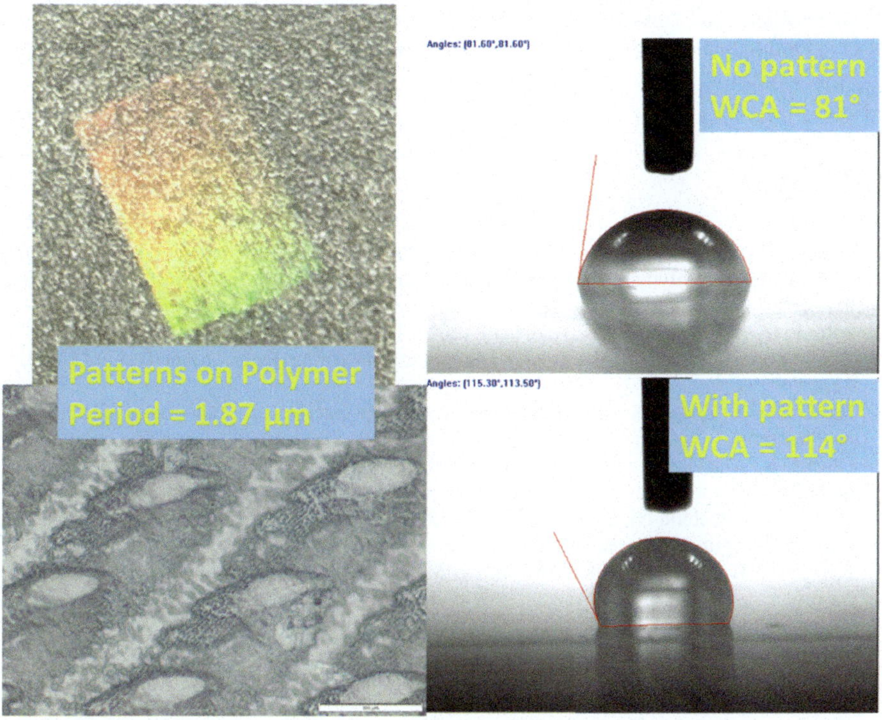

Fig. 8 The pattern transferred from stainless steel to polymer (PMMA) to increase its hydrophobicity

stainless steel to transfer the patterns on to the polymer (PMMA) surface. The temperature of imprinting is 120C°. The pressure used is 2.5 ton and the time duration is 3 min. The result shows that these patterns transferred to the polymer will increase its hydrophobicity drastically. Therefore, our proposition is feasible in future applications. Figure 8 shows the result of hot embossing and the change to its hydrophobicity.

4 Conclusions

In summary, laser interference patterning process is an excellent way to create patterns on material surface to modify its wetting property. It can greatly increase metal materials' hydrophobicity. Through transferring the patterns to polymer surface, it can also increase its hydrophobicity. Therefore, laser interference patterning process is a useful tool on microfluidic applications.

Acknowledgements This work is supported by the project "Laser surface texturing & Interference process for microfluidic applications" from Singapore Institute of Manufacturing Technology.

References

1. Lasagni A, Acevedo D, Barbero C, M¨ucklich F (2007) One-step production of organized surface architectures on polymeric materials by direct laser interference patterning. Adv Eng Mater 9(1–2):99–103
2. Alfredo IA-M, Sabri A, Tim K, Andrés FL (2018) Influence of processing parameters on surface texture homogeneity using direct laser interference patterning. Optics Laser Technol 107:216–227
3. He J, Lin Y, Zhang X (2014) Fiber-based flexible interference lithography for photonic nanopatterning. Opt Exp 22(21):26386–26391
4. Oliveira V, Vilar R, Serra R, Oliveira JC, Polushkin NI, Conde O (2013) Sub-micron structuring of silicon using femtosecond laser interferometry. Optics Laser Technol V54:428–431
5. Oliveira V, Polushkin NI, Conde O, Vilar R (2012) Laser surface patterning using a Michelson interferometer and femtosecond laser radiation. Opt Laser Technol 44:2072–2075
6. Kawamura K-I, Ito N, Sarukura N, Hirano M, Hosono H (2002) New adjustment technique for time coincidence of femtosecond laser pulses using third harmonic generation in air and its application to holograph encoding system. Rev Sci Instrum 73:1711

Investigation of Shot Peening on Residual Stress Distribution in Common Aerospace Materials

Muhammad Azrul Shukri Azmi, Teo Xian Zhong Augustine, Yu Han Aldrich Chua, and Koh Hui Fang

Abstract Intensity is an indirect quantification of the kinetic energy in the air-media stream through the usage of Almen strips in shot peening. However, intensity does not give an insight to the energy transferred into the material nor the compressive residual stress (CRS) on different materials. Previous studies have been done on individual materials but few serve to investigate and compare across materials. In this study, the effect of Shot Peening on common aerospace materials Al7075-T6, IN718, SUS304 and Ti-6Al-4 V were investigated to assess residual stress distribution and other material properties such as hardness across different material properties for the same peening intensity. Through this work, the authors aim to elucidate the mechanism of differences in residual stress distribution by establishing a relationship with the different material properties studied. Specimens were shot peened using ASR70 media, while residual stress and hardness measurements were conducted at various depths thereafter. It was found that the variation in pre-peened material hardness from all materials does not correlate to obtaining higher work hardening and deeper residual stresses under the same shot peening intensity and coverage. Among the materials studied, Al7075-T6, the softest, exhibited the highest post-peening surface roughness, but showed limited work hardening and induced residual stresses. In contrast, Ti-6Al-4 V, the hardest material, demonstrated the lowest post-peening surface roughness, with relatively minor work hardening and residual stresses compared to Inconel 718 and SS304. These differences can be attributed to the distinctive material properties and microstructure, leading to varying deformation behavior when subjected to cold work processes like shot peening.

Keywords Al7075-T6 · SUS304 · Ti-6Al-4 V · Inconel 718 · Shot peening · Hardness · Compressive residual stress · Surface roughness

---

M. A. S. Azmi (✉) · T. X. Z. Augustine · Y. H. A. Chua
Advanced Remanufacturing & Technology Centre (ARTC), Agency for Science, Technology and Research (A*STAR), 3 CleanTech Loop, #01-01 CleanTech Two, Singapore 637143, Singapore
e-mail: azrul_azmi@artc.a-star.edu.sg

K. H. Fang
School of Mechanical and Aerospace Engineering, Nanyang Technological University, Singapore, Singapore

N. Maharjan and W. He (eds.), *Proceedings of the 3rd International Conference on Advanced Surface Enhancement (INCASE) 2023*, Lecture Notes in Mechanical Engineering, https://doi.org/10.1007/978-981-99-8643-9_13

1 Introduction

Shot peening is a surface enhancement process, widely used in aerospace and transport industries to increase component fatigue life by imparting compressive residual stress (CRS). In an air blast shot peening machine, countless spherical media (shot, glass, or ceramic beads) accelerates to velocities between 50 and 150 m/s by compressed air. The bombardment of spherical media onto the surface acts as tiny hammer impacts, deforming the surface over time and inducing CRS. Areas under compressively stressed state delay the onset of cracks initiation and propagation, increasing fatigue life [1–4]. Therefore, critical aerospace components such as engine blades, discs and stators often undergo shot peening in production or during repair.

Shot peening requirements are usually stated by the type of media, Almen intensity and coverage required. Almen intensity is an indirect quantification of the kinetic energy in the air-media stream through the peening of Almen strips mounted securely in an Almen holder. Residual stress induced causes the Almen strips to increasingly bend with peening time, forming an arc. These arcs are measured and input into a saturation curve solver which is usually employed to simplify the process of determining Almen intensity [5]. Coverage is defined as the percentage of surface deformed by media impacts, and the process of determining coverage is visual and highly susceptible to inaccuracy. Coverage development involves peening on a rigid specimen of the same material as the component of interest, until 98% coverage is attained. The peening time for 98% coverage is the baseline for coverage requirement over 100%; to achieve 200% coverage requires twice the exposure time for 98% coverage [6].

While intensity provides a means for process control, it does not give an insight to the energy transferred from the media to the component. Several factors such as material properties, thickness and geometry may contribute to difference in induced CRS on different components or parts of a component. For a peening intensity, the maximum CRS and the depth achieved differs with material as it highly dependent on the material properties.

Although there are many studies regarding residual stress from shot peening on specific materials, there are little comparing multiple materials through the same experimental procedures. Therefore, this study aims to evaluate residual stress from shot peening of commonly used aerospace materials such as Al7075-T6, SUS304, Inconel 718 and Ti-6Al-4V [7].

2 Methodology

2.1 Material Preparation

Al7075-T6, SUS304, Inconel 718 (UNS N07718), Ti-6Al-4V specimens were fabricated from raw materials in accordance with ASTM standards B209M-14, A240, B670 and B265 respectively. The chemical compositions of each material are shown

Table 1 Al7075-T6 chemical composition as per manufacturer datasheet (weight %)

Elements	Zn	Mg	Cu	Fe	Si	Mn	Ni	Ti	Others
Composition	6.1	2.9	2	0.5	0.4	0.3	0.28	0.2	0.15

Table 2 SUS304 chemical composition as per manufacturer datasheet (weight %)

Elements	Cr	Ni	Mn	Si	C	N	P	S
Composition	18.1	8	1.17	0.32	0.24	0.06	0.025	0.001

Table 3 Inconel 718 chemical composition as per manufacturer datasheet (weight %)

Elements	C	Mn	Si	P	S	Cr	Co	Mo
Composition	0.008	0.07	< 0.05	< 0.005	0.0005	17.68	0.15	2.94
Elements	Nb	Ta	Ti	Al	Fe	Cu	Ni	B
Composition	4.96	< 0.02	0.95	0.52	18.42	0.03	54.2	0.0028

Table 4 Ti-6Al-4 V chemical composition as per manufacturer datasheet (weight %)

Elements	Al	V	H	Fe	C	O	N	Other
Composition	5.8	3.9	0.12	0.07	0.02	0.02	0.001	< 0.4

in Tables 1, 2, 3 and 4. All specimens were fabricated with dimensions of 76 × 19 × 5 mm (L × W × H), with the arithmetic average roughness (R_a) kept below 0.5 μm.

2.2 *Shot Peening*

Shot peening was conducted with a pressure type robotic shot peening machine, equipped with ASR70 cast steel shots with a diameter of 0.1778 mm and a hardness range of 45–52 HRC. All samples (Al7075-T6, Inconel 718, SUS304 and Ti-6Al-4V) were treated at an Almen intensity of 0.16 mmA and a coverage of 200%, in accordance to SAE J443 [5] and J2277 [6] respectively.

2.3 *Material Characterisation*

The roughness evaluation chosen for this study is the mean arithmetic average roughness (R_a). Measurements were conducted using the Taylor Hobson Form Talysurf, Phase Grating Interformeter (PGI), and calculated using its own software. The measurement was conducted on 3 different locations, with 3 repetitions per location for each specimen before and after the shot peening treatment.

Microhardness measurements were conducted on the surface of all samples before and after the shot peening treatment to study the amount of strain hardening from the shot peening process. As it would be difficult to study the microhardness on the surface after shot peening, the shot peened samples were polished lightly till there were no visible dimples before microhardness measurements were conducted. Two microhardness profile was also obtained for each shot peened samples to a depth of 1 mm to study the depth of strain hardening. All microhardness measurements were conducted using the Innovatest FALCON 500 Vickers Hardness Tester with a load of 200 g and dwell time of 10 s per measurement.

Lastly, residual stress profiles were measured by Stresstech's Prism® system, an advanced hole drilling technique that measures the surface distortion using electronic speckle pattern interferometry (ESPI), instead of the typical usage of strain gauges. Upon each step of hole drilling, the surface is illuminated with coherent laser light and a high-resolution camera captures the distortion by analyzing the speckle pattern. Speckle pattern occurs when there is interference from the diffusely reflected light due to the rough surface. Therefore, all specimens were coated with a thin coat of spray-paint to reduce the reflectivity before hole drilling was conducted. Both pre-peened and peened specimens had measurements taken at hole depth of 0.04, 0.08, 0.12, 0.16, 0.20, 0.24, 0.32, 0.40 mm. It should be noted that surface residual stress is not possible with such hole drilling method.

3 Results and Discussion

3.1 Surface Hardness and Surface Roughness

Figure 1 shows the measured values of surface roughness and microhardness, before and after shot peening treatment. Shot peening has not only induced work hardening on the surface of all specimens, but also increased the surface roughness during the process. Al7075-T6 achieved the highest surface roughness (8.34 μm), followed by SUS304 (3.45 μm), IN718 (3.37 μm) and Ti-6Al-4V (2.51 μm) respectively. Highest surface roughness from Al7075-T6 is likely due to the lowest pre-peened surface hardness amongst the 4 materials, allowing greater deformation on the surface. Higher pre-peened surface hardness from SUS304, IN718 and Ti-6Al-6V exhibited better resistance to plastic deformation incurred from the shot peening process, resulting in lower surface roughness as compared to Al7075-T6. However, it should be noted that pre-peened surface hardness alone cannot fully determine the resultant surface roughness across materials after shot peening treatment is applied. Existing

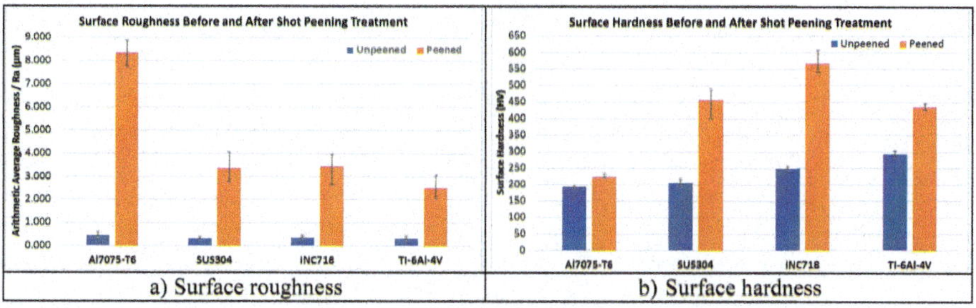

Fig. 1 **a** Surface roughness, **b** surface hardness of unpeened and peened specimens

material properties and microstructure can exhibit different levels of plastic deformation upon application of shot peening treatment, resulting in various mechanical properties (i.e. hardness, roughness, residual stresses etc.) on the surface and subsurface level.

This can be observed for the shot peened Al7075-T6 specimens, where the amount of work hardening from the shot peened specimens is significantly lower than that of Inconel 718, SUS304, and Ti-6Al-4V. This could be likely due to higher dislocation mobility, lower initial dislocation densities and finer grain structures in the Al7075-T6 alloy that reduces the opportunity to form dislocations, resulting in lesser work hardening. Zagar et al. showed that the different content of alloying elements in Al7075 alloy obtained through various aging treatments can also result in different microhardness, residual stresses, and surface roughness [7].

3.2 *Cross-Sectional Hardness*

Figure 2 shows the microhardness depth profiles for Al7075-T6, SUS304, Inconel 718, and Ti-6Al-4V respectively. Measurements below 0.4 mm can be safely assumed as the unpeened hardness of the material since the shot size used is ASR70 with a diameter of 0.1778 mm. It is unlikely that work hardening will occur at a depth twice the diameter of the shot. Hardness profile graph for Al7075-T6 was gradual and observed minimal change as its unpeened condition (194.1–217.5 HV/0.2), with a peak hardness at 228.17 HV/0.2 after peening. Hence, indicating a small amount of work hardening occurring after shot peening, due to the low content alloying elements, Mg and Cu which are the main contributors of work hardening in AL7075-T6 [8].

SUS304 specimens have significant work hardening as seen from the significant increase in hardness from unpeened condition (199.20–215.70 HV/0.2) to a peak hardness at 487.69 HV/0.2 after peening. This is due to phase change from austenite to martensite in SUS304, due to the rearrangement of the crystal structure and the introduction of carbon and dislocations [9–12].

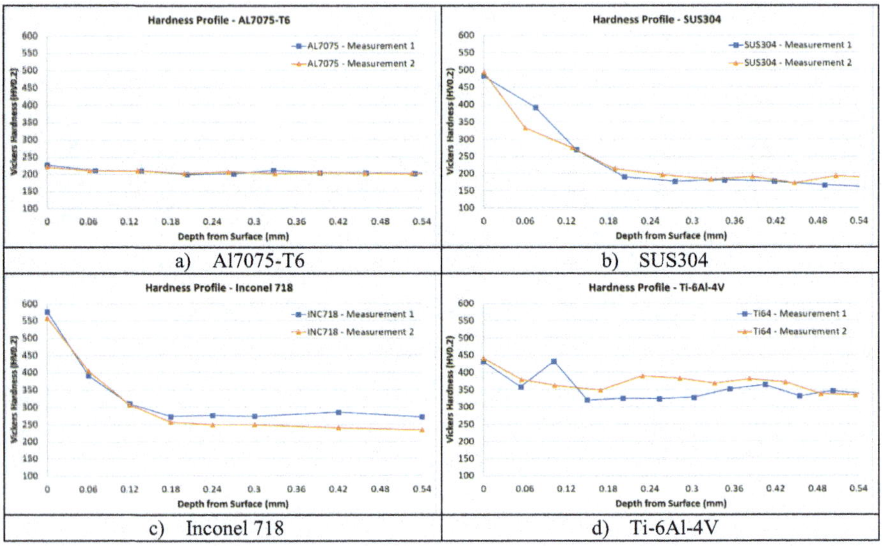

Fig. 2 Hardness profiles obtained from cross-sectional measurements on peened specimens of **a** AL7075-T6, **b** SUS304, **c** Inconel 718, **d** Ti-6Al-4 V

Inconel 718 had the steepest gradient increasing in hardness from unpeened condition (245.10–254.60 HV/0.2) to an average peak hardness of 568.36 HV/0.2. The unpeened specimens were observed to have lower hardness than the value stated in the supplier provided datasheet, resulting in the observed steep gradient. The increase in hardness is likely due to grain refinement in surface layer after shot peening [13]. The hardness values after peening are in line with findings from several other studies [4, 13]

For Ti-6Al-4V, hardness increased from unpeened condition (294–356 HV/0.2) to peak hardness at 441.8 HV/0.2 after peening. The increase in hardness is likely due to the formation of nanograins and the high dislocation density in the near surface layer [14].

3.3 Residual Stress

The residual stress profiles for each material up to 0.4 mm are shown in Fig. 3. AL7075-T6 achieved maximum CRS of approximately 500 MPa at a depth of 0.125 mm from the surface. The zero-crossing at which residual stress changes from compressive to tensile is approximately 0.25 mm. Measurements near surface show almost no induced CRS while at subsequent depths below zero-crossing does not show any tensile stress beneath the compressive region. Ti-6Al-4V achieved maximum CRS of 769 MPa at 0.08 mm from the surface and zero-crossing depth of 0.186 mm. In both specimens, tensile stress was observed below zero-crossing depth, ranging from 50 to 150 MPa. The results obtained for Ti-6Al-4V is similar to other literature conducted previously [15]. SUS304 and Inconel 718 had significantly higher maximum CRS of 1380 MPa at 0.075 mm and 1409 MPa at 0.08–0.09 mm respectively, as compared to AL7075-T6 and Ti-6Al-4V. The zero-crossing for the two SUS304 specimens are different (between 0.21 and 0.31 mm), and for Inconel 718 specimens are 0.24–0.26 mm. In both SUS304 specimens, there was a sudden change in gradient in decreasing CRS. In addition, for the second SUS304 specimen, the longitudinal CRS was greater than transverse CRS. This could be due to existing stresses that were present on the specimen before peening.

Residual stress profiles for all materials indicate a general observation that the harder the base material, the shallower is the depth of max CRS after peening. However, factors such as the phase change in SUS304 from shot peening increases the hardness too and reduces the depth of max CRS implying the decreasing magnitude of zero-crossing depth occurring with increasing hardness. Overall, all materials have induced CRS after shot peening, though the extent of the region where surface or internal crack growths and propagations are slowed, varies from one material to another. Results from AL7075-T6 indicate that near surface cracks will not be slowed as the induced CRS is minimal.

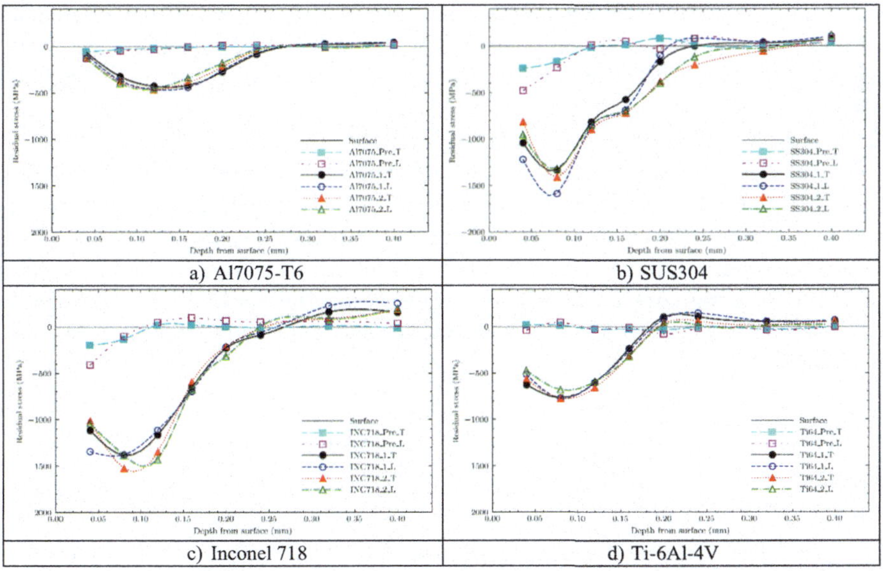

Fig. 3 Compressive residual stress profiles of unpeened and peened specimens of **a** AL7075-T6, **b** SUS304, **c** Inconel 718, **d** Ti-6Al-4 V

4 Conclusion

Shot peening have been carried out on AL7075-T6, SUS304, Inconel 718, Ti-6Al-4V specimens with Almen intensity of 0.16 mmA and coverage of 200%.

Surface roughness of all specimens increased substantially after shot peening, especially for AL7075-T6, which has the lowest surface hardness amongst the tested materials. Surface hardness greatly increased after shot peening except AL7075-T6 where work hardening was observed to be minimal. The hardness profiles from cross-sectional measurements at increasing depth indicates the same behavior where AL7075-T6 specimens have barely any increase in hardness, while SUS304, Inconel 718 and Ti-6Al-4V specimens have steep increase in hardness from the base material. This was likely due to the higher dislocation mobility, lower initial dislocation densities and finer grain structures in the Al7075-T6 alloy that reduces the opportunity to form dislocations, resulting in lesser work hardening.

Residual stress measurements from PRISM indicate an inverse relationship between the hardness of the base material and the depth of max CRS after shot peening, with harder materials exhibiting shallower depths. However, the induced compressive residual stress profiles differ greatly for each material. SUS304 and Inconel 718 specimens were induced with the most CRS, while AL7075-T6 and Ti-6Al-4V specimens were generally induced with lower CRS. Cross referencing various literature on shot peening of SUS304, the surface layer of the material would have undergone phase change from austenite to martensite, attributing to the high CRS.

Overall, shot peening of AL7075-T6, SUS304, Inconel 718, Ti-6Al-4V in this study resulted in induced CRS which will delay crack growth and propagation. Although the material hardness may offer some insights into post-peened surface roughness and depth of maximum CRS, a comprehensive analysis of the individual material properties, characteristics, and microstructure is essential to gain a profound understanding of the effects of work hardening, residual stresses imparted and eventually fatigue life improvements through cold work processes like shot peening.

References

1. Torres MAS, Voorwald HJC (2002) An evaluation of shot peening, residual stress and stress relaxation on the fatigue life of AISI 4340 steel. Int J Fatigue 24(8):877–886
2. De los Rios ER, Walley A, Milan MT, Hammersley G (1995) Fatigue crack initiation and propagation on shot-peened surfaces in A316 stainless steel. Int J Fatigue 17(7):493–499
3. Jiang XP, Man CS, Shepard MJ, Zhai T (2007) Effects of shot-peening and re-shot-peening on four-point bend fatigue behavior of Ti–6Al–4V. Mater Sci Eng A 468:137–143
4. Klotz T, Delbergue D, Bocher P, Lévesque M, Brochu M (2018) Surface characteristics and fatigue behavior of shot peened Inconel 718. Int J Fatigue 110:10–21
5. Standard SAE (2013) J2277 shot peening coverage determination. In: Developed by surface enhancement committee, publication date: Apr, 1

6. Standard, S. A. E. (2017). J443-Procedures for using standard shot peening test strip. SAE International.
7. Mouritz AP (2012) Introduction to aerospace materials. Elsevier
8. Žagar S, Markoli B, Naglič I, Šturm R (2021) The Influence of Age hardening and shot peening on the surface properties of 7075 aluminium alloy. Materials 14(9):2220
9. Zhou L, He W, Luo S, Long C, Wang C, Nie X, Li Y (2016) Laser shock peening induced surface nanocrystallization and martensite transformation in austenitic stainless steel. J Alloys Compounds 655:66–70
10. Liu H, Wei Y, Tan CKI, Ardi DT, Tan DC, Lee CJ (2020) XRD and EBSD studies of severe shot peening induced martensite transformation and grain refinements in austenitic stainless steel. Mater Charact 168:110574
11. Jebaraj AV, Kumar TS, Kumar LA, Deepak CR (2017) Influence of shot peening on surface quality of austenitic and duplex stainless steel. In: IOP conference series: materials science and engineering, vol 263, no 6, IOP Publishing, p 062057
12. Kirk D, Payne NJ (1999) Transformations induced in austenitic stainless steels by shot peening. In: ICSP7: 7th international conference on shot peening, Warsaw, Poland
13. Maleki E, Unal O, Guagliano M, Bagherifard S (2021) The effects of shot peening, laser shock peening and ultrasonic nanocrystal surface modification on the fatigue strength of Inconel 718. Mater Sci Eng A 810:141029
14. Wen Y, Wu Y, Hua L, Xie L, Wang L, Zhang LC, Lu W (2021) Effects of shot peening on microstructure evolution and mechanical properties of surface nanocrystal layer on titanium matrix composite. Mater Des 206:109760
15. Lainé SJ, Knowles KM, Doorbar PJ, Cutts RD, Rugg D (2017) Microstructural characterisation of metallic shot peened and laser shock peened Ti–6Al–4V. Acta Mater 123:350–361

Construction of a Controllable Hydrophobic and Hydrophilic Shot Blasting Technique for Aluminum

Satoru Ikeda and Koichiro Nambu

Abstract In recent years, many problems have arisen, such as increased power consumption in commercial refrigerators due to water droplets on aluminum alloy surfaces, increased frost formation due to bacteria and mold, and increased transportation costs for bullet trains. To solve these problems, we turned our attention to surface texturing technology and shot blasting, a new surface component that can switch between hydrophobic and hydrophilic properties. Shot blasting can be applied to large areas at a lower cost than laser processing and can be used for surface texturing. This study reports the results of an evaluation of the effect of shot blasting on the surface wettability of aluminum alloys. Shot blasting was performed on aluminum alloy A5052 using various particles, and the wettability of the specimens was evaluated. It was found that the wettability after shot blasting varied with the particle material and particle size as in previous studies. Using the aspect ratio (dent depth/dent width) calculated from the surface roughness, it was found that the specimens became hydrophilic to hydrophobic up to an aspect ratio of 7 or less, and that the contact angle gradually increased to hydrophobic at an aspect ratio of 7 or greater.

Keywords Shot blast process · Aluminum alloy · Hydrophobic

1 Introduction

In recent years, water droplets on the surface of aluminum alloys have caused various problems, such as increased power consumption in commercial refrigerators, frost formation caused by bacteria and mold, and increased transportation costs on bullet trains. In addition, aluminum alloys are expected to become lighter and stronger as the weight of cars and motorbikes is reduced. Attention is therefore focused on

S. Ikeda · K. Nambu (✉)
Osaka Sangyo University, 3-1-1 Nakagaito, Osaka Prefacture, Daito City, Japan
e-mail: knambu.mech@ge.osaka-sandai.ac.jp

N. Maharjan and W. He (eds.), *Proceedings of the 3rd International Conference on Advanced Surface Enhancement (INCASE) 2023*, Lecture Notes in Mechanical Engineering, https://doi.org/10.1007/978-981-99-8643-9_14

wettability, such as the water repellency of lotus leaves and the hydrophilic properties of snail shells. The wettability of these organisms has the function of keeping surfaces clean. Wettability is achieved by the microstructure of the surface, and research is underway to impart biological wettability to material surfaces as well [1–3]. It is believed that by intentionally imparting wettability to material surfaces, water droplets and dirt that cause corrosion can be removed. If corrosion resistance and antifouling properties are improved by wettability, this will lead to a longer service life of materials and a resource- and energy-saving society. Laser processing has attracted attention and research as a surface modification technique to improve wettability. Laser processing is excellent for forming the intended surface shape, but it is not suitable for processing large materials because the area that can be processed at one time is small and the processing cost is high.

Therefore, this study focused on shot blasting, which has a larger processing area and lower processing costs than laser processing. Shot blasting is mainly used for surface roughening and cleaning as a surface preparation before deburring and painting. In addition to this, it has recently been shown that new properties can be imparted to the surface [4].

Previous studies have shown that wettability control by shot blasting is possible on stainless steel [3, 5]. On the other hand, the effect on aluminum alloys has not yet been determined. The aim of this study is to determine the effect of shot blasting on the wettability of aluminum alloys.

2 Experimental Procedure

An aluminum alloy (A5052) was used for the test specimens. The chemical composition is shown in Table 1. The specimens were 20 × 20 × 5 mm plate specimens. The shot blasting process was performed using a pneumatic injection device made by Fuji. In this study, we focused on particles of natural materials from the viewpoint of environmental issues. Resin particles and hard particles were also used as comparison materials. The types of resin particles used were polyethylene, walnut grid, AP grid (apricot), and PS grid (peach). Zirconia 205, Zirconia 505, glass beads, and SG20 (Steel) were used as hard particles. A list of the particles used is shown in Table 2. Shot blasting treatment conditions were 90°, 10 s, and 0.5 MPa pressure. The surface profile of the treated specimens was measured with a laser microscope. The surface roughness parameters measured were maximum height roughness and root mean square roughness. The contact angle and surface free energy of the treated samples were measured using Excimer's smart Contact PRO II.

Table 1 Chemical compositions

	Si	Fe	Cu	Mn	Mg	Cr	Zn
A5052	0.25	0.40	0.10	0.10	2.2–2.8	0.15–0.35	0.10

Table 2 List of particles

	Particle diameter (μm)
Walnut	80
PE (Polyethylene)	180
PS (Peach)	150
AP (Apricot)	110
Glass beads	100
Zirconia 205	50
Zirconia 505	10
SG20 (steel)	100

3 Result and Discussion

3.1 Wettability of Shot-Blasted Specimens

The results of contact angle measurements are shown in Fig. 1. The figure shows that the contact angles vary greatly depending on the particles used, which is similar to the results obtained in the past with stainless steel materials [5]. On the other hand, the specimens treated with apricots and walnuts, which are particles of natural materials, were found to be hydrophilic. Next, a comparison by particle size shows that the contact angle is higher for zirconia 205, which has a larger particle size. These results indicate that the wettability of aluminum alloys also changes with shot blasting treatment, and that the particle material and particle size are influential factors.

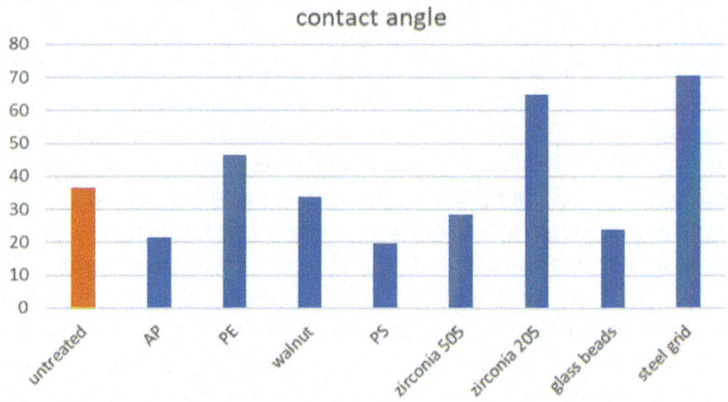

Fig. 1 Contact angle measurement result of shot blast samples

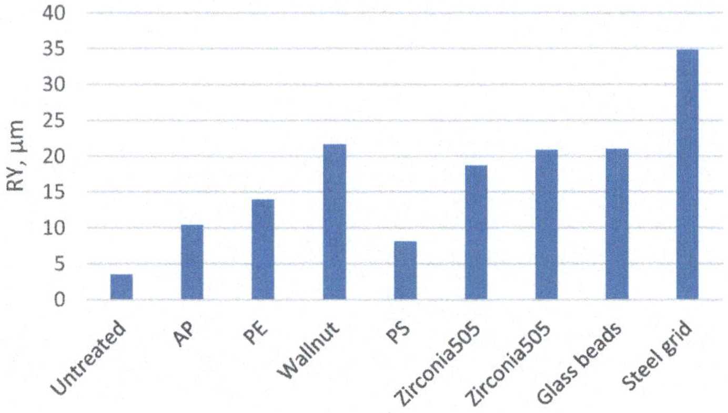

Fig. 2 Maximum height roughness (RY) of shot blast samples

3.2 Surface Roughness Measurement Results

The results of the surface roughness measurements are then shown in Figs. 2 and 3. The figures show that the surface roughness of the shot-blasted specimens is higher than that of the untreated material in both cases. It can be seen that the surface roughness of the natural material particles is also higher than that of the hard particles, although it is smaller. Focusing on the zirconia with a contact angle that varied significantly with grain size, it can be seen that the maximum height roughness (RY) was similar, but the root mean square roughness (RMS) was greater for the 205 with a larger grain size. The steel beads with the largest contact angle have a larger RY, but the RMS is not significantly different compared to the other particles. Focusing on the zirconia with a contact angle that varied significantly with grain size, RY was similar, but RMS was greater for the 205 with a larger grain size. The steel beads with the largest contact angle have a larger RY, but the RMS is not significantly different compared to the other particles.

Previous studies have shown that for stainless steel, surface roughness is important in terms of whether the surface wettability is a Cassie-Baxter state, which indicates water repellency, or a hydrophilic state due to increased surface area [5]. Therefore, we examined the aspect ratio of the indentation calculated from the surface roughness. The aspect ratio here is the value calculated from (indentation depth (RY/2)/width (RSM/2)). A larger aspect ratio indicates that narrower grooves and deeper indentations are formed. When the grooves are narrow and deep, air is trapped inside the grooves (Cassie-Baxter condition), which is considered to indicate water repellency.

The results are shown in Fig. 4. The leftmost plot shows untreated material. The results show that the contact angle decreases until the aspect ratio is about 7, after which the contact angle increases, indicating water repellency. This result suggests that the hydrophilicity increases up to an aspect ratio of 7 due to the increase in surface area, after which the material becomes water-repellent due to the Cassie-Baxter state. These results suggest that surface microstructure control by changing

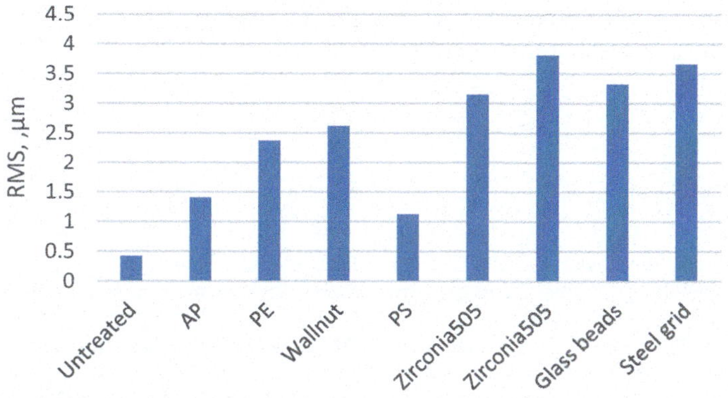

Fig. 3 Root means square roughness (RMS) of shot blast samples

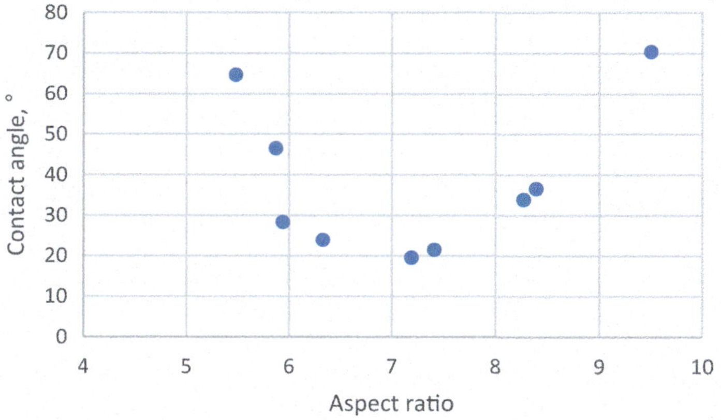

Fig. 4 Relationship between contact angle and aspect ratio of shot blast samples

particle material and shape is effective in controlling the wettability of aluminum alloys.

4 Conclusions

The effect of shot blasting on the wettability of aluminum alloys was evaluated, with the following conclusions.

1. wettability of aluminum alloys is affected by shot blasting
2. particle material, shape, and size are factors that change wettability.
3. aspect ratio calculated from surface roughness up to 7 shows hydrophilicity as surface area increases.

4. for aspect ratios greater than 7, the particles gradually become Cassie-Baxter, indicating hydrophobic properties.

References

1. Cabezudo N, Sun J, Andi B, Ding F, Wang D, Chang W, Luo X, Xu BB (2019) Enhancement of surface wettability via micro- and nanostructures by single point diamond turning. Nanotechnol Precis Eng 2(1):8–14
2. Cai Y, Chang W, Luo X, Sousa AML, Lau KHA, Qin Y (2018) Superhydrophobic structures on 316L stainless steel surfaces machined by nanosecond pulsed laser. Precis Eng 52(1):266–275
3. Chen T, Liu H, Yang H, Yan W, Zhu W, Liu H (2016) Biomimetic fabrication of robust self-assembly superhydrophobic surfaces with corrosion resistance properties on stainless steel substrate. RSC Adv 50(6):43937–43949
4. Kameyama Y, Komotori J (2009) Effect of micro ploughing during fine particle peening process on the microstructure of metallic materials. J Mater Process Technol 209(20):6146–6155
5. Sano F, Nambu K, Uchiumi Y, Okumiya M (2020) Control of hydrophobic and hydrophilic characteristics of metallic materials using shot blasting. Int J Abras Technol 10(2):96

Recent Development of Low Damage Laser Microprocessing on Surface and Subsurface Treatment of Silicon for Industrial Applications

XinXin Li and Yingchun Guan

Abstract Silicon is one of the most commonly used semiconductor materials. However, working with silicon presents challenges due to its high brittleness and hardness. Drilling, milling, dicing, and surface treatment require careful handling. Fortunately, laser processing technology offers a solution. With its high energy density and smaller heat-affected zone, laser processing is well-suited for working with silicon materials. This paper reviews the interaction mechanism between laser and silicon material and its applications. The surface and subsurface treatments are introduced. The surface roughness, microstructure, residual stress, and electrical properties of silicon material after laser microprocessing are investigated. Based on the current research status, we summarize some of the challenges that must be overcome to allow a wider application of laser in the new generation of semiconductor materials.

Keywords Laser microprocessing technology · Surface treatment · Silicon material · Surface quality · Properties

1 Introduction

Silicon is a widely used semiconductor material in microelectronics and photovoltaics due to its excellent electrical and mechanical properties. However, the high hardness and high brittleness make it difficult to precisely and efficiently process by

X. Li · Y. Guan (✉)
School of Mechanical Engineering and Automation, Beihang University, Beijing 100083, China
e-mail: guanyingchun@buaa.edu.cn

Y. Guan
National Engineering Laboratory of Additive Manufacturing for Large Metallic Components, Beihang University, 37 Xueyuan Road, Beijing 100083, China

Ningbo Innovation Research Institute, Beihang University, Beilun District, Ningbo 315800, China

N. Maharjan and W. He (eds.), *Proceedings of the 3rd International Conference on Advanced Surface Enhancement (INCASE) 2023*, Lecture Notes in Mechanical Engineering, https://doi.org/10.1007/978-981-99-8643-9_15

traditional mechanical processing techniques. For example, after grinding, subsurface damage, such as dislocations and microcracks, will be generated in silicon wafers [1, 2]. Laser processing, such as nanosecond and picosecond laser processing, is an advanced technique that enables rapid and precise silicon materials machining without generating extra mechanical loads. Thereby the mechanical damage is minimized. As a result, the pulsed laser was widely employed in silicon and silicon-based materials processing. Some of these applications include dicing [3, 4], grooving [5], drilling [6, 7], annealing [8–10], and surface treatment [11, 12]. Moreover, pulsed laser technology is also employed in the fabrication of silicon devices, such as the production of free-standing membranes of SOI-CMOS RF circuits/functions on Silicon-on-Insulator (SOI) wafers [13] and magnetic field sensors [14]. To further enhance processing accuracy, research has been conducted on techniques such as water jet-guided laser micromachining and ultrasonic-assisted water-confined laser micromachining [15].

This paper primarily focuses on applying pulsed lasers for silicon material microprocessing and surface treatment. The existing theoretical and experimental research works in this field are reviewed. An overview of our recent work on laser processing of silicon material was presented. Finally, laser technology's challenges and future works on new-generation semiconductor materials were briefly discussed.

2 Laser Microprocessing of Silicon Materials

In silicon wafer manufacturing, the thinning process of silicon wafer is an important step. The conventional thinning method is mechanical grinding. To avoid surface damage resulting from mechanical processes, our research group introduced laser microprocessing to thin silicon material [16]. The laser thinning process includes rapid materials removal and surface quality improvement. When the high fluence picosecond laser irradiates the silicon material surface, the photon energy is absorbed by free electrons and extra-nuclear electrons of silicon material. The high-energy free electrons become more active, and the extra-nuclear electrons absorb additional photon energy to overcome the Coulombic force [17]. The active electrons also transmit energy to other free and extra-nuclear electrons via electron collisions. Due to the large mass difference between electrons and lattice, only a small fraction of energy is transferred in each electron-lattice collision. The electron-phonon relaxation time exceeds the electron or ion collision times, and thermodynamic equilibrium between the electron system and the lattice is reached after multiple electron-phonon relaxation times. As a result, the lattice temperature remains low during picosecond laser microprocessing [18]. With continuous picosecond laser radiation, the lattices eventually heat up through collisions with energetic electrons. Simultaneously, the number of free electrons increases, and avalanche ionization occurs. Moreover, the high laser fluence directly breaks bonds between electrons in the irradiated silicon material [19]. The material is removed through avalanche ionization and bond breakage. Consequently, the amount of material removed increases with the laser energy absorbed by the silicon.

After the laser treatment, laser scanning traces were observed on the treated surface (Fig. 1a), and the depth of the heat-affected zone (HAZ) was found to be less than 1 μm (Fig. 1b). The minimal HAZ was achieved by a highly focused picosecond laser with a small laser power density at the focal point. After laser thinning, the residual stress of silicon material declined from −0.63 to −0.34 Gpa (compressive stress), as shown in Fig. 1c. The accumulation of laser energy leads to a thermal effect that generates tensile stress on the surface, while laser shock induces compressive stress during laser thinning. It is well-known that the thickness of silicon material influences the residual stress, increasing as the thickness of the silicon material decreases. A combination of thermal effects, laser shock, and silicon material thickness determines the overall residual stress on the laser-processed surface. As the laser energy decreases, the residual stress transitions from initial compressive stress to tensile stress and then back to compressive stress. The compressive stress at the laser microprocessing surface will benefit the silicon material's mechanical properties and service life. After laser thinning, as shown in Fig. 1d, the slope of the I-V curves decreases, and the resistivity of the silicon material decreases from 1.8 to 1.767 Ω.cm. Thus, the proposed laser thinning process of the silicon material has minimal impact on the resistivity of silicon material.

3 Subsurface Repairing

Subsurface damages are generated on silicon material during the mechanical grinding process. To remove subsurface damage, the laser has been explored as an available strategy for improving the silicon material surface quality. The mechanism of this strategy is presented in Fig. 2a. After the laser irradiates the silicon material surface, a melted layer is formed, starting from the top and gradually thickening as the laser irradiation continues until it reaches the defect-free bulk region. Following laser irradiation, bottom-up epitaxial growth initiates from the defect-free bulk. This microprocessing forms a defect-free single-crystalline structure within the laser-irradiated region [20]. Consequently, the defects in the subsurface layer are completely eliminated, and the surface quality is improved.

Due to its simplicity and precise controllability, this method has garnered significant attention and has been the subject of study by various researchers. Yan et al. [21, 22] employed a nanosecond laser with different energy to irradiate diamond-grinded single-crystal silicon wafers. Fig. 2(**b**-2) shows a transmission electron microscope (TEM) micrograph of a sample irradiated at a laser energy density of 0.30 J/cm^2. Despite the nearly complete damage repair in the adjacent region, a residual dislocation is observed beneath the surface. On the other hand, when the laser energy was increased to 0.65 J/cm^2, the subsurface damage was entirely repaired (Fig. 2(**b**-3)). The results revealed that the amorphous layers transformed into single-crystal silicon and dislocations and microcracks were completely eliminated after laser irradiation.

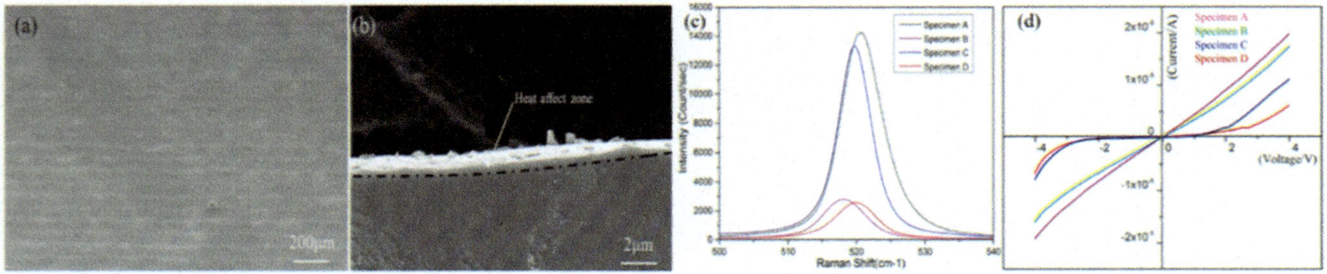

Fig. 1 Laser thinning silicon material [16]. **a** SEM images of thinned silicon; **b** heat-affected zone of laser-thinned silicon; **c** Raman spectrums of as-received and laser-thinned silicon surfaces; **d** I-V characteristic curves of as-received and laser-thinned silicon samples

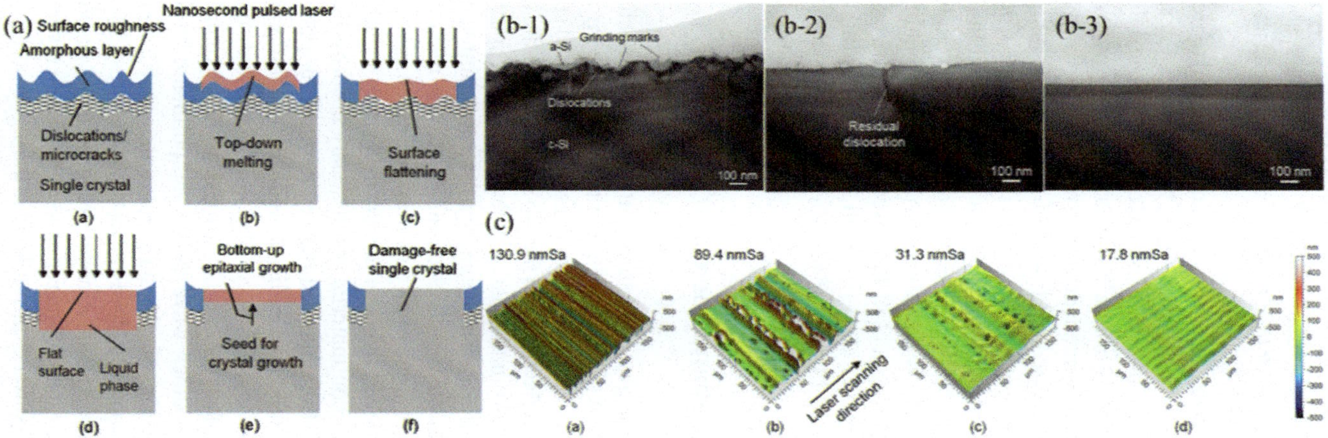

Fig. 2 Laser repairing silicon material. **a** Schematic diagram of laser repairing mechanism [22]; **b** Cross-sectional TEM micrographs of a silicon material [22], (**b**-1) unirradiated surface and laser irradiated surface at energy of (**b**-2) 0.30 J/cm^2 and (**b**-3) 0.65J/cm^2 [25]; **c** local three-dimensional surface topography of the notch irradiated at different laser fluences [23]

The surface roughness of the edges and notches was significantly improved after nanosecond laser irradiation (Fig. 2c) [23]. A waviness perpendicular to the scanning direction was observed in this particular condition. This waviness could be attributed to the convection of liquid silicon or the pressure exerted by the laser-induced plasma during laser irradiation.

Additionally, two-step nanosecond laser technique was used to treat sliced single-crystal silicon material [24]. In the first step, higher laser energy density was employed to address deeper surface defects, which unfortunately led to increasing surface roughness. The flow was primarily governed by Marangoni convection in the molten pool, which gradually intensified with higher energy density. The surface tension gradient exceeded the viscosity forces of the liquid, resulting in the region with the highest surface tension located at the edge exerting forces towards the outer perimeter. This surface flow during solidification resulted in the formation of high-frequency features, contributing to the increased surface roughness. The same surface was then exposed to a lower laser energy density, eliminating residual high-frequency features. The improvement of high-frequency features was attributed to surface tension and capillary action. As a result, the surface roughness of silicon material was reduced from 1.057 μm to 26 nm after two-step laser microprocessing.

4 Surface Property Enhancement

To efficiently remove the damage on the silicon material after mechanical grinding, our research group has studied the laser grinding process in the Ar air environment [26]. As shown in Fig. 3a, when the nanosecond laser irradiates the silicon material, a portion of the laser energy is reflected from the top surface while the material absorbs other energy. The free electrons in the material absorb the energy from the photons, leading to the excitation of electrons. The energy from the excited electrons is then transferred to the lattice and other free electrons. Since the relaxation time of the electron-photon interaction is significantly shorter than the nanosecond laser pulse width, an equilibrium is achieved between the electrons and the lattice. As the laser continues to irradiate, the silicon material continuously absorbs energy, increasing the temperature. When the temperature surpasses the melting point of silicon, the material melts. The molten silicon is redistributed due to the combined effects of surface tension and the Marangoni effect. The laser with high energy density is absorbed by the surface oxide layer, which forms a rapidly expanding plasma and produces a shock wave. The shock wave causes the pollutants to fragment and be removed. Thus, as shown in Fig. 3b, the surface roughness (Ra and Rz) is reduced from 0.4 and 2.47 to 0.075 and 0.34 μm after laser ground, respectively.

Additionally, the embossments and grooves are eliminated, resulting in a significant improvement in surface quality. The removal of the oxide layer and the absence of oxygen injection into the surface during laser grinding in an argon environment, as shown in Fig. 3c. Figure 3d indicates that the polycrystalline silicon is completely transformed into single-crystal silicon. After laser irradiation, the polycrystalline

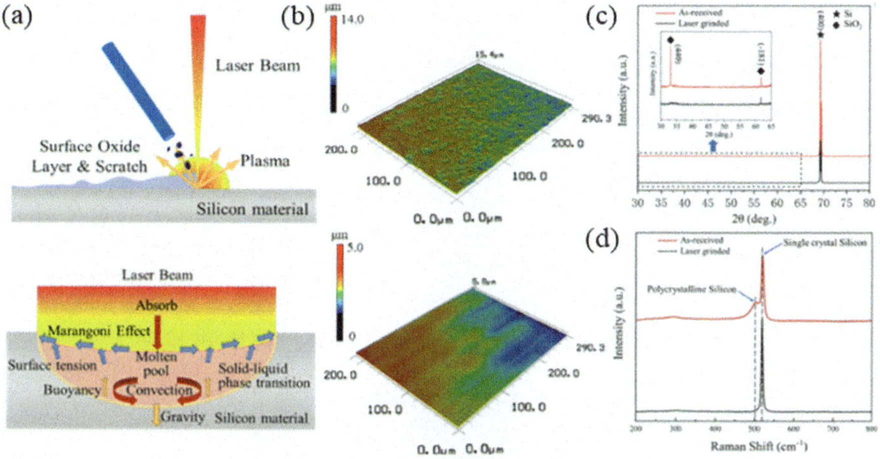

Fig. 3 Laser grinding single crystal silicon [26]. **a** Schematic diagram of laser grinding mechanism; **b** three-dimensional topographic images of as-received and laser ground single crystal silicon surfaces; **c** X-ray diffraction XRD spectra of as-received and laser ground single crystal silicon surfaces; **d** Raman spectra of as-received and laser ground single crystal silicon surfaces

silicon layers undergo melting. As the laser irradiation continues, the melted layer gradually thickens. Once the laser pulse ceases, the cooling process leads to bottom-up epitaxial regrowth from the single-crystal silicon substrate. Additionally, the resistivity of the laser-grinded silicon was measured and found to be lower than that of the as-received silicon. This decrease in resistivity can be attributed to the absence of polycrystalline silicon, which has been removed during the laser grinding process.

Laser grinding efficiently reduced the damage to mechanical ground silicon. Thus, our research group utilized a hybrid laser approach to enhance the surface quality of the sliced silicon [27]. Based on the interaction between laser and materials, the hybrid laser process includes two steps, as shown in Fig. 4a. Firstly, the sliced silicon was irradiated by a picosecond laser in ambient air. Subsequently, the sample that underwent picosecond laser processing was subjected to a nanosecond laser treatment within an Ar gas environment.

After picosecond laser treatment, all saw-mark defects were effectively eliminated. Consequently, the quality of the silicon was significantly enhanced after the hybrid laser treatment, as shown in Fig. 4b. The volume fraction of SiO_2 was increased after picosecond laser treatment in air, while it decreased after nanosecond laser treatment with Ar gas (Fig. 4c). During picosecond laser treatment, silicon reacted with oxygen to convert it into silicon dioxide. However, when silicon was irradiated with a nanosecond laser in Ar gas shielding, the reaction between silicon and oxygen did not occur. Additionally, the oxide layer is broken down into oxygen and silicon during nanosecond laser irradiation and disappears. However, the hybrid laser treatment led to the residual stress of silicon material change from +0.13 (tensile stress) to −

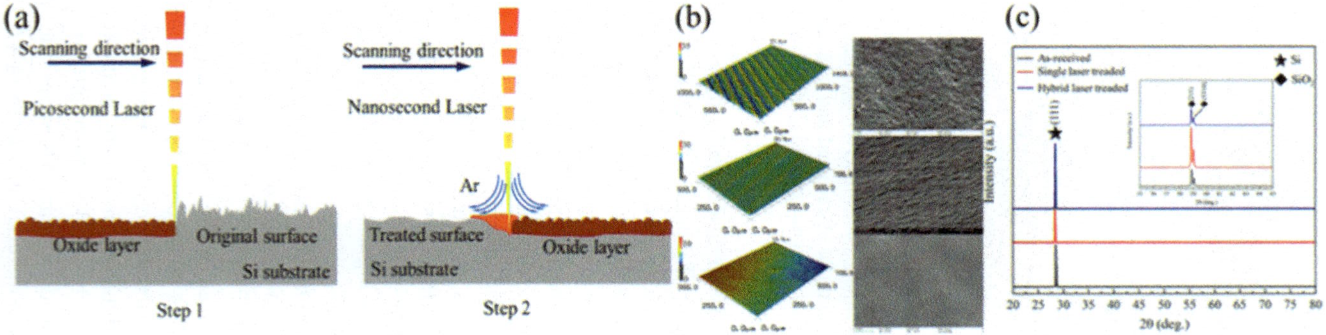

Fig. 4 Hybrid laser treating single crystal silicon material [27]. **a** Schematic representation of hybrid laser treating; **b** surface topography of as-received and hybrid laser treated single crystal silicon samples; **c** XRD spectra of as-received and hybrid laser treated single crystal silicon surfaces

0.19 GPa (compressive stress). This phenomenon is mainly attributed to the laser-induced shock force during laser machining. Compressive stress has the advantage of suppressing crack propagation [28].

One of the critical requirements for successful semiconductor applications is achieving a high concentration of impurity doping, which can significantly enhance the electrical, structural, and optical properties. Silicon materials doped with phosphorus have gained significant attention as promising candidates for various applications, including photovoltaic, electrochemical, and quantum computing devices. Traditionally, ion implantation has been the method for phosphorus doping. However, this technique can damage the substrate's ion radiation, causing atoms displacement from their lattice sites. Therefore, post-implantation annealing is necessary to repair the damaged lattice.

Laser annealing technology has emerged as an effective post-processing technique for P-doped epitaxial layers owing to high energy and flexibility. Extensive research has been conducted to investigate the effects of laser annealing on P-doped epitaxial layers. Shin et al. [9] found that the defects in the recrystallized layer decreased with an increase in pulse frequency and power density. When single-pulse annealing was performed at 600 mJ/cm^2, the number of defects in the recrystallized layer was reduced, but some end-of-range defects were still present (Fig. 5a). However, when 10-pulse annealing was performed at 600 mJ/cm^2, both the defects in the recrystallized layer and the end-of-range defects disappeared (Fig. 5**a**). Following laser annealing, the thickness of the phosphorus concentration region increased from 40 to 60 nm. This can be attributed to the higher diffusivity of liquid silicon than single-crystalline silicon, resulting in the phosphorus distribution in the melted silicon during laser annealing. Lee et al. [10] observed no changes in the P concentration of highly P-doped Si thin films before or after a short annealing time. This suggests that P atoms were incorporated into the substitutional lattice sites, and the physical states remained unchanged without any dopant diffusion, as shown in Fig. 5b. The room-temperature Hall measurement of the silicon film was conducted before and after Dynamic surface annealing (DSA). As shown in Fig. 5c, it was observed that the free-carrier concentrations in the Si:P samples increased with higher P concentrations after DSA. However, the free-carrier concentration decreased with further increases in P concentration (Fig. 5(**c**-1)). The increased active P concentration from the DSA process reduced Hall mobility due to increased ionized impurity scattering (Fig.5(**c**-2)). Consequently, the resistivity of all the annealed Si:P samples decreased due to the increased active P concentration (Fig. 5(**c**-3)). Ryu et al. [11] observed that P atoms were slightly accumulated in the near-surface region of the Si:P film after laser annealing. However, there was no significant diffusion of P atoms towards the Si (100) substrate, and the macroscopic P concentration was maintained throughout the Si:P film, as shown in Fig. 5d. After HF cleaning, a significant decrease in P concentration was observed in the near-surface region of the laser-annealed film (Fig. 5d). This decrease was attributed to the loss of dopants during the native oxide removal.

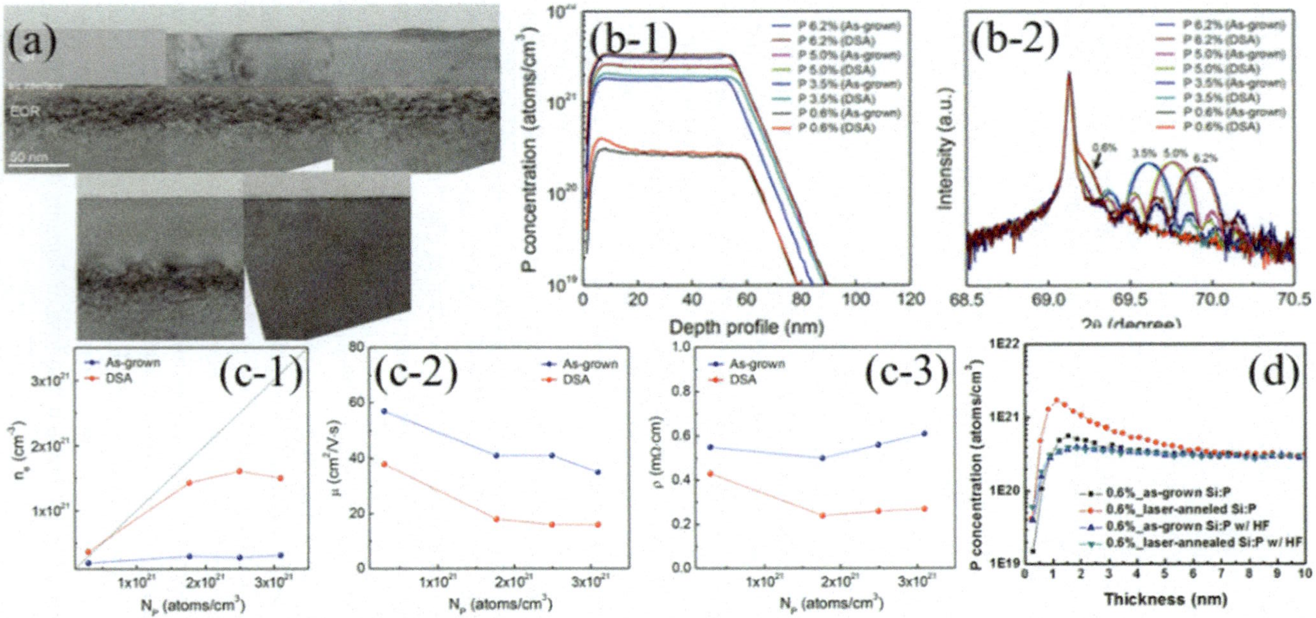

Fig. 5 Laser annealing P-doped epitaxial Si (Si:P) layers. **a** Cross-sectional TEM images in different laser energy density and pulse number [9]; **b**-1 P depth profile measured by SIMS and **b**-2 HR-XRD ω/2θ (004) rocking curves of Si:P samples before and after laser annealing [10]; **c** electrical properties of Si:P before and after DSA: **c**-1 free-carrier concentration, **c**-2 hall mobility, **c**-3 resistivity as a function of P concentration. **d** SIMS profiles of Si:P films with the P concentration [11]

5 Conclusion and Outlook

This mini-review primarily focuses on applying laser microprocessing and surface treatment of silicon materials. The utilization of lasers in silicon material processing has gained significant attention. The advantages of laser microprocessing technology have been extensively demonstrated and discussed. However, as the demand for the microelectronics industry continues to grow, second, third, and even fourth-generation semiconductor materials have emerged and are gradually being applied. One such example is single-crystal silicon carbide (SiC). SiC is the most mature and widely used third-generation semiconductor material due to its special mechanical, electrical, and thermal properties. Single-crystal silicon carbide possesses different physical and chemical properties compared to single-crystal silicon. Therefore, the application of laser technology in the new generation of semiconductor materials needs to be further developed.

Acknowledgements This work was supported by the Ningbo Science and Technology Major Project [Grant number 2023Z012].

References

1. Pei Z, Kassir S, Bhagavat M, Fisher GR (2004) An experimental investigation into soft-pad grinding of wire-sawn silicon wafers. Int J Mach Tools Manuf 44(2–3):299–306
2. Li Z, Pei Z, Fisher GR (2006) Simultaneous double side grinding of silicon wafers: a literature review. Int J Mach Tools Manuf 46(12–13):1449–1458
3. Sikora A, Grojo D, Sentis M (2017) Wavelength scaling of silicon laser ablation in picosecond regime. J Appl Phys 122(4)
4. Marks MR, Cheong KY, Hassan Z (2018) An improved three-point bending test method for the investigation of nanosecond laser dicing of ultrathin Si dies with Cu stabilization layer. Mater Charact 136:29–40
5. Moutin1 C, Bouillard B, Cowache C, Robin O, Porto M, Noraz E (2022) Laser grooving process modifications to avoid recast growth. In: Micro/nano-electronics packaging & assembly, design and manufacturing forum, Grenoble, France
6. Jiao L, Zheng H, Zhang Y, Ng EYK (2018) Picosecond laser drilling of silicon with applied voltage. SN Appl Sci 1(1)
7. Shaheen ME, Gagnon JE, Fryer BJ (2019) Studies on laser ablation of silicon using near IR picosecond and deep UV nanosecond lasers. Opt Lasers Eng 119:18–25
8. Kim B, Han KN, Ryu JC, Kim YJ, Kim SW (2021) Non-periodic nanoscale structuring of crystalline silicon surface by using ultrashort laser pulses. Appl Surf Sci 565
9. Shin HS, Lee JH, Lee MH, Ryu HY, Park S, Park HG, Ko DH (2020) Recrystallization and activation of ultra-high-dose phosphorus-implanted silicon using multi-pulse nanosecond laser annealing. Japanese J Appl Phy 59(SG)
10. Lee MH, Ryu HY, Ko EJ, Ko DH (2019) Effects of phosphorus doping and postgrowth laser annealing on the structural, electrical, and chemical properties of phosphorus-doped silicon films. ACS Appl Electron Mater 1(3):288–301
11. Ryu HY, Lee M, Park H, Ko DH (2020) Chemical bonding states and dopant redistribution of heavily phosphorus-doped epitaxial silicon films: Effects of millisecond laser annealing and doping concentration. Appl Surf Sci 504

12. Shalimi S, Samuel GL (2021) Ultrafast pulse laser inscription and surface quality characterization of micro-structured silicon wafer. J Manuf Process 62:323–336
13. Maia JM, Amorim VA, Viveiros D, Marques PVS (2020) Magnetic field sensors in fused silica fabricated by femtosecond laser micromachining. J Phys Photon 2(1)
14. Zhou J, Xu RW, Jiao H, Bao JD, Liu QY, Long YH (2020) Study on the mechanism of ultrasonic-assisted water confined laser micromachining of silicon. Optics Lasers Eng 132
15. Guo HM, Cao SB, Li L, Zhang XF (2022) A review on the mainstream through-silicon via etching methods. Mater Sci Semiconduct Process 137
16. Fang ZH, Chen LF, Guan YC, Zheng HY (2019) Picosecond laser micromachining of silicon wafer: characterizations and electrical properties. Surf Rev Lett 27(05):1950142
17. Keller WJ, Shen N, Rubenchik AM, Ly S, Negres R, Raman RN, Yoo J-H, Guss G, Stolken JS, Matthews MJ, Bude JD (2019) Physics of picosecond pulse laser ablation. J Appl Phys 125(8):085103
18. Li XX, Guan YC (2020) Theoretical fundamentals of short pulse laser–metal interaction: a review. Nanotechnol Precis Eng 3(3):105–125
19. Phillips KC, Gandhi HH, Mazur E, Sundaram SK (2015) Ultrafast laser processing of materials: a review. Adv Optics Photon 7(4):684–712
20. Niitsu K, Tayama Y, Kato T, Maehara H, Yan JW (2018) Characterization of recrystallized depth and dopant distribution in laser recovery of grinding damage in single-crystal silicon. Mater Sci Semicond Process 82:54–61
21. Yan JW, Muto SY, Kuriyagawa T (2009) Processing grinding-damaged silicon wafers by high-frequency nano-second laser irradiation. Adv Mater Res 76–78:451–456
22. Yan JW, Kobayashi F (2013) Laser recovery of machining damage under curved silicon surface. CIRP Ann 62(1):199–202
23. Niitsu K, Tayama Y, Kato T, Yan JW (2019) Laser recovery of grinding-induced subsurface damage in the edge and notch of a single-crystal silicon wafer. Surf Topogr Metrol Prop 7(1):015013
24. Ren MY, ZY Z (2022) Surface of nanosecond laser polished single-crystal silicon improved by two-step laser irradiation. Acta Optica Sinica 42(7):0714004
25. Yan J, Sakai S, Isogai H, Izunome K (2009) Recovery of microstructure and surface topography of grinding-damaged silicon wafers by nanosecond-pulsed laser irradiation. Semicond Sci Technol 24(10):105018
26. Li XX, Wang YM, Guan YC (2021) Laser grinding of single-crystal silicon wafer for surface finishing and electrical properties. Micromachines 12(3):262–271
27. Li XX, Cui ZQ, Guan YC (2022) Surface modification of single-crystal silicon by hybrid laser treatment. In: 13th international photonics and optoelectronics meetings (POEM 2021)
28. Liu TY, Ge PQ, Bi WB, Gao YF (2017) Subsurface crack damage in silicon wafers induced by resin bonded diamond wire sawing. Mater Sci Semicond Process 57:147–156

Electrochemical Polishing of Ti-6Al-4V 3D Printed Internal Channels with Customized Functionalized Cathode Using HF and Perchlorate Free Electrolyte

Adri Abu Bakar and Hirotaka Sato

Abstract This study aimed to achieve a polished surface on the internal channel of Ti-6Al-4V, a popular titanium alloy in the 3D printing industry, using electrochemical polishing (ECP). We used a custom functionalised cathode and less toxic electrolyte to achieve this goal. We found that a smooth surface with a minimum roughness of Ra value 9.9 μm could be achieved, and localised polishing of channels could be done without masking the workpiece itself. The study investigated the interaction between the workpiece, cathode, and electrolyte within the internal channel and found that the formation and removal of a viscous layer during ECP affected the polishing effect, as well as the bubbling produced at the cathode. The optimal process parameters are expected to vary for different geometries and could be determined through a linear sweep voltammetry test followed by ECP at a constant voltage. This research provides useful insights and methods for achieving ECP on the internal surface of Ti-6Al-4 V components which could have significant implications for the design of more complex components for applications of 3D printing in the aerospace, marine or med-tech industries.

Keywords Electrochemical polishing · Internal finishing · Additive manufacturing

A. A. Bakar (✉)
Advanced Remanufacturing and Technology Centre, 3 Cleantech Loop, #01/01, CleanTech 2, Singapore 637143, Singapore
e-mail: muhammad_adri@artc.a-star.edu.sg

H. Sato
Nanyang Technological University, 50 Nanyang Avenue, Singapore 639798, Singapore

N. Maharjan and W. He (eds.), *Proceedings of the 3rd International Conference on Advanced Surface Enhancement (INCASE) 2023*, Lecture Notes in Mechanical Engineering, https://doi.org/10.1007/978-981-99-8643-9_16

1 Introduction

1.1 Background and Problem Statement

Metal additive manufacturing exhibited exceptional growth in recent years [1, 2]. Strong interest is focused on the aerospace [3] and biomedical [4, 5] industries to produce complex and valued parts, which are usually customised or produced in smaller quantities. Metal 3D printing requires the melting of metal powder or wire using a laser or electron beam and the melted metal will construct the solid part layer-by-layer to the desired structure [6–8]. Powder bed fusion AM (PBF-AM) is generally used for fabricating metallic structures due to its capability in producing complex structures with high accuracy and good mechanical properties [9]. Some AM Laser-PBF techniques, for example, include direct metal laser sintering (DMLS) and selective laser melting (SLM), and electron beam melting PBF (EBM) [10].

Despite the advantages of metal AM, the surface quality of the printed parts is vastly uneven and not appropriate for immediate use in most applications [11–13]. The printed parts have high surface roughness which made the part vulnerable to fatigue, crack generation, corrosion, and poor fitting during assembly [14–17]. The printed surfaces are mostly coarse due to the presence of asperities, such as loosely attached powder particles, partial melts, and surface undulations existing on the surface during the process of melting/sintering powder particles. At present, there is strong interest in developing AM to fabricate components such as conformal cooling channels for gas and fluid flow applications in aerospace and automotive industries [18, 19], but the poor surface finish of inner surface poses a huge challenge for such applications. Traditional post-processing surface finishing techniques (e.g. turning and milling) are not applicable for these complex internal surfaces [20] due to geometric limitation, thus there a need to develop reliable methods to achieve fine internal surface finish for industrial application.

1.2 State of the Art for Internal Surface Finishing

Numerous surface finishing techniques have been developed and tailored specifically for the post-processing of additive manufacturing (AM) components. However, achieving the desired surface requirements remains challenging, particularly for the intricate internal channels that are difficult to access. Consequently, only a limited number of processes have proven successful in attaining low roughness for AM parts, as indicated by the existing literature. Notably, the following processes have demonstrated efficacy in achieving the desired surface quality: (i) abrasive flow machining (AFM) [21, 22], (ii) magnetic abrasive finishing (MAF) [23], (iii) fluidized bed machining (FBM) [24], and (iv) hydrodynamic cavitation abrasive finishing (HCAF) [19]. While each of these processes has advantages and limitations, the complex nature of internal surface finishing for AM structures still poses a significant challenge yet to be fully overcome. The industrialization of these techniques is impeded by their inherent limitations. Therefore, it is crucial to explore alternative approaches in order to achieve the desired fine surface finishing of AM workpieces, thereby making them truly functional in various industries.

1.3 Application of Electrochemical Polishing for Ti-6Al-4 V

In this study, the authors propose a novel electrochemical-based solution with a customized counter electrode to achieve the desired finishing requirements within the internal channels of additive manufacturing (AM) components. Electrochemical polishing (ECP) is a well-established finishing method in various industries, but its application for internal surface finishing remains underexplored. The primary challenge lies in placing the counter electrode within the workpiece to ensure a consistent and controlled rate of material removal. Since the target workpiece is conductive, any inadvertent contact between the counter electrode and the workpiece can cause a short-circuit, disrupting the intended chemical reactions. Thus, a novel system that allows the insertion of the counter electrode into the target channels is necessary to achieve surface finishing for these complex AM components with internal features as shown in Fig. 1.

Furthermore, the chemical etching and polishing of titanium (Ti) based alloys typically involve hazardous acids such as sulfuric acid, hydrofluoric acid, and perchloric

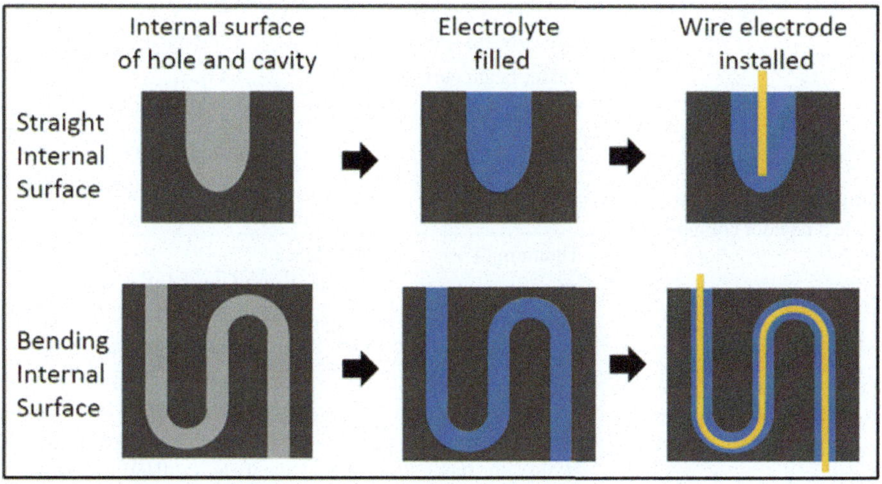

Fig. 1 Graphical explanation of positioning of counter electrode (yellow) in internal features of structures for electropolishing

acid as electrolytes. This study explores the utilization of less toxic electrolytes to mitigate the environmental impact associated with the surface finishing process.

2 Methodology

2.1 AM Workpiece

To investigate the capability of the ECP process for AM Ti-6Al-4 V, straight channel coupons were printed by EBM with an A2XX, Arcam printer using EOS Titanium Ti64 Grade 5 with the following parameters given in Table 1 and elemental composition given in Table 2.

These coupons were printed with a length of 60 mm, internal diameter of 10 mm or 20 mm, and wall thickness of 5 mm, as shown in Fig. 2.The EBM printed coupons exhibit one of the highest pristine average surface roughness with Ra values within the range of 40–60 μm while typical powder bed fusion (PBF) components have an average Ra value of 5–10 μm.

Table 1 EBM parameters for printing of straight channel coupons

Parameter	Value
Preheating temperature	730 °C
Preheating beam current	30 mA
Preheating beam speed	13,000 mm/s
Number of contours	2
Offset-beam speed	340 mm/s
Offset-beam current	4 mA
Hatching speed	4530 mm/s

Table 2 EOS titanium Ti64 grade 5 powder composition wt %

Element	Minimum	Maximum
Titanium (Ti)	Balance	
Aluminum (Al)	5.50	6.75
Vanadium (V)	3.50	4.50
Oxygen (O)		0.20
Nitrogen (N)		0.05
Carbon (C)		0.08
Hydrogen (H)		0.015
Iron (Fe)		0.30
Yitrium (Y)		0.005
Other elements		0.40

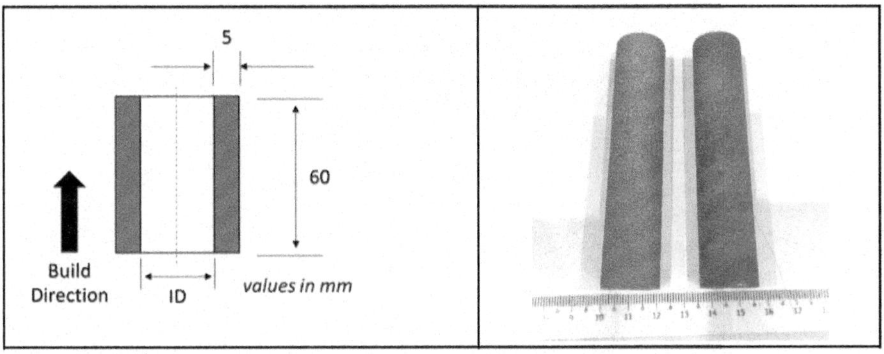

Fig. 2 **a** Build geometry for straight channel coupons **b** Image of EBM printed Ti-6AL-4 V coupon

2.2 Electrochemical Experimental Setup

The schematic of the basic electrochemical setup is shown in Fig. 3 with variations of the setup orientations indicated in subsequent figures. A Keithley source meter 2601B was used to control the two-electrode cell system for electrochemical experiment such as linear scan voltammetry and chronoamperometry. Kickstart software was used as the platform to control the source meter.

$$\text{Anode} : M \rightarrow M^{n+} + ne^- \tag{1}$$

$$\text{Cathode} : 2H^+ + 2e^- \rightarrow H_2 \tag{2}$$

$$\text{Cathode} : TiCl_4 + C_2H_5OH \rightarrow TiO + C_2H_5Cl(complex) + H_2 \tag{3}$$

When power supply is applied during ECP process, electrical current passes from the anode to the cathode and will cause oxidation and reduction reactions at the anode and cathode respectively. The oxidized metal at the anodic surface turns into metal ions (Eq. 1) and diffuse to the cathode at a controlled rate. Typically, hydrogen generation occurs at the cathode due to the reduction reaction (Eq. 2). However, the electrolyte consists of a 1 M ethylene glycol (EG) sodium chloride (NaCl) mixture diluted with ethanol as the polar solvent instead of an aqueous electrolyte, with an optimal ratio of 80–20% as determined in a study conducted by Kim et al. [25]. Therefore, the expected consequent reaction at the cathode is shown in (Eq. 3) is due to formation of a stable $TiCl_4$ compound in presence of ethanol to form a Ti oxide, ethanol—chloride complex and hydrogen gas. Compared to the typical acid electrolytes employed in electrochemical polishing (ECP) and etching of titanium (Ti), such as perchloric acid and hydrofluoric acid, the electrolyte used in this study, based on ethylene glycol and sodium chloride, is considerably safer to handle. Moreover, the utilization of acid electrolytes in surface finishing processes can also pose risks

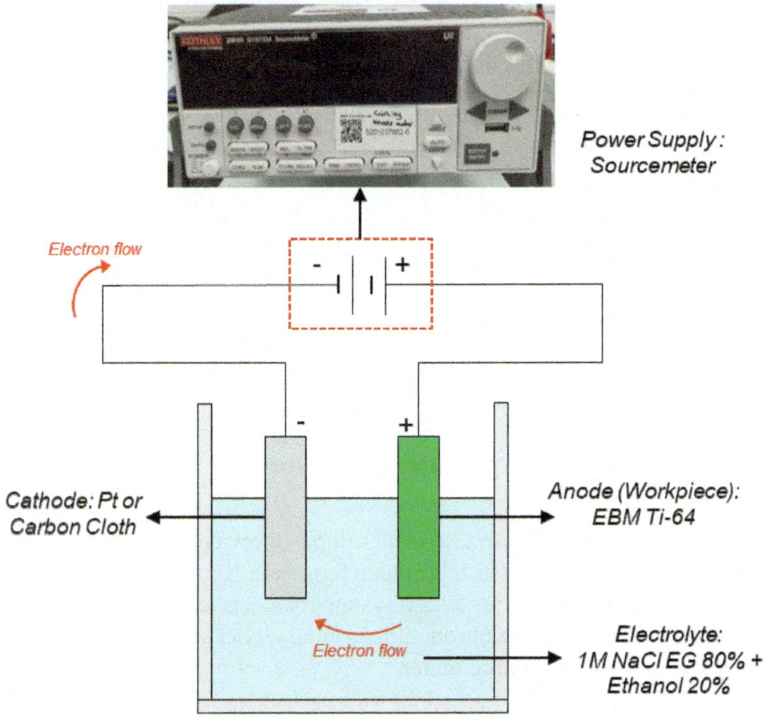

Fig. 3 Schematic diagram of electropolishing experimental setup

to the environment if not properly managed. The improper disposal of these acids can contaminate water sources, harm aquatic life, and contribute to environmental degradation.

In this study, a novel concept for the cathode or counter electrode (CE) was developed, and its implementation is currently in progress. The CE design incorporates a carbon cloth loaded with platinum particles, which is coiled around a flexible core and then encased with a polymer mesh. This unique design offers two important features: flexibility and bendability, allowing it to be inserted into complex channels, while mitigating the risk of short-circuiting when the CE comes into contact with the workpiece. By utilizing a carbon cloth loaded with platinum particles, the CE achieves enhanced conductivity and efficiency during the electrochemical polishing process. The flexibility and bendability of the CE enable it to adapt to the intricate internal channels of additive manufacturing components, ensuring thorough surface finishing. The addition of a polymer mesh provides an extra layer of protection, minimizing the likelihood of unintended contact and subsequent short-circuits between the CE and the workpiece. The details of the sectioned EBM printed Ti-6Al-4 V samples and the counter electrode are shown in Fig. 4.

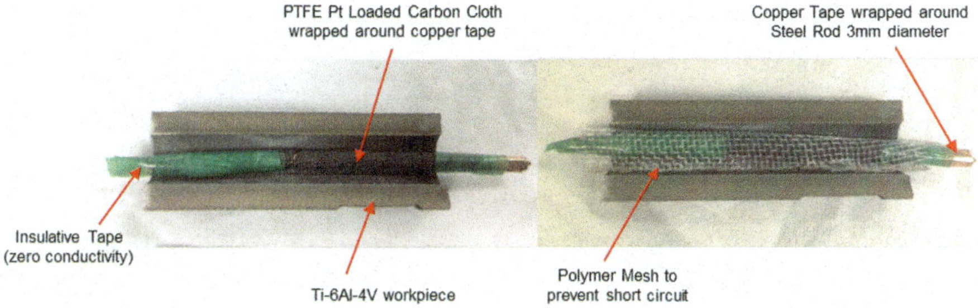

Fig. 4 Sectioned EBM Ti-6Al-4 V workpieces and custom fabricated counter electrode

3 Results and Discussion

3.1 Polarisation Curve (I-V) Through Linear Sweep Voltammetry

The combination of the parameters mentioned above are used to further investigate the mechanism of ECP within constricted channels. The workpiece was first sectioned vertically and a glass panel was introduced to constrict the channel while still enabling direct observation of the process. In order to understand the anodic dissolution behaviour, a polarization curve (I-V curve) was measured. This involved sweeping the voltage across a predetermined range while measuring the corresponding current. This technique, known as linear voltammetry sweep (LSV), provides valuable information about the electrochemical behaviour and performance of the system. The details of the LSV sweep experiment are shown in Table 3 and the results shown in Fig. 5.

The experiment was repeated 3 times and the results are split into 3 voltage regions 1 through 3. In region 1 (15–21 V), as the voltage increases, there was little to no change in the current indicating that the resistance in the system is increasing. This is likely due to the formation of a yellow viscous layer of $TiCl_4$ as during the ECP of Ti-6Al-4V in NaCl EG solution as discussed by Zhang et al. [26]. This $TiCl_4$

Table 3 Experimental parameters for linear voltammetry sweep (LSV) for Ti-6Al-4 V workpiece with custom counter electrode

Parameter	Value
Electrolyte	1 M NaCl + EG (80%) + Ethanol (20%)
Cathode	Pt Loaded PTFE Carbon Cloth
Anode	EBM Ti64 (internal surface 10 mm)
Voltage range (DC)	LSV: 15–35 V sweep @ 0.1 V/s
Current	Measured
Temperature	25 °C
Agitation/stirring	None

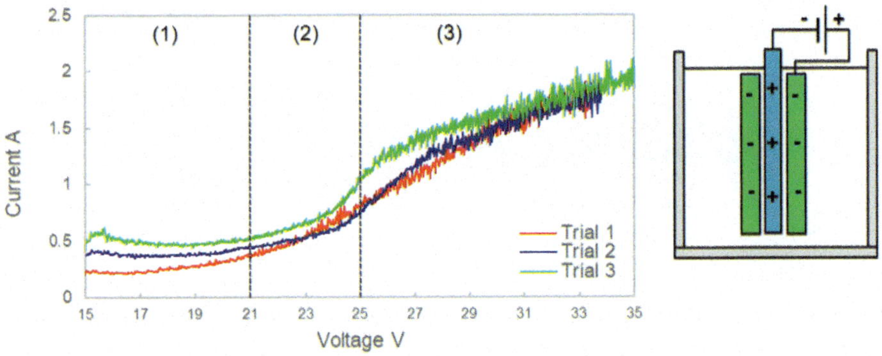

Fig. 5 LSV sweep results for three repeats sweeping from 15 to 35 V and schematic of experimental setup

complex compound has low electrical conductivity accounting for the increase in resistance but is also able to dissolve into the bulk EG electrolyte. There are two key insights from this region: (i) There is formation of a $TiCl_4$ viscous layer and the rate of formation is increasing as shown by the increasing resistance and stable current despite increasing voltage. (ii) When the experiment is repeated, the overall resistance is lower as evidenced by the higher current likely due to lower surface area due to the removal of asperities from the anodic dissolution during the full LSV sweeps.

In the next voltage region 2 (21–25 V), the current starts to increase in proportion with the voltage indicating that the electrical resistance is stabilizing. When considering the mechanism within the constrained channels, this indicates that the amount of $TiCl_4$ is also stabilizing. Apart from the formation of the viscous complex compound, there is also formation of hydrogen gas in the form of bubbling that travels up the channel inducing a flow out of the constrained region as can be seen in Fig. 6b. As the voltage increases, the rate of bubbling increases as well thus pulling the $TiCl_4$ out of the region and dissolving into the bulk electrolyte. Therefore, it is likely that within region 2, the rate of removal and formation of $TiCl_4$ is at equilibrium. This is favourable conditions for ECP as the forward reaction is enabled as the $TiCl_4$ by-product can be continuously replaced and not be over concentrated in the constrained region. Although this is a specific case, the optimal voltage range for electrochemical polishing typically occurs within region 2 for most cases.

Region 3 (>21 V) still shows an increase in current as the voltage increases but the current starts to show fluctuations and instability in the current. This can be directly attributed to the increasing rate of gaseous hydrogen product formed as the rate of reaction is increased from the higher potential difference applied. This can be seen in Fig. 6c where the yellow $TiCl_4$ is visible and large hydrogen bubble forming. At this voltage region the gaseous products are interfering with the reaction and may result in an uneven ECP reaction across the internal walls of the channel. For typical ECP reactions, if the voltage is too high it is believed that the passivated layer breaks

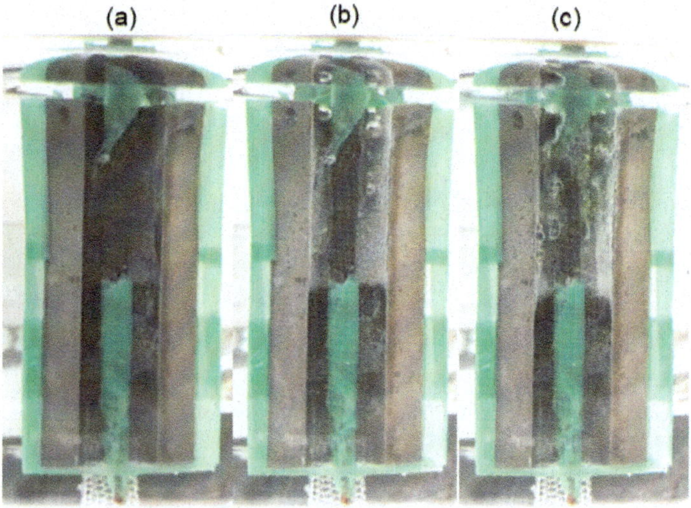

Fig. 6 Images from LSV sweep experiment at varying voltages. **a** 0 V **b** 25 V **c** 30 V

down and gas evolution occurs when the applied voltage is further increased above the gas evolution potential [27]. During the experiment, severe bubbling was also observed in the voltage ranges above 30 V.

The results indicate that there is a key mechanism of viscous layer ($TiCl_4$) transport and there is an optimal voltage region for which ECP can be conducted. However, it is expected that the voltage region will vary based on the experimental setup. For example, if the setup were not vertical as shown in Fig. 4, but horizontal instead, the transport of $TiCl_4$ would be expected to be different. Within the scope of this study, the vertical setup is further explored with electropolishing conducted at region 2 with the surface finishing results also shared for the next set of experiments.

3.2 *Selective Region Polishing*

In addition to the LSV sweep, additional experiments were also conducted with variation in the exposed counter electrode. Similar parameters were used as in Table 3 but instead of LSV, the voltage was held at 25 V for 20 min. The results and the schematic of the experiment is shown in Fig. 7. From the images there is a visible improvement in brightness and surface finish of the wall regions that are close to the exposed regions of the counter electrode. The surface roughness results measured using a PGI profilometer are shown in Fig. 8.

The initial surface roughness of the EBM workpiece was expectedly high and resulted in large variation in surface roughness when measure with a stylus profilometer. However, the corresponding sections that are not directly exposed to

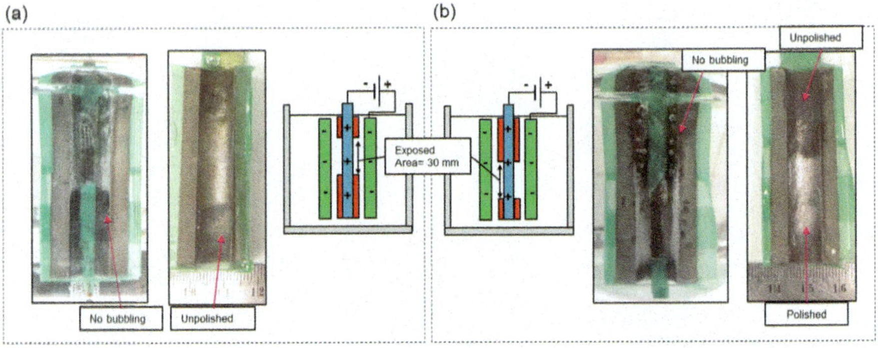

Fig. 7 Experimental Schematic and results for **a** upper section exposed and **b** lower section exposed with before and after polishing (25 V, 20 min) showing regions of unpolished workpiece where the counter electrode is not exposed

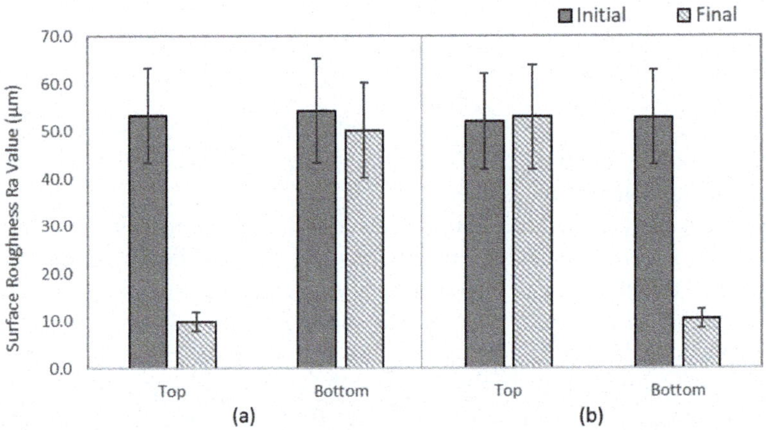

Fig. 8 Surface roughness results from ECP trials measure via stylus profilometer for lower and upper sections for two configurations as shown in Fig. 7: **a** upper section exposed and **b** lower section exposed

the counter electrode show no significant variation in surface roughness Ra value before and after the ECP trials. The areas that are exposed, Top for (a) and Bottom for (b) show up to 80% roughness improvement for both trials with a minimum surface roughness of Ra value of 9.9 μm. The range of effective ECP is slightly beyond the range the exposed counter electrode at approximately 10 mm. There is also evidence of uneven and minimal polishing for setup (b) where the lower section of the counter electrode is exposed. This indicates that the mechanism of the transport of the viscous layer is vital to the effectiveness of the ECP within constrained channels. Additional study would be required for further optimization of the ECP parameters for internal polishing of Ti-6Al-4V components with non-corrosive electrolyte to bring the capabilities inline with the performance of acidic electrolytes that are able to achieve surface finish with Ra values < 1 μm.

The ability to mask the counter electrode instead of the workpiece to enable polishing of targeted areas is also an important finding as selective internal polishing is difficult to achieve with any other post processing method. This is especially important for post processing of AM parts where there is varying surface roughness at various build angles resulting in different levels of post processing required for different regions.

4 Conclusion

The mechanism of ECP of Ti-6Al-4 V using 1 M NaCl + EG (80%) + Ethanol (20%) was studied for internal channels and demonstrated the capability of achieving targeted surface finishing with an Ra value of at least 9.9 μm. The key findings of the study are highlighted as follows:

1. The novel counter electrode developed for this study successfully demonstrated the capability to polishing the internal surface of AM printed components while minimizing the risk of short circuiting.
2. There is an optimal voltage range for ECP of internal components and it will vary based on the setup and is highly dependent on the transport mechanism of the $TiCl_4$. Being able to manipulate the transport of the $TiCl_4$ and enable an equilibrium of the formation and removal of the viscous layer is vital for optimal ECP.
3. The range of effective ECP is approximately 10 mm from the exposed counter electrode and this can be taken advantage of via masking of the counter electrode to enable selective polishing of different regions of internal sections of AM components.
4. The less toxic electrolyte has demonstrated capability in ECP but has yet to achieve similar performance to its acidic counterparts.

Post processing of AM components remains a large roadblock for the full utilization of AM in possibly limitless applications. Specifically, controlled and targeted internal feature finishing is vital towards this goal. This study provides a proof of concept of using a novel counter electrode for selective internal channel finishing that can be applied to different materials with a combination of different materials as counter electrode and electrolyte. The further development of this technology is in progress by the authors.

References

1. Thompson et al (2016) Design for additive manufacturing: trends, opportunities, considerations, and constraints. CIRP Annals 65(2):737–760, 2016/01/01/ 2016. https://doi.org/10.1016/j.cirp.2016.05.004
2. Wohlers Associates I (2017) Wohlers report 2017: 3D printing and additive manufacturing state of the industry : annual worldwide progress report. Wohlers Associates
3. Kumar LK, Krishnadas Nair CG (2017) Current trends of additive manufacturing in the aerospace industry. In: Wimpenny DI, Pandey PM, Kumar LJ (eds) Advances in 3D printing & additive manufacturing technologies. Springer Singapore, Singapore, pp 39–54
4. Ackland DC, Robinson D, Redhead M, Lee PVS, Moskaljuk A, Dimitroulis G (2017) A personalized 3D-printed prosthetic joint replacement for the human temporomandibular joint: from implant design to implantation. J Mech Behav Biomed Mater 69:404–411, 2017/05/01/. https://doi.org/10.1016/j.jmbbm.2017.01.048
5. Chen RK, Jin YA, Wensman J, Shih A Additive manufacturing of custom orthoses and prostheses—a review. Addit Manufact 12:77–89
6. Tian Z et al (2020) A review on laser powder bed fusion of Inconel 625 nickel-based alloy. Appl Sci 10(1):81. Retrieved from https://www.mdpi.com/2076-3417/10/1/81
7. Ding D, Pan Z, Cuiuri D, Li H (2015) Wire-feed additive manufacturing of metal components: technologies, developments and future interests. Int J Adv Manufact Technol 81(1):465–481. https://doi.org/10.1007/s00170-015-7077-3
8. Yap CY et al (2015) Review of selective laser melting: materials and applications. Appl Phys Rev 2(4):041101. https://doi.org/10.1063/1.4935926
9. Gu DD, Meiners W, Wissenbach K, Poprawe R (2012) Laser additive manufacturing of metallic components: materials, processes and mechanisms. Int Mater Rev 57(3):133–164. https://doi.org/10.1179/1743280411Y.0000000014
10. Ngo TD, Kashani A, Imbalzano G, Nguyen KTQ, Hui D (2018) Additive manufacturing (3D printing): a review of materials, methods, applications and challenges. Comp Part B: Eng 143:172–196, 2018/06/15/ 2018. https://doi.org/10.1016/j.compositesb.2018.02.012
11. Hebert RJ (2016) Viewpoint: metallurgical aspects of powder bed metal additive manufacturing. J Mater Sci 51(3):1165–1175, 2016/02/01 2016. https://doi.org/10.1007/s10853-015-9479-x
12. Townsend A, Senin N, Blunt K, Leach RK, Taylor JS (2016) Surface texture metrology for metal additive manufacturing: a review. Precis Eng 46:34–47, 2016/10/01/ 2016. https://doi.org/10.1016/j.precisioneng.2016.06.001
13. Strano G, Hao L, Everson RM, Evans KE (2013) Surface roughness analysis, modelling and prediction in selective laser melting. J Mater Process Technol 213(4):589–597, 2013/04/01/ 2013. https://doi.org/10.1016/j.jmatprotec.2012.11.011
14. Poulin JR, Brailovski V, Terriault P (2018) Long fatigue crack propagation behavior of Inconel 625 processed by laser powder bed fusion: Influence of build orientation and post-processing conditions. Int J Fatigue 116:pp. 634–647, 2018/11/01/ 2018. https://doi.org/10.1016/j.ijfatigue.2018.07.008
15. Brandl E, Heckenberger U, Holzinger V, Buchbinder D (2012) Additive manufactured AlSi10Mg samples using selective laser melting (SLM): microstructure, high cycle fatigue, and fracture behavior (in English). Mater Des 34:159–169. https://doi.org/10.1016/j.matdes.2011.07.067
16. Qian GA, Jian ZM, Pan XN, Berto F (2020) In-situ investigation on fatigue behaviors of Ti-6Al-4V manufactured by selective laser melting," (in English). Int J Fatigue 133:9, 105424. https://doi.org/10.1016/j.ijfatigue.2019.105424
17. Sander G, Babu AP, Gao X, Jiang D, Birbilis N (2021) On the effect of build orientation and residual stress on the corrosion of 316L stainless steel prepared by selective laser melting (in English). Corr Sci 179:9. https://doi.org/10.1016/j.corsci.2020.109149
18. Labonnote N, Rønnquist A, Manum B, Rüther P (2016) Additive construction: State-of-the-art, challenges and opportunities. Automat Constr 72:347–366, 2016/12/01/ 2016. https://doi.org/10.1016/j.autcon.2016.08.026

19. Nagalingam AP, Yeo SH (2020) Surface finishing of additively manufactured Inconel 625 complex internal channels: A case study using a multi-jet hydrodynamic approach. Addit Manufact 36:101428, 2020/12/01/ 2020. https://doi.org/10.1016/j.addma.2020.101428
20. Polishetty A, Shunmugavel M, Goldberg M, Littlefair G, Singh RK (2016) Cutting force and surface finish analysis of machining additive manufactured titanium alloy Ti-6Al-4V. Procedia Manufact 7:284–289, 2017/01/01/ 2017. https://doi.org/10.1016/j.promfg.2016.12.071
21. Bouland C, Urlea V, Beaubier K, Samoilenko M, Brailovski V (2019) Abrasive flow machining of laser powder bed-fused parts: numerical modeling and experimental validation. J Mater Process Technol 273:116262, 2019/11/01/ 2019. https://doi.org/10.1016/j.jmatprotec.2019.116262
22. Han S, Salvatore F, Rech J, Bajolet J (2019) Abrasive flow machining (AFM) finishing of conformal cooling channels created by selective laser melting (SLM). Precis Eng 64:20–33, 2020/07/01/ 2020. https://doi.org/10.1016/j.precisioneng.2020.03.006
23. Guo J et al (2019) Novel rotating-vibrating magnetic abrasive polishing method for double-layered internal surface finishing. J Mater Process Technol 264:422–437, 2019/02/01/ 2019. https://doi.org/10.1016/j.jmatprotec.2018.09.024
24. El Hassanin et al A (2019) Fluidised bed machining of metal additive manufactured parts. In: Galdos L et al (eds) Proceedings of the 22nd international Esaform conference on material forming, vol 2113, AIP Conference Proceedings. Melville: Amer Inst Physics
25. Kim D, Son K, Sung D, Kim Y, Chung W (2015) Effect of added ethanol in ethylene glycol-NaCl electrolyte on titanium electropolishing. Corros Sci 98:494–499. https://doi.org/10.1016/j.corsci.2015.05.057
26. Zhang Y, Li J, Che S, Tian Y (2020) Electrochemical polishing of additively manufactured Ti–6Al–4V alloy. Met Mater Int 26(6):783–792. https://doi.org/10.1007/s12540-019-00556-0
27. Mendez J, Akolkar R, Andryushchenko T, Landau U (2008) A mechanistic model for copper electropolishing in phosphoric acid," (in English). J Electrochem Soc 155(1):D27–D34. https://doi.org/10.1149/1.2801396

Effect of Laser Shock Peening on Fatigue Performance of Fracturing Pump

Ping Liu, Guojie Li, Yazhou Wang, Liangshuyi Zhang, Nan Jiang, and Xiaoyin Li

Abstract Hydraulic fracturing is widely accepted and has become the most useful method to increase exploiting production of shale oil and gas in more complex geological environment. As the key component of the hydraulic fracturing, fracturing pump bears high cyclic pressure and is prone to fatigue failure around intersection line of the fluid cylinder, which is closely associated with the material surface state. In this paper, residual stress and axial tension fatigue test are investigated on the specimens with and without Laser Shock Peening (LSP) treatment. Numerical simulation is conducted to identify stress distribution and LSP region of fracturing pump, which is then subjected to LSP and fatigue performance test. Results showed that LSP produced a maximum compressive residual stress (CRS) of up to 730 MPa on the specimen surface and a CRS layer with a thickness of 1.5 mm under the upper surface. Tension-tension fatigue life was increased by 141.08% at 800 MPa after LSP treatment. Stress concentration of the fracturing pump was located at intersection line according to finite element analysis. And then, LSP region was designated as the area where Mises stress larger than 80% of Max. The service life was increased by 68.3% with subsequent LSP which was attributed to the compressive residual stress.

P. Liu (✉) · G. Li · L. Zhang · N. Jiang
Xi'an Tyrida Optical Electric Technology Co., Ltd, Xi'an, China
e-mail: 604341474@qq.com

G. Li
e-mail: ligj@tyrida.com

L. Zhang
e-mail: zhangliangshuyi@126.com

N. Jiang
e-mail: 2652529840@qq.com

Y. Wang
Air Force Engineering University, Xi'an, China
e-mail: wangyz@tyrida.com

X. Li
Xi'an Sool Photoelectric Technology Co., Ltd, Xi'an, China
e-mail: 63316618@qq.com

N. Maharjan and W. He (eds.), *Proceedings of the 3rd International Conference on Advanced Surface Enhancement (INCASE) 2023*, Lecture Notes in Mechanical Engineering, https://doi.org/10.1007/978-981-99-8643-9_17

Keywords Laser shock peening · Fracturing pump · finite element analysis · Residual stress · Fatigue life

1 Introduction

Fracturing pump is an important component being used for supplying fluid to oil and gas in the hydraulic fracturing process to increase exploit production. However, interior cracks occurring at certain location results in less than 300 h of service life and great financial losses [1, 2]. The main failure mechanism of fracturing pump is fatigue failure under alternating high loads with the source of cracks locating at the intersection line of inner cavity. In addition to improving the structure and material of pump, surface enhancement technology could also be adopted to improve fatigue performance [3–6].

The fatigue crack initiation region of fracturing pump is located in inner cavity and conventional surface enhancement techniques, such as shot peening, cold rolling, electroplating, etc., are difficult to implemented due to the complex structure. Hua Jian et al. investigated the autofrettage technology through finite element analysis and showed that the residual compressive stress generated by the autofrettage technology can effectively improve the fatigue life of the fracturing pump [7, 8]. Zhou Sizhu et al. concluded that the maximum autofrettage pressure of the fracturing pump is 400–450 MPa by finite element analysis for different types of fracturing pumps [9, 10]. Currently fracturing pump studies based on finite element analysis have found that resolving stress concentrations at the intersection line can improve mechanical damage. But it is difficult to achieve autofrettage pressure for fracturing pump with multiple sets of cross-hole cavities in practice [11, 12].

Laser shock peening is a new surface enhancement technique, which can generate plastic deformation on metals by the mechanical effect of high-energy pulsed laser-induced shock waves. Thereby, the fatigue performance can be improved due to implanted compressive residual stresses and altered surface microstructure [13, 14]. Zhou Wang et al. [15] carried out LSP investigation for 17-4 PH stainless steel. It was found that LSP provided -700 MPa residual stress on surface and elongated the fatigue life by a factor of around 5. Ramadas et al. [16] experimentally conducted LSP on 15-5 PH stainless steels. The fatigue test results showed LSP increased the fatigue life by 2.5 times compared to the none-LSP specimens at an intermediate stress value of 650 MPa. The aforementioned research results showed that LSP can significantly improve the fatigue resistance of stainless steels materials. However, no studies involved LSP region of fracturing pump and fatigue performance of fracturing pump after LSP.

In this paper, 17-4 PH stainless steel is subjected to LSP treatment, residual stress characterization and fatigue test. Preferable LSP parameter is obtained and applied to fatigue crack initiation region of fracturing pump acquired by finite element analysis. The effect of LSP on the fatigue performance of fracturing pump is evaluated by a trial under actual working conditions.

Table 1 Chemical compositions of 17-4 Ph

Elements	Cr	Ni	Cu	Si	Mn	C	Fe
Composition (wt%)	15.5–17.5	3–5	3–5	0–1	0–1	0–0.07	Bal

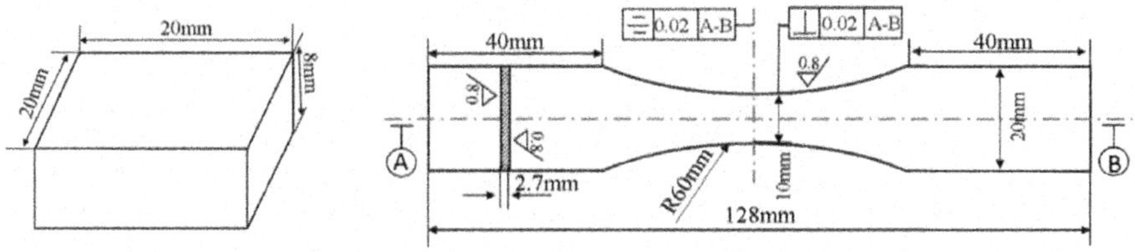

(a) Residual stress test sample (b) Fatigue test sample

Fig. 1 Sample preparation

2 Material and Methods

2.1 Material and Sample

The 17-4 PH stainless steel material is taken from the pump, which use the conventional heat treatment process: solution treatment at 1030–1050 °C, air cooled or water-quenched. The chemical composition is listed in Table 1.

The fatigue test of the sample mode actual working conditions and ignore some influencing factors. So, the result of fatigue test and pump service life will have Slight difference. The plate samples for residual stress test is machined from round bar material with the geometry of 20 × 20 × 8 mm shown in Fig. 1a. One of 20 × 20 mm surface is mechanically polished to a roughness of Ra 0.8. The dimensions and schematic diagrams of the specimen for fatigue test is shown in Fig. 1b.

2.2 LSP Experiments

LSP is performed by using a YAG laser apparatus (TYRIDA, China) with wavelength 1064 nm, pulse duration 20 ns, laser power density 5.6 GW/cm^2, and 1 time laser shock impact. Two types of LSP parameters are implemented on both residual stress and fatigue test sample. The detailed parameters are shown in Table 2

The LSP region of the residual stress test sample is 20 × 20 mm. Fatigue test sample is LSPed on double sides with Z-path in 50 × 20 mm zone, as shown in Fig. 2. For fracturing pump, LSP region is determine according to simulation results in Sect. 3.2 and LSP parameter is selected as LSP1 based on the residual stress and fatigue performance in Sect. 3.1.

Table 2 LSP parameters design of different specimens

No	Wavelength	Pulse duration	Ablative layer	Overlapping ratio
None-LSP	/	/	/	/
LSP1	1064 nm	20 ns	Black tape	50%
LSP2	1064 nm	20 ns	/	50%

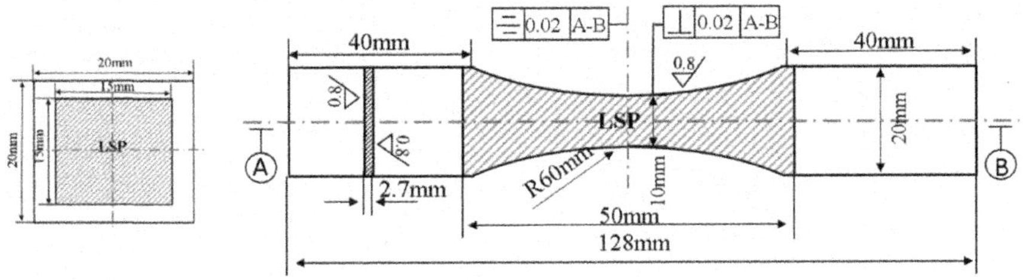

(a) residual stress specimen (b) fatigue specimen

Fig. 2 LSP region

2.3 *Residual Stress Characterization*

The residual stress test is carried out using the Proto-LXRD X-ray diffractometer (Proto, Canada) as shown in Fig. 3. via the sin2φ method. A single axis goniometer with Cr–Kα X-rays is used in the tests. Measurement points are shown in Fig. 3. X-ray source is Cr-Kα and the X-ray is 1 mm. The diffraction plane and angle are {211} and 156°, respectively. The PROTO's POLISHER 8818 V-3 electropolishing machine (Shanghai, China) is used to remove the extra surface material layer by layer. The corrosion liquid is Saturated NaCl solution, and the diameter of the polishing head is 6 mm.

2.4 *Fatigue Test and Fracture Morphology Observation*

The fatigue life test is performed on a QBG high-frequency fatigue equipment (QBG-100, China), as shown in Fig. 4. The fatigue performance of 17-4 PH samples with and without LSP is evaluated at room temperature with maximum load of 800 MPa and stress ratio of 0.1. Loading frequency is automatically determined by the testing control system and specimen based on the principle of resonance [17–19]. During the test, the load frequency is quickly and automatically searched and stable at 98 Hz in this work. After fatigue test, No. 1, 4 and 7 specimens with the state of none-LSP, LSP1 and LSP2 respectively are selected for fracture morphology observation to characterize fracture behavior by SEM.

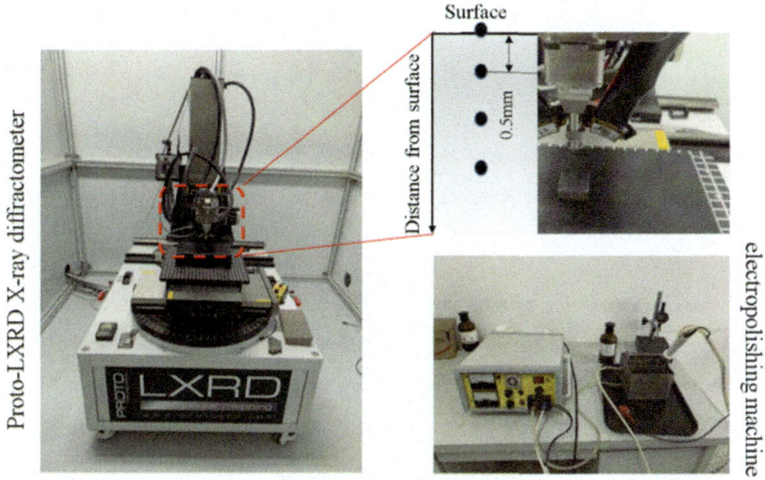

Fig. 3 Machine and measurement points of the residual stress

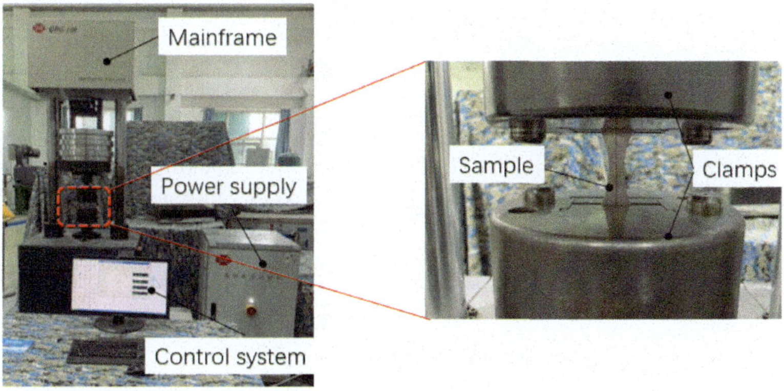

Fig. 4 QBG-100 high frequency fatigue testing equipment

2.5 *Finite Element Modeling*

The numerical simulation is carried out to extract the stress distribution under working conditions and then, LSP region can be determined according to working stress distribution. As shown in Fig. 5., a simplified pump model is setup and meshed with C3D4 (a 4-node linear tetrahedron, three dimensional) elements whose number is 779,000. Uneven density mesh strategy is applied with 1 mm minimum grid size on intersection line and 10 mm maximum grid size on other area. Material is 17-4 PH steel, elastic modulus is 210 GPa and Poisson's ratio is 0.3. A pressure of 80 MPa is set on inner cavity of fracturing pump.

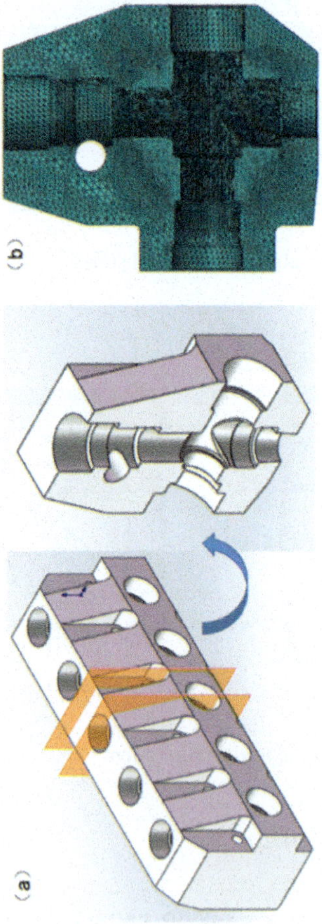

Fig. 5 Finite element model

3 Results and Discussion

3.1 Residual Stress and Fatigue Performance of Specimen

With LSP treatment, a compressive residual stress layer with a gradient trend is generated below the surface of the sample (Fig. 6). Similar surface residual stress distributions are observed in LSPed Ti and Al alloys [17]. The maximum compressive residual stress is 730 and 702 MPa for LSP1 and LSP2, respectively. Compressive residual stress depth is deeper than 1.5 mm both for LSP1 and LSP2.

The fatigue life of samples: none-LSP, LSP1 and LSP2 are shown in Table 3. The comparison of average fatigue life is displayed in Fig. 7. The fatigue life under 800 MPa of none-LSP, LSP1 and LSP2 are 5.1×10^5 cycles, 12.3×10^5 cycles and 9.22×10^5 cycles respectively. Compared with none-LSP state, the fatigue life of 17-4 PH is increased by 141.08% and 80.6% for LSP1 and LSP2. Based on the results, LSP1 is much better and selected to peen the pump.

Fatigue crack source of none-LSP is on the top surface and about 250 μm away from side surface (Fig. 8a). As for LSP1 and LSP2, fatigue crack sources are located on the side surface and are around 600 and 400 μm from bottom and top surface separately, where undergone LSP processing. Such findings could be attributed to the LSP being able to produce a compressive residual stress layer on the bottom and top surface of the material, which forces the fatigue crack source to shift. Furthermore, LSP1 has greater residual compressive stress, resulting in a deeper crack source from LSP surface than LSP2. The propagation path is more tortuous after LSP (Fig. 8b). Compressive residual stress induced by LSP hinder the main crack growth, which then turn the direction to relatively weak strength region, resulting in tortuous crack growth path. Crack propagation path changing retards fatigue crack propagation and reduces the fatigue crack propagation rate. In addition to the fatigue striations during crack propagation (Fig. 8c), secondary cracks could be obviously observed in crack

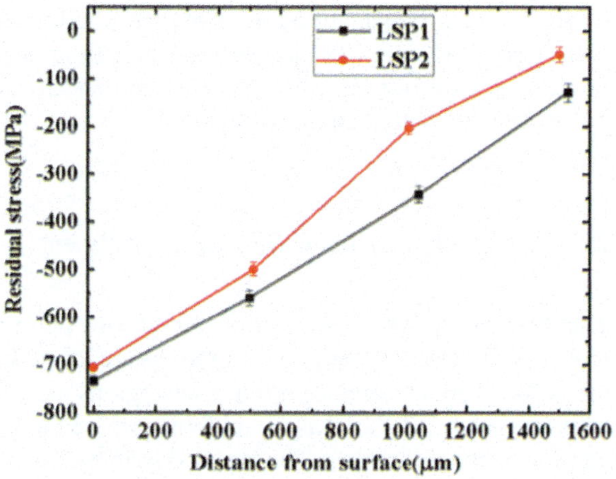

Fig. 6 Residual stress distributions

Table 3 Fatigue life of none-LSP, LSP1 and LSP2

State	No	σ (MPa)	R	N_f ($\times 10^5$)	Average($\times 10^5$)
None-LSP	1	800	0.1	5.47	5.1
	2			4.99	
	3			4.85	
LSP1	4			16.13	12.3
	5			9.30	
	6			11.48	
LSP2	7			10.27	9.22
	8			8.57	
	9			8.81	

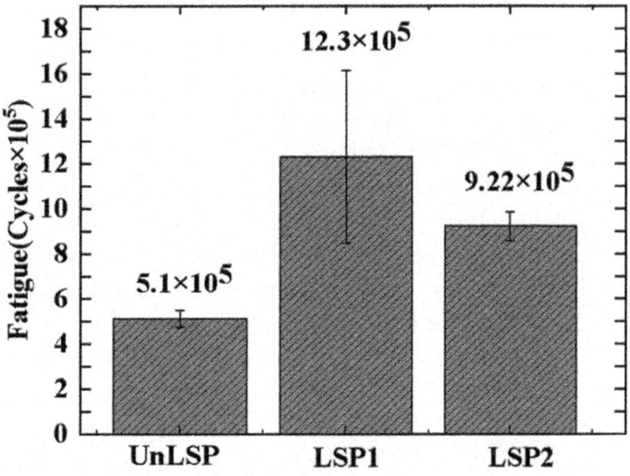

Fig. 7 Average fatigue life of none-LSP, LSP1 and LSP2

propagation area (Fig. 8d), due to local material tearing brought by the crack tip when the crack propagation rate is higher. The higher number of secondary cracks in the none-LSP indicates that the crack propagation rate is higher and the cracks are prone to destabilization, whereas LSP1 and LSP2 have fewer number of secondary cracks and a lower crack expansion rate.

3.2 LSP Region Identification of Pump

The inner cavity stress distribution of fracturing pump under work condition is shown in Fig. 9. Significant stress concentrations are located along the intersection line, and the order of Mises stresses at the four regions are A > B > C > D. The maximum stress region A is located at the midpoint where the intersection of plunger chamber and the suction chamber. The stress distributions in two paths on each region are

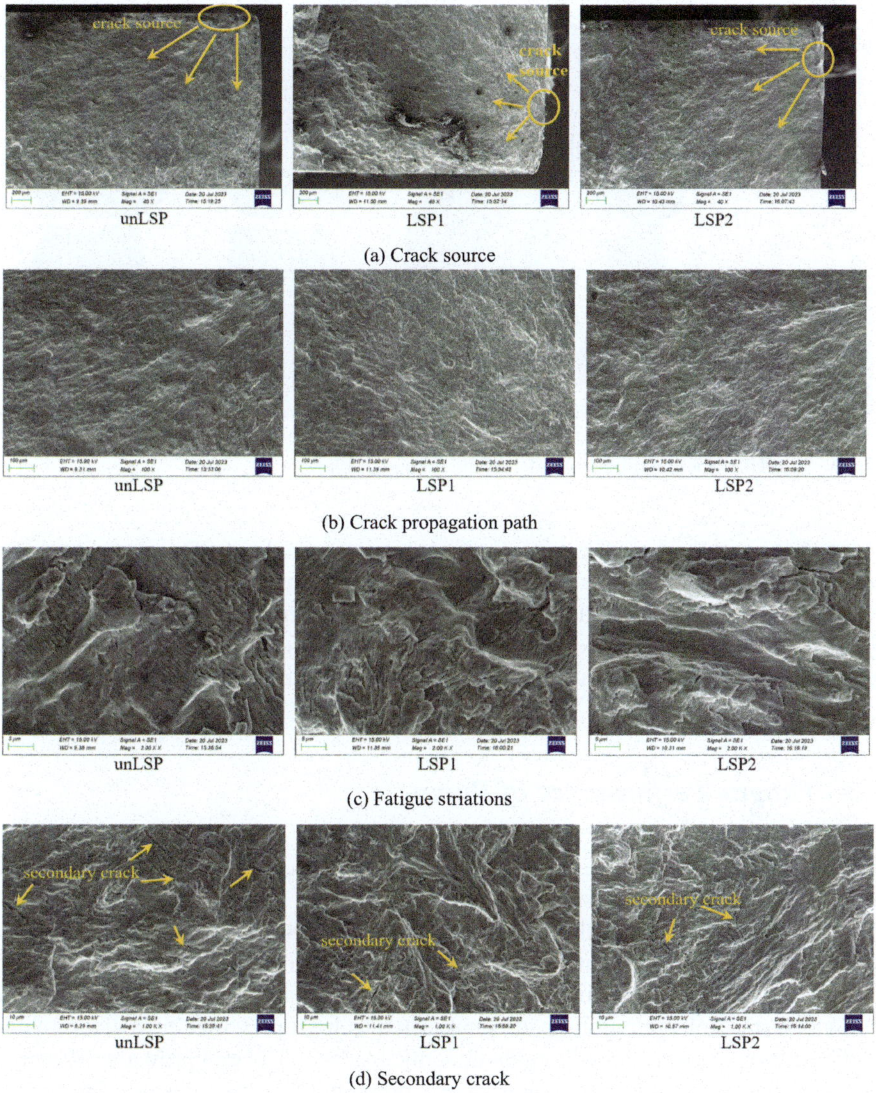

Fig. 8 Fracture morphologies of none-LSP, LSP1 and LSP2

extracted to analyze the LSP region. L1 path is along the intersection line and L2 is along the axial extension of the chamber boundary.

As shown in Fig. 10, the maximum stress concentration level σ_{max}, the safety stress level σ_0 (selected as $0.8\sigma_{max}$) and the location safe stress level region is extracted. In region A, the safety stress is 345 MPa and the region with stress level from 345 to 441 MPa is identified as LSP region. In inner cavity, 198 mm away from the edge of the plunger chamber and 127 mm away from the edge of the suction chamber, 31 mm

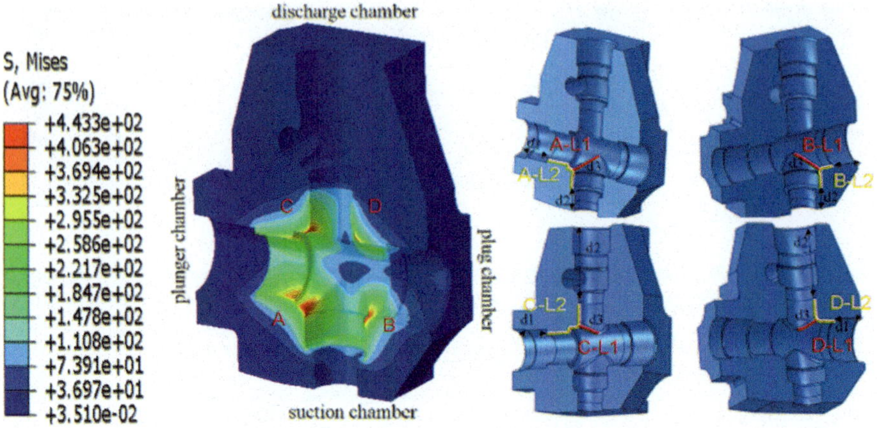

Fig. 9 Stress distribution of the cylinder of the fracturing pump

along the circumference of the intersection line is covered in LSP region. Similar to region A, LSP region in area B to D are determined with parameters d1-d3 shown in Fig. 7. As a results, LSP region is identified as area where working stress exceeding the selected safety stress level to meet the requirement for extending service life and lower LSP costs of the fracturing pump at the same time.

3.3 Fatigue Performance of Pump

The fracturing pump service life (Anton) trial is carried on Anton Wei Yuan 204 H40 and 204 H51 platform. Results are shown in Table 4. The average service life of fracturing pump without and with LSPed are 347 and 584 h, separately. The average service life of fracturing pump is increased by 68.3% after LSP.

The working conditions of the pump head are complex and non-constant, and the internal surface of the pump head is subjected to high pressure, corrosion/erosion of the fracturing medium and fatigue dynamic loads, and a large pressure difference between the suction pressure and discharge pressure of the pump head. The fatigue life of the sample is conducted under constant stress levels and testing frequency conditions, with a relatively single environment and working conditions, ignoring some influencing factors, and cannot fully comply with the actual working conditions. Therefore, the fatigue life of the sample is higher than the service life, it can be used as a reference factor for predicting the fatigue life of the pump head and has certain engineering value.

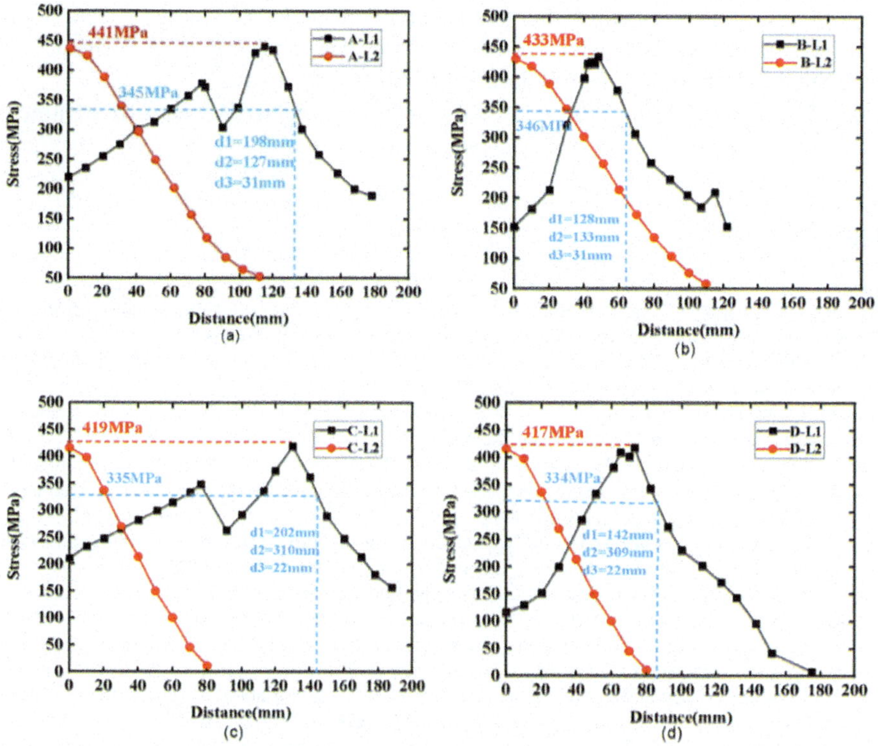

Fig. 10 Stress distribution in inner cavity of fracturing pump

Table 4 Service life test paraments

Fracturing pump	Pressure (MP)	Gear	Liquid volume (m^3/h)	Liquid volume statistics (m^3)	Usage time (h)
None-LSP	65.5	2.45	34.32	13,984	347
LSP	65.5	2.45	34.32	14,540	584

4 Conclusions

In this work, LSP process parameters and region of 17-4 PH fracturing pump are identified based on the residual stress, fatigue test and established numerical simulation model. The fatigue performance of fracturing pump is investigated with/without LSP treatment. The main research results can be summarized as follows:

(1) For 17-4 PH, LSP could introduce the residual compressive stress of 730 MPa and the depth of compressive stress layer exceeds 1.5 mm.

(2) Stress concentration of the fracturing pump is located at intersection line according to finite element analysis. And then, LSP region is designated as the area whose Mises stress larger than 80% of max stress.
(3) The service life of fracturing pump is increased by 68.3% after LSP treatment.

References

1. Jiang FG, Liang Z, Zhao GH et al (2012) Research on hydraulic self-enhancement technology of fracturing pump's valve box[J]. Adv Mater Res 479–481:2076–2081
2. Wang GR, Chen LY, Zhao M et al (2014) The research on failure analysis of fluid cylinder and fatigue life prediction[J]. Eng Fail Anal 40:48–57
3. Xu L, Wu G, Si T (2018) Analysis and countermeasures of cross hole cracking of fracturing pump[J]. Jinshu Rechuli/Heat Treat Metals 43(3):233–236
4. Hoehbusch J, Bucher S, Haensel H et al (2015) Evaluation of the fatigue strength of multi-phase pump components under corrosive conditions using a short crack fracture mechanics approach[J]. Eng Fail Anal 47:265–272
5. Wei Y, Li Y, Lai J et al (2020) Analysis on corrosion fatigue cracking mechanism of 17–4PH blade of low pressure rotor of steam turbine[J]. Eng Fail Anal 118:104925
6. Fu JD, Wan S, Yang Y et al (2021) Accelerated corrosion behavior of weathering steel Q345qDNH for bridge in industrial atmosphere[J]. Constr Build Mater 306:124864
7. Fang HJ, Du DJ, Zhou HJ et al (2013) Local mesh refinement-based finite element analysis of pump head strength [J]. China Petrol Mach 41:98–101
8. Hua J, Liao JM, Zhou SZ et al (2015) Research on secondary autofrettage treatment of head body of ultrahigh pressure fracturing pump[J]. Sci Technol Eng 15:170–174
9. Zhou SZ, Yan FL, Li L et al (2013) Application of the Bauschinger effect on pump head autofrettage, China Petroleum Machinery [J]. China Petrol Mach 41:62–66
10. He X, Zhao M, Chen LX et al (2014) Fatigue life analysis for fluid cylinder based on autofrettage technology [J]. Mech Sci Technol Aerospace Eng 33:208–211
11. Yao ZM, Yi XZM, Zhou YH et al (2019) Mechanical analysis of the pump head of the CDJY2500 five-cylinder fracturing pump [J]. China Petrol Mach 47(10):7
12. Zhang S, Zhou YH, Li N et al (2017) Research on stress distribution and stress intensity factor of crack tip on interesting line of fracturing pump head [J]. Sci Technol Eng 17(5):7
13. Peyre P, Fabbro R, Merrien P et al (1996) Laser shock processing of aluminium alloys. application to high cycle fatigue behaviour[J]. Mater Sci Eng A 210(1–2):102–113
14. Clauer AH, Lahrman DF (2001) Laser shock processing as a surface enhancement process[J]. Key Eng Mater 197:121–142
15. Zhou W, Jiang C, Gan X et al (2011) Influence of shot peening on the fatigue life of laser hardened 17–4PH steel[J]. Int J Fatigue 33(4):549–556
16. Ramadas H, Sarkar S, Ganesh P et al (2023) Enhancing the static and dynamic mechanical properties of laser powder bed fusion process built 15–5 precipitation hardening stainless steel specimens by laser shock peening. Mater Sci Eng A 866:144657
17. Gao T, Ji C, Zhan Z et al (2023) A novel defect-based fatigue damage model coupled with an optimized neural network for high-cycle fatigue analysis of casting alloys with surface defect [J]. Int J Fatigue 170:107538
18. Zhang Y, Zhang H, Xue J et al (2023) Microstructure transformed by heat treatment to improve fatigue property of laser solid formed Ti6Al4V titanium alloy [J]. Mater Sci Eng A 865:144363
19. Huang C, Hai L, Jiang J et al (2023) High-cycle fatigue properties of explosion bonded titanium-clad bimetallic steel [J]. Int J Fatigue 169:107499

Effect of Plasma Electrolytic Polishing Technology on Stainless Steel Surface Composition and Cr/Fe

Mei-Yi Liu, Tzu-Hong Chen, Wen-Chieh Wu, and Chen hui Chang

Abstract Plasma electrolytic polishing (PEP) is a technology with high efficiency, environmental friendly, simple manufacturing process, high quality and ability to improve corrosion resistance. Plasma electrolytic polishing improves corrosion resistance by forming oxides on the surface of workpieces and achieving passivation. In this study, the SUS 316 was polished with different plasma electrolytic polishing electrolytes. We observed the influence of electrolyte on the ratio of chromium to iron to understand the effect of plasma electrolytic polishing technology on improving the corrosion resistance of materials. After plasma electropolishing of SUS 316 using different electrolytes (voltage: 320 V; time: 180 s; temperature: 80 °C), it was found that the surface components were mainly oxides of iron and oxides of chromium. The oxide is concentrated on the surface of the workpiece with a thickness of about 3–750 nm. Plasma electrolytic polishing with different electrolytes will produce oxide layers with different thicknesses. After plasma electrolytic polishing, the stainless steel workpiece has a surface chromium-iron ratio of 1.64 (optimal condition). This study shows that plasma electrolytic polishing technology can optimize the oxide layer on the surface of stainless steel and achieve the effect of improving corrosion resistance.

Keywords Plasma electrolytic polishing (PEP) · 316 Stainless steel · Corrosion resistance · Cr/Fe ratio · Electrolyte

M.-Y. Liu · T.-H. Chen (✉) · W.-C. Wu · C. Chang
Metal Industries Research and Development Centre (MIRDC), Industrial Area, 25, 37 Road, Taichung, Taiwan, Republic of China
e-mail: ancel.chen@mail.mirdc.org.tw

M.-Y. Liu
e-mail: mei-yi@mail.mirdc.org.tw

W.-C. Wu
e-mail: wcwu@mail.mirdc.org.tw

N. Maharjan and W. He (eds.), *Proceedings of the 3rd International Conference on Advanced Surface Enhancement (INCASE) 2023*, Lecture Notes in Mechanical Engineering, https://doi.org/10.1007/978-981-99-8643-9_18

1 Introduction

When plasma electrolytic polishing (PEP), the workpiece is connected to the anode, the metal plate is connected to the cathode, and the both of them are immersed in the electrolyte. Provide a specific voltage to lead to the cathode and anode occur violent electrolytic reactions and release great amount of thermal energy.

A layer of gas film is formed around the workpiece (anode). The main component of the gas film is water vapor and some oxygen. This layer of gas film isolates the workpiece from the electrolyte.

However, due to the presence of a strong electric field, the gas film is punctured, forming discharge channels that remove material. There is a higher probability of discharge channels forming and material removal at the micro-protrusions on the surface of the workpiece, thereby achieving the polishing effect [1, 4, 5].

Plasma electrolytic polishing offers advantages such as high polishing efficiency, easy to operate, excellent polishing quality, environmentally friendly, and improved corrosion resistance. One method to enhance corrosion resistance is through the passivation phenomenon that occurs on the surface during the plasma electrolytic polishing process. Taking plasma electropolishing of stainless steel as an example, during the plasma electrolytic polishing process, oxidation occurs on the workpiece surface, forming iron oxide and chromium oxide. This iron oxide and chromium oxide rich film will protect the workpiece and reduce more material corrosion, achieve passivation effect and improve corrosion resistance. Of which, chromium is more easily oxidized and forms a dense oxide film compared to iron. Therefore, the chromium content on the surface of stainless steel significantly influences its corrosion properties, and the Cr/Fe ratio serves as a metrics of corrosion resistance [2].

In this study, different electrolytes were used to perform plasma electrolytic polishing on SUS316 stainless steel samples. The effects of the electrolytes on the surface Cr/Fe ratio and oxide film thickness of stainless steel were investigated to understand their influence. Furthermore, a preliminary evaluation of the plasma electrolytic polishing technique's effectiveness in enhancing material corrosion resistance was conducted.

2 Methods

The SUS316 plates were cut into sizes of $10 \times 10 \times 2$ mm and $55 \times 30 \times 2$ mm, respectively. A test samples with a size of $10 \times 10 \times 2$ mm was used for surface composition analysis, and a test samples with a size of $55 \times 30 \times 2$ mm was used for surface roughness and gloss measurement.

The SUS316 samples were connected to the anode and fixed in a beaker containing the electrolyte and cathode. The size of the cathode is 3 times that of the anode. Plasma electrolytic polishing was performed at a polishing voltage of 320 V for a duration of

Table 1 Plasma electrolytic polishing electrolytes and codes

Electrolyte	Code
5 wt% Sulfur-containing electrolyte	S-5
3 wt% Sulfur-containing electrolyte	S
3 wt% Sulfur-containing electrolyte with a small amount of citric acid	S-Citric acid
3 wt% nitrogen-containing Electrolyte (A)	N(A)
3 wt% nitrogen-containing Electrolyte (B)	N(B)
3 wt% nitrogen-containing Electrolyte (B) with a small amount of citric acid	N(B)-Citric acid

3 min. The electrolyte temperature was maintained at 80 °C. The electrolyte formula is listed in Table 1.

The surface roughness (Ra) of the SUS316 samples was measured with a surface roughness meter (Accretech/E-35A/Japan) before and after polishing, and then the rate of reduction in surface roughness (Ra) was calculated. The calculation formula is shown in formula 1.

$$\text{Rate of reduction in surface roughness (Ra)} = Ra_{(after)}/Ra_{(before)} * 100 \quad (1)$$

The gloss of the SUS316 samples was measured using a gloss meter (3nh/YG268/China) at measurement angles of 20°, 60°, and 85°. Except for two specific cases, all measurements were conducted at a 60° angle. The first exception is when the data obtained at a 60° angle exceeds 70 GU (gloss units), in which case the data obtained at a 20° angle is additionally recorded. The second exception is when the data obtained at a 60° angle is less than 10 GU, in which case the data obtained at an 85° angle is used instead.

The surface composition of the SUS316 samples was analyzed using an X-ray Photoelectron Spectroscope (XPS) (ULVAC-PHI/PHI 5000 VersaProbe/Japan). Both surface composition analysis and depth profile were performed on the samples. The scanning speed for the depth profile analysis was set at 30 nm/min. The measured elements included O1s, C1s, Fe2p3, and Cr2p3, with the additional measurement of N1s under certain conditions.

3 Results and Discussion

After polishing the SUS316 samples with S-5, S, S-Citric acid, N(A), N(B), and N(B)-Citric acid, it was observed that the surfaces had significant glossiness with S-5, S, S-Citric acid and N(B)-Citric acid, while there was no glossiness observed with N(A) and N(B). The surface roughness data is shown in Table 2, which demonstrates the reduction in surface roughness of the SUS316 samples after polishing with different electrolytes. Specifically, after polishing with the sulfur-containing electrolyte, the surface roughness is reduced to 65.63–68.64% of its original value, demonstrating an effective improvement in surface roughness. On the other hand, polishing with the nitrogen-containing electrolyte showed the surface roughness is reduced to 72.73–90.91% of its original value.

By measuring the gloss at a 60° angle, it was found that the gloss of the SUS316 samples polished with S-5, S, and S-Citric acid electrolytes achieved 70 GU, indicating a high gloss material. The data obtained at a 20° angle, the best gloss of 71.5 GU and 72.0 GU was achieved with S-5 and S polishing, respectively, while the SUS316 samples polished with S-Citric acid electrolytes, the gloss was 56.7 GU. When N(A) and N(B) electrolytes were used for polishing, the gloss measured at a 60° angle was 11.4 GU and 7.7 GU, respectively, with N(B) having a gloss of less than 10 GU, indicating a low gloss material. The gloss measured at an 85° angle for N(B) was 10.7 GU. When N(B)-Citric acid electrolytes were used for polishing, the gloss measured at a 60° angle was 46.2 GU, which was better than other nitrogen-containing electrolytes.

The surface composition of the SUS316 samples polished with different electrolytes is shown in Fig. 1. The surface composition of all conditions contains carbon, oxygen, iron and chromium elements, which may be oxides (or hydroxides) formed by oxidation of iron and chromium [3]. After polishing the SUS316 samples using N(A) and N(B) electrolytes, the presence of nitrogen was observed on the surface. However, when N(B)-Citric acid was used for polishing, nitrogen was not observed on the surface of the samples, indicating a possible uneven distribution of nitrogen.

The chromium-to-iron ratio (Cr/Fe) for different conditions is compared in Table 3. Comparing S, N(A), and N(B), it was found that S had a higher Cr/Fe ratio (0.74). Additionally, when comparing S and S-5, it was observed that increasing

Table 2 Surface roughness and gloss

Electrolyte	Rate of reduction in surface roughness (Ra) (%)	Surface roughness (Ra) (pre/post polishing (μm))	Gloss (GU) (Measurement Angle: 60°)	Gloss (GU) (Measurement Angle: 20°)	Gloss (GU) (Measurement Angle: 85°)
S-5	65.63	1.1000/ 0.7219	97.9	71.5	
S	66.59	1.1000/ 0.7325	98.3	72.0	
S-Citric acid	68.64	1.1000/ 0.7550	79.0	56.7	
N(A)	72.73	1.1000/ 0.8000	11.4		
N(B)	90.91	1.1000/ 1.0000	7.7		10.7
N(B)-Citric acid	84.55	1.1000/ 0.9300	46.2		

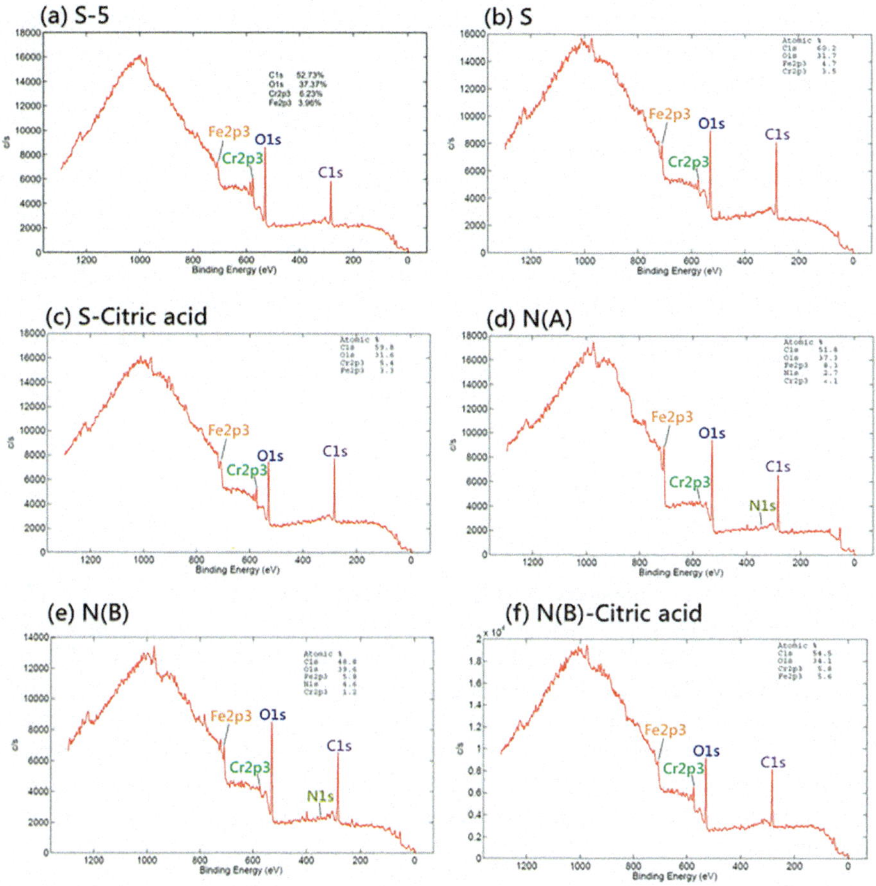

Fig. 1 XPS surface composition analysis of SUS316 samples polished with different electrolytes

the concentration contributed to an improvement in the Cr/Fe. The addition of citric acid also contributed to an increase in the surface Cr/Fe, possibly due to the acidic environment provided by citric acid favoring the formation of chromium oxides.

The results of XPS depth profile analysis for SUS316 samples polished with different electrolytes are shown in Fig. 2. The analysis provides insights into the thickness of iron and chromium oxides (or hydroxides) from the surface to the interior of the substrate. The horizontal axis represents the sputter time, which increases with increasing measurement depth. It can be observed that after polishing SUS316

Table 3 Chromium-to-iron ratio

Electrolyte	S-5	S	S-Citric acid	N(A)	N(B)	N(B)-Citric acid
Cr/Fe	1.57	0.74	1.64	< 0.01	0.21	1.04

samples with S-5, S and S-Citric acid, the thickness of iron and chromium oxides (or hydroxides) layer on the surface is approximately 3 nm. However, when polishing with N(A), the thickness of iron and chromium oxides (or hydroxides) layer on the surface exceeds 600 nm, and the most of the oxides of iron and chromium are concentrated on the surface of the test piece with a thickness of 153 nm. After polishing with N(B), the thickness of iron and chromium oxides (or hydroxides) layer on the surface exceeds 750 nm, and the most of the oxides of iron and chromium are concentrated on the surface of the test piece with a thickness of 93 nm.

These results differ significantly from those obtained with N(B)-Citric acid, where the thickness of iron and chromium oxides (or hydroxides) layer on the surface is only 3 nm. In addition, trace amounts of nitrogen were observed on the surfaces of N(A), N(B), and N(B)-Citric acid samples, with a thickness of at least 60 nm and possibly exceeding 750 nm.

4 Conclusion

This study conducted plasma electrolytic polishing on 316 stainless steel (SUS316) samples using different electrolytes. The results showed that sulfur-containing electrolyte (S-5, S, and S-Citric acid) had better polishing effects and improved the Cr/Fe ratio. Among them, S-Citric acid had a Cr/Fe ratio of 1.64, which was the optimal condition for Cr/Fe enhancement. The thickness of the passivation layer with a higher Cr/Fe ratio is approximately 3 nm. This study investigates the effect of electrolytes used in plasma electrolytic polishing on the surface Cr/Fe ratio, beneficial to improve the selection of electrolytes for polishing workpieces with high corrosion resistance requirements.

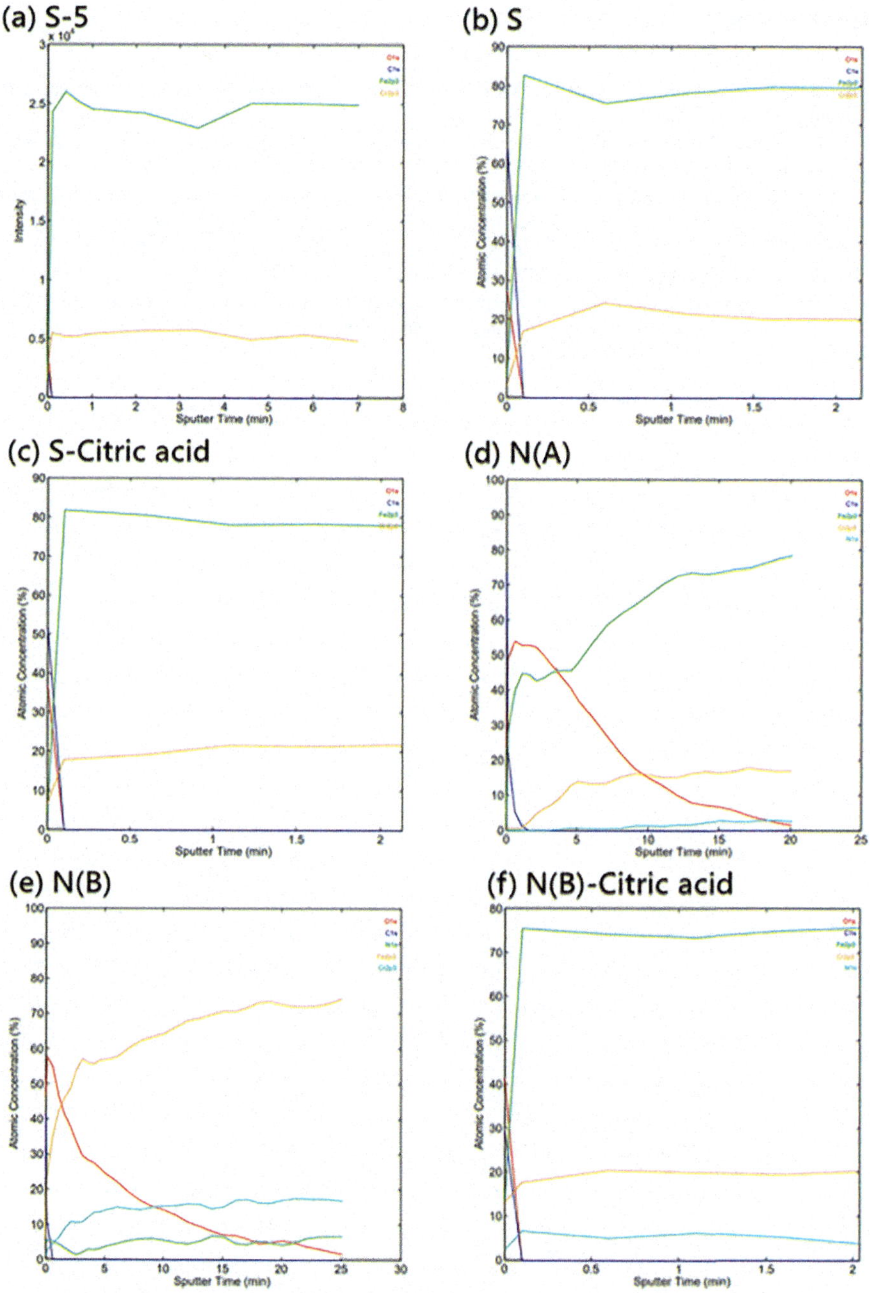

Fig. 2 The XPS depth profile analysis of SUS316 samples polished with different electrolytes

References

1. Cornelsen M, Deutsch C, Seitz H (2017) Electrolytic plasma polishing of pipe inner surfaces. Metals 8(1):1–12
2. Han W, Fang F (2019) Electropolishing of 316L stainless steel using sulfuric acid-free electrolyte. J Manuf Sci Eng 141:1–12
3. Seo B, Park HK, Kim HG, Kim WR, Park K (2021) Corrosion behavior of additive manufactured CoCr parts polished with plasma electrolytic polishing. Surf Coat Technol 406:126640
4. Wang J, Suo LC, Guan LL, Fu YL (2012) Analytical study on mechanism of electrolysis and plasma polishing. Adv Mater Res 472–475:350–353
5. Zeidler H, Böttger T, Schröder S, Schneider M, Lämmel C, Sahr F, Tardelli J, Exbrayat L (2022) Analysis of plasma-electrolytic polishing process initiation. Procedia CIRP 108:782–786

Characteristics of Residual Stress Profiles in Spring Steel Induced by Stress Rolling

Eckehard Mueller

Abstract Stress peening is a modern technique to enhance the durability of spring. The springs are shot peened under prestress, and after unloading, the residual stress will increase significantly. The same procedure can also be done by deep rolling. The difference is that deep rolling has two different residual stress directions (amounts) if you look along or perpendicular to the rolling track. Therefore, you get in sum four totally different residual stress distributions if you take into consideration that the prestress (before rolling) is along or perpendicular to the track. The residual stress distributions in the dependence of the above-explained facts, and different prestresses are investigated. The different depths of the interaction zones are shown. Also, the limit of the theoretical inducible residual stresses is compared with the experimental results. Last, but not least, a practical application of a torsion bar is presented.

Keywords Deep rolling · Stress rolling · Residual stress · Mechanical surface treatment

1 Introduction

Mechanical surface treatment such as shot peening and deep rolling is now standard practice. Improving the process is difficult. Shot peening parameters such as shot size or pressure can be change. A big step forward is stress peening which is mainly used for (automotive) springs [1]. The spring is pre-stress during the shot peening process, and then unloaded with the effect of a significant increase in the compressive residual stress. Deep rolling does not a wide range of applications because the geometry is important to get a good performance of the rolling process, and subsequently an

E. Mueller (✉)
Bochum University of Applied Sciences, Bochum, Germany
e-mail: eckehard.mueller@hs-bochum.de

Steinbeis-Transfercenter for Spring Technology, Component Behavior and Process, Iserlohn, Germany

N. Maharjan and W. He (eds.), *Proceedings of the 3rd International Conference on Advanced Surface Enhancement (INCASE) 2023*, Lecture Notes in Mechanical Engineering, https://doi.org/10.1007/978-981-99-8643-9_19

increase in durability. There is no fundamental obstacle to applying the same process to deep rolling. Let's call it stress rolling.

2 Basics

2.1 Deep Rolling

Altenberger defines deep rolling: "In all further elaborations, the terminology "deep rolling" refers to a surface rolling treatment using rolls or ball-point tools with the purpose of inducing deep plastic deformations and residual stresses in near-surface layers in contrast to "roller burnishing" which is usually applied with much lower forces or pressures and mostly aims to obtain a certain surface quality especially in terms of roughness" [2]. It is from this definition that the understanding of deep rolling is taken in this paper.

2.2 Stress Peening

Shot peening is often a process as described in [3]. Shot peening induces compressive residual stresses in the surface layers. If the process (also stress peening) is well done, the compressive residual stresses are independent of the residual stress state before [4].

Stress peening means that there are three steps in the shot peening process: loading the component in the direction of the later tensile stress. Peening the component and then unload it. The last step resulted an increase of the compressive residual stress and a greater depth of interaction of the induced residual stress (Fig. 1).

2.3 Stress Rolling

Stress rolling is an equivalent to stress peening. The peening process is replaced by a rolling process. The main difference is that a deep rolling process produces two different states of compressive residual stress. Along the rolling track, the compressive residual stress at the surface is approximately half as high or less then that perpendicular to the track for normal deep rolling [5]. Therefore, there are two cases: the rolling track along the direction of loading and the rolling track perpendicular to the direction of loading (see Fig. 3) As a result of this fact, four different residual stress states are possible for stress rolling, measured along and perpendicular to the track. Because this is a new field of investigation no literature as used are available.

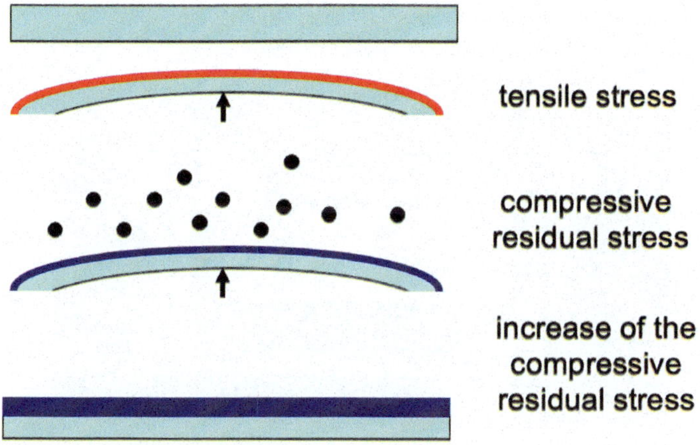

Fig. 1 Different steps of stress peening

3 Experimental Setup

3.1 Specimen

The specimen used was a small leaf of spring steel 50 Cr V 4 (1.8159) with dimensions of 350 mm * 80 mm * 10 mm (see Fig. 2). The tensile strength was 1500 MPa. The surface was ground to have no decarburization. Due to the varying depths of decarburization, sometimes a small amount of up to 50 μm may still still be in some areas of the specimen.

3.2 Rolling

A four-point-bending jig was used to obtain an area of constant stress (see Fig. 2). Rolling was carried out using tool HG 6 from the company Ecoroll with a hydrostatically controlled of a 6 mm diameter ball [6]. The pressure was 100 bar ± 10%. The distance between the tracks was 0.15 mm. The pattern is a meandering shape (see Fig. 3). In Fig. 2 you can see two areas in the middle that have been deep rolled in this way.

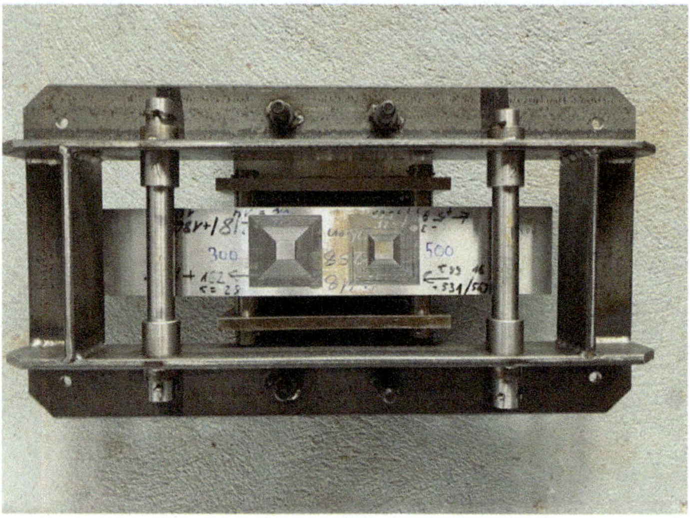

Fig. 2 Loading tool and specimen

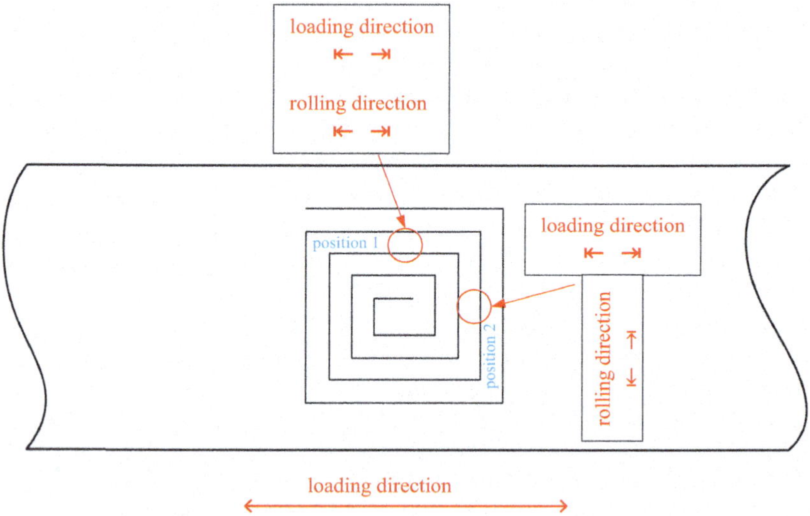

Fig. 3 Pattern of the tracks and the different combinations of the rolling and loading directions

3.3 X-Ray Determination

Residual stresses were determined by X-ray diffraction. The instrument used was a Rigaku Strainflex MSF-2 M with the $\sin^2\psi - 2\theta$ method. The errors of the residual stress results are ± 30 MPa or 7% (whichever is greater). Further explanations can be found e.g. in [7].

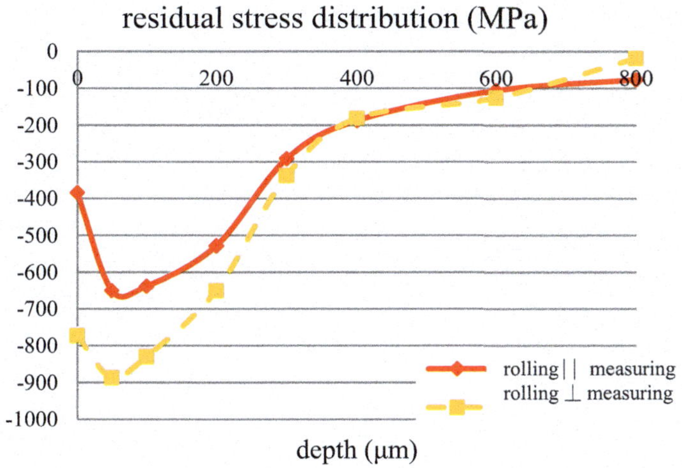

Fig. 4 Residual stress distribution in dependence of the depth for normal rolling (‖ = parallel, ⊥ perpendicular)

4 Results

4.1 Baseline—Normal Rolling

One specimen was deep rolled without any prestress to provide a baseline against which the increase of the compressive residual stress can be seen. Figure 4 shows the residual stress distribution as a function of the depth in the material for normal rolling. Near the surface the compressive residual stress parallel to the rolling track is significantly lower than perpendicular. The surface values are somewhat affected by decarburization and are not reliable. Around a depth of 400 μm, the same stress level is reached for both directions. The highest compressive residual stress is close to 900 MPa perpendicular to the track.

4.2 Stress Rolling

The prestress (preload) before the stress rolling process was 260, 370 and 710 MPa. The following figures show the residual stress distribution for the different preloads.

For a better understanding of the keys in the diagrams, the following explanations are given. The different positions (pos) are shown Fig. 3 (Table 1).

The residual stress states are (at a pre-stress of 710 MPa, the pressure of the rolling ball was approximately 10% less then to the other) (Fig. 5).

Table 1 Different measurement possibilities for the residual stresses

Key	Track	Measuring direction in relation to the track
pos1 ‖ measuring	Along the loading direction	Parallel
pos1 ⊥ measuring	Along the loading direction	Perpendicular
pos2 ‖ measuring	Perpendicular to the loading direction	Parallel
pos2 ⊥ measuring	Perpendicular to the loading direction	Perpendicular

4.3 Analysis

For further analysis, the residual stresses at the surface were not taken into account to avoid surface effects fin the measurements and possible decarburization.

Figure 6 shows the increase of the maximum compressive residual stress as a function of the pre-stress during rolling. The unloading stress difference is smaller than the loading stress because of the need to compensate for the induced residual stress during rolling and the enhancement. The red line (along the track) shows this effect (Fig. 6). Compare this with the green line, which shows the effect perpendicular to the track where the higher compressive residual stresses are present before unloading. The increase is less because microplastification starts. At a stress $\sigma = 2/3$ R_m (tensile strength of the spring steel) a limit is reached where no higher compressive residual stresses can be induced [8]. This explains the large difference in the prestress between 710 MPa and the increase obtained. This effect can also be seen at stress peened specimens or leaf springs [9]. However, a high prestress is not useless because the depth of interaction becomes deeper. Locking at Fig. 5c, at a depth of 400 μm, the compressive residual stress is significantly higher than at a prestress of 270 MPa Fig. 5a.

5 Conclusions

Stress rolling, like stress peening, is a way to significantly increasing the compressive residual stresses compared to normal rolling significantly. It also increasing the interaction depth of the rolling process is increased. The roughness is less than 10% compared to shot or stress peening. The exact value depends on the pressure on the rolling tool. Deep rolling is not possible at whole leaf spring to increase the durability, because the edges cannot be deep rolled sufficiently. An example of a component where stress rolling can be applied on torsion bar springs. Initial results show that you can be increased compressive residual stresses in such a way that the amount under 45° where the maximum shear stresses occur, is increased over the stresses level perpendicular to the rolling track. This improves the durability of the component.

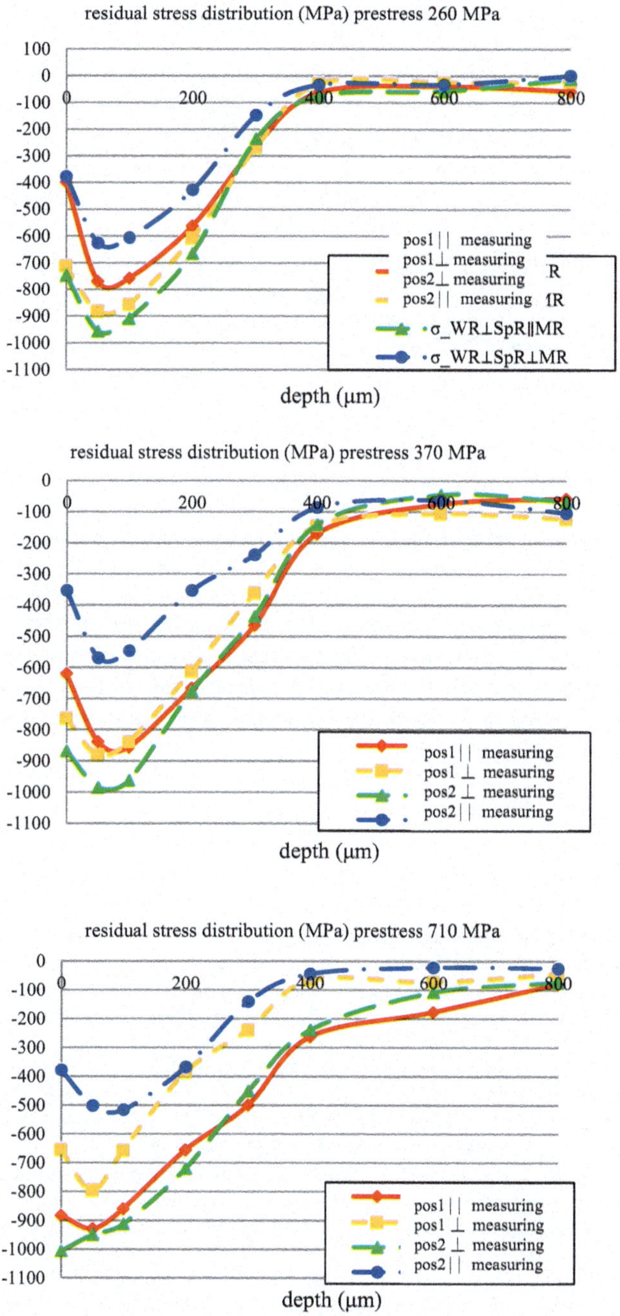

Fig. 5 a–c Different residual stress distributions for the performed preloads

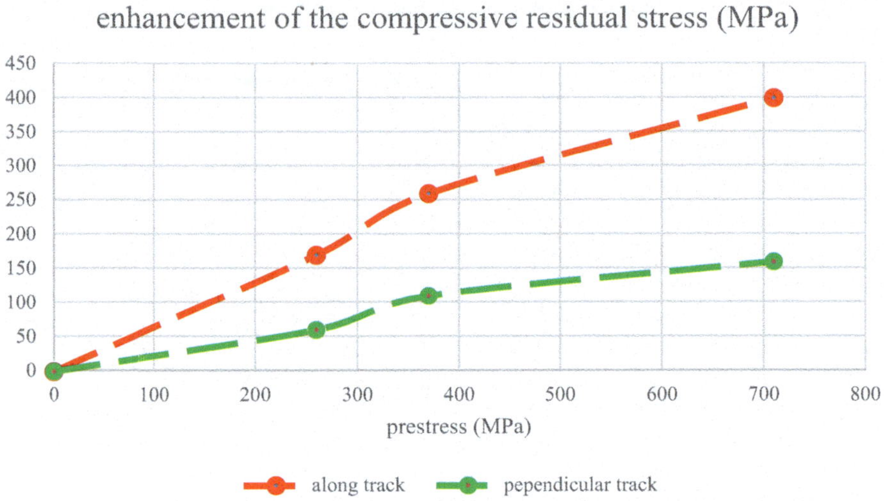

Fig. 6 The enhancement of the residual stresses in dependence on the prestress

The difficulty is to keep the torsion bar under prestress during the rolling process and must be investigated further [10]. The results are the first one to be obtained by anyone. A patent is made for the application at torsion bars [11] that shows the innovation of this subject. In the future it will give a increase of the durability which has to be investigate.

References

1. Müller E (2019) Stress peening-a sophisticated way of normal shot peening. J Mater Sci Eng A & J Mater Sci Eng B 9(2):56–63
2. Altenberger I (2005) Deep rolling–the past, the present and the future. In: Niku-Lari A, Schulze V (eds) 9th international conference on shot peening, Shot Peening and other mechanical surface treatments, I.I.T.T. International, Noisy-le-Grand, pp 144–155
3. Baiker St (ed) (2022) Shot peening, 6th edn. MFN publishing house, Wetzikon
4. Müller E (1999) The development of residual stress at bending specimens under the influence of setting and stress peening. In: Nakonieczny A (ed) 7th international conference on shot peening. Warzsaw, Institute of Precision Mechanics, Warzsaw, pp 88–95
5. Denkena B, Grove T, Breidenstein B, Abrão A, Meyer K (2018) Correlation between process load and deep rolling induced residual stress profiles. Procedia CIRP 78:161–165
6. Ecoroll Homepage: https://www.ecoroll.de/produkte/hydrostatische-werkzeuge.html. Accessed 5 May 2023
7. He BB (2009) Two-dimensional X-Ray diffraction. Wiley, Hoboken
8. Müller E (2005) Evolution of the residual stresses by stress rolling. In: Niku-Lari A, Schulze V (eds) 9th international conference on shot peening, shot peening and other mechanical surface treatments, I.I.T.T. International, Noisy-le-Grand, pp 436–441
9. Hilaire NKJ (2015) Untersuchung der maximalen Druckeigenspannung im Federelementen im Vergleich zur Zugfestigkeit. BA-Thesis, University of Applied Sciences Bochum, Germany

10. Halemeyer J (2017) Eigenspannungsentwicklung beim Festwalzen unter Vorspannung an Drehstäben. BA-Thesis, University of Applied Sciences Bochum, Germany
11. Müller E (2018) Verfahren zum Einbauen einer Druckeigenspannung in ein rotationssymmetrisches Metallbauteil, Drehstabfeder sowie Verwendung einer solchen, Patent DE 10 2017 102 447 A1

Control of Residual Stress by Cutting Condition on the Surface of Aluminum and Steel

Yoshitaka Yokogawa, Kazuyuki Oguri, Kazuki Takai, Makoto Mori, Toshiya Nakagami, and Takahiro Nozaki

Abstract Distortion of machined parts is still a big issue not only in the aircraft parts industry but also in any machine industry. Compressive residual stress is favorable compared with tensile residual stress after machining for fatigue and stress corrosion properties. However, it is not easy to get compressive residual stress on the surface after machining. Cutting is the main machining process for various machine parts including aircraft parts. Therefore, in this research, we try to find out the appropriate cutting condition to obtain compressive residual stress after cutting. Especially cutting conditions of up-cut and down-cut methods are compared from the viewpoint of residual stress after the cutting. The residual stress was measured by X-ray diffraction method using the μ-X360 device of Pulstec Industrial Co., Ltd. The results confirm that up-cut tends to generate compressive residual stress. On the other hand, down-cut tends to generate tensile residual stress. The details including the effect of repetitive processing are also discussed.

Keywords Residual stress · Cutting · X-ray · Up-cut · Down-cut

1 Introduction

Today, most metal parts are formed by cutting. Residual stress is generated on the surface of parts formed by cutting due to plastic deformation and heat caused by machining. This residual stress can cause parts to distort, which is a major problem. There are two types of residual stress, namely compressive and tensile residual stress.

Y. Yokogawa (✉)
Graduate School, Kanazawa Institute of Technology, Ishikawa 921-8501, Japan
e-mail: b1904897@planet.kanazawa-it.ac.jp

K. Oguri
Kanazawa Institute of Technology, Ishikawa 921-8501, Japan
e-mail: kaz_oguri@neptune.kanazawa-it.ac.jp

K. Takai · M. Mori · T. Nakagami · T. Nozaki
Asahi-Seiki Manufacturing, Co., Ltd, Aichi 488-0852, Japan

N. Maharjan and W. He (eds.), *Proceedings of the 3rd International Conference on Advanced Surface Enhancement (INCASE) 2023*, Lecture Notes in Mechanical Engineering, https://doi.org/10.1007/978-981-99-8643-9_20

When compressive residual stress is generated, it is a factor that inhibits crack initiation and propagation. Therefore, it can improve fatigue strength and fatigue life, which is a great advantage. On the other hand, when tensile residual stress is generated, it is a factor that propagates cracks and leads to component failure.

Relationship between cutting conditions and residual stress is not fully understood yet. The purpose of this study is to find the appropriate cutting conditions to obtain compressive residual stress after side cutting method as a first step.

2 Specimen

Aluminum alloy A7050 and carbon steel S50C were used as specimens. The specimens were rectangular, approximately 200 mm × 40 mm × 40 mm, and the 200 mm × 40 mm face was machined.

3 Equipment

The μ-X360n [1] from Pulstec Industrial Co., Ltd. was used to measure the residual stress. The measurement equipment used is shown in Fig. 1.

When evaluating residual stress in the depth direction after cutting, electropolishing and measurement are repeated. This is because surface removal by sanding or machining will impart residual stress. In this study, 10% sodium chloride solution was used as the electrolytic solution, and the polishing range was φ8[mm].

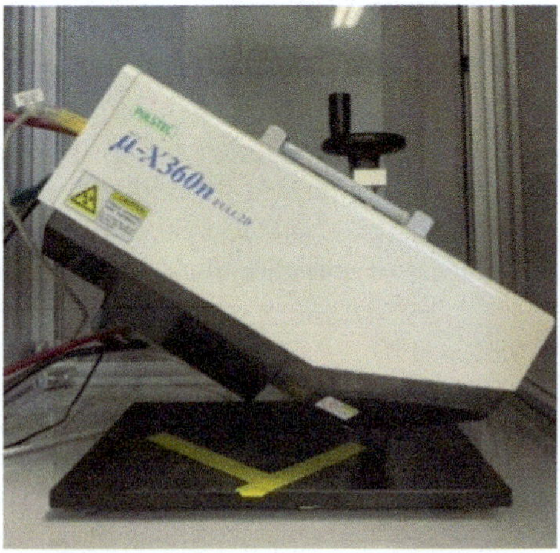

Fig. 1 Measuring instruments, μ-X360

4 Processing Conditions

There are two types of side-cutting methods, namely up-cut and down-cut. First, machining was performed on A7050 to see how the residual stress changes with the difference. The machining speed was varied as shown in Table 1. The tool was a two-flute end mill made of cemented carbide.

Next, to investigate how the residual stress changes by additional cutting after machining once, machining was performed under the conditions shown in Tables 2 and 3. The work material was A7050. The tool was a two-flute end mill made of high-speed tool steel.

The specimens to measure the residual stress depth distribution by electropolishing were machined under the conditions shown in Table 4. The work material was A7050 and the tool was a two-flute carbide end mill. Specimens of S50C was also machined as a comparison.

Table 5 shows the conditions to check the effect of cutting fluid on the residual stress. The work material is S50C, and the tool is a two-flute end mill made of high-speed tool steel.

Table 6 shows the machining conditions to check material is S50C, and the tool is a two-flute end mill made of high-speed tool steel.

Table 1 Processing conditions of down-cut and up-cut machining methods for A7050

Processing method	Rotational speed [min^{-1}]	Feed speed [min/mm]
Up-cut Down-cut	500	150
	1000	300
	3000	900
	5000	1500
	7000	2100
	10,000	3000
	20,000	6000
	30,000	9000

Table 2 Processing sequence for A7050

No.	Processing details (Infeed amount [mm])
1	Up-cut(0.2)
2	Up-cut(0.2) → up-cut(0.1)
3	Down-cut(0.2) → up-cut(0.1)
4	Down-cut(0.2)
5	Up-cut(0.2) → down-cut(0.1)
6	Down-cut(0.2) → down-cut(0.1)

Table 3 Processing conditions of Table 2

No.	Rotational speed [min^{-1}]	Feed speed [min/mm]
1,2,3,4, 5,6	750	150

Table 4 Processing conditions of residual stress depth profile measurement for A7050

Processing method	Rotational speed [min^{-1}]	Feed speed [min/mm]
Up-cut for A7050	7000	3000
	10,000	3000
Down-cut for A7050	7000	2100
Up-cut, Down-cut for S50C	1330	266

Table 5 Processing conditions with and without cutting fluid for S50C

Processing method	Rotational speed [min^{-1}]	Feed speed [min/mm]
Up-cut, Down-cut	280	70
	320	70
	460	100
	730	150
	900	180
	1280	300
Up-cut, Down-cut with cutting fluid	280	70
	320	70
	460	100
	730	150
	900	180
	1280	300

Table 6 Processing conditions by the tool with and without TiN coating for S50C

Processing method	Rotational speed [min^{-1}]	Feed speed [min/mm]
Up-cut, Down-cut	180	33
	380	85
	530	108
	1100	243
Up-cut, Down-cut with TiN coating for tools	180	33
	380	85
	530	108
	1100	243

Table 7 Processing conditions on the effect of cutting depth for S50C

Processing method	Rotational speed [min^{-1}]	Feed speed [min/mm]	Depth of cut [mm]
Up-cut, Down-cut	311	62	2.0
	580	116	
	885	177	
	1330	266	
	311	62	0.2
	580	116	
	885	177	
	1330	266	

Finally, we investigated the effect of the depth of cut on the residual stress by the machining conditions in Table 7.

The work material was S50C, and the tool was a two-flute carbide end mill.

The changes of the residual stress under various processing conditions shown above were measured by X-ray residual stress measurement devise, μ-X360n [1] from Pulstec Industrial Co., Ltd., and evaluated.

5 Results and Discussion

First, Fig. 2 shows the results of residual stress measurements on work material machined under the conditions in Table 1. It can be seen that the residual stress imparted to the work material is significantly different between up-cut and down-cut. The compressive residual stress was generated higher by the up-cut, reaching about − 340 MPa at 7000 rpm. Stress values increased on the tensile residual stress side with increasing speed in all speed ranges for the down-cut and above 7000 rpm for the up-cut. One possible cause of the change to tensile stress is the increase in surface temperature during processing. In other words, surface heating generates thermal stress from inside the material and imparts compressive residual stress. After machining, the cooling of the machined surface is thought to generate tensile residual stress that tries to undo the deformation that occurred during machining [2, 3].

Figure 3 shows the results of residual stress changes due to additional processing under the conditions of Tables 2 and 3.

The results of Fig. 2 show that the up-cutting process is effective to provide compressive residual stress. And the results in Fig. 3 show that regardless of the initial machining method, if the final machining is an up-cut, a large compressive residual stress can be applied. In other words, it was found that residual stress due to machining can be overridden by the final machining process.

The results of residual stress measurements in the depth direction for specimens processed under the conditions of Table 4 are shown in Fig. 4.

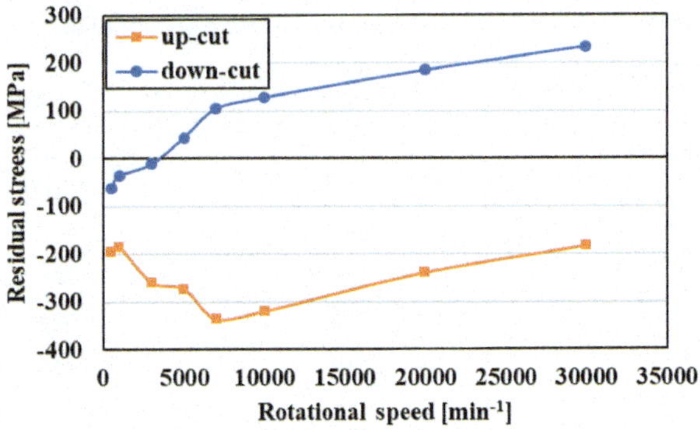

Fig. 2 Relationship between rotational speed and residual stress processing methods for down-cut and up-cut for A7050

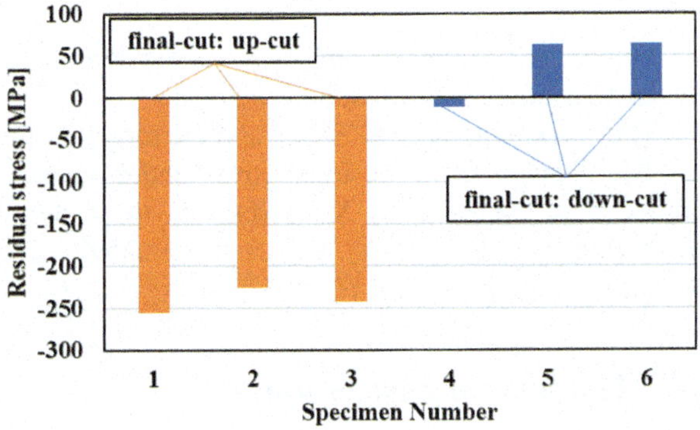

Fig. 3 Effect of combination processing methods on residual stress for A7050

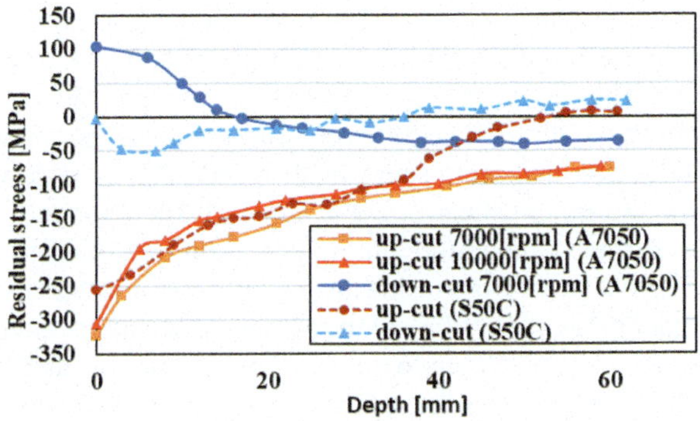

Fig. 4 Depth profile of residual stress by down-cut and up-cut processing method

The up-cut imparts compressive residual stress, while the down-cut imparts tensile residual stress in the surface area. The up-cut can impart compressive residual stress to a depth of 50 μm or more depending on the conditions. However, it can be seen that there is little change in residual stress in the depth direction with the rotational speed. Residual stress trends in A7050 and S50C were similar.

Fig. 5 Effect of the presence of cutting fluid on residual stress for S50C

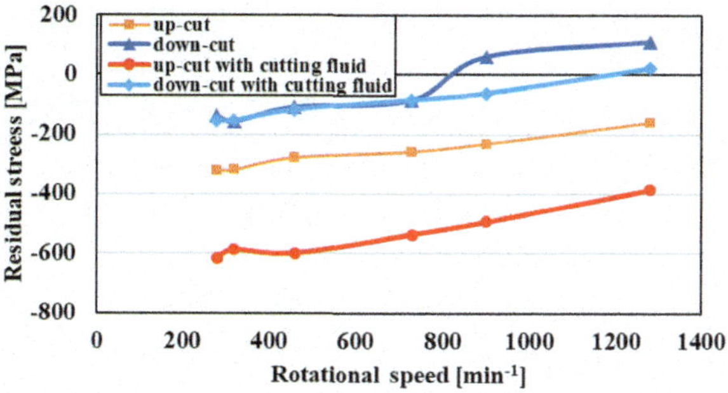

Figure 5 shows the results of the effect of cutting fluid by the conditions in Table 5. It can be seen that the down-cut is not affected by the cutting fluid. However, the up-cut is significantly affected by cutting fluid. In the up-cut process without cutting fluid, the maximum compressive stress was about − 300 MPa. On the other hand, when the up-cutting process was performed using cutting fluid, the maximum compressive stress reached about − 600 MPa, a two-fold increase. The results indicate that it is better to use cutting fluid to impart compressive residual stress by up-cut.

Figure 6 shows the results of evaluating the effect of residual stresses with and without tool coating. The machining conditions are those shown in Table 6. It can be seen that there is little difference in the trend of residual stress change for both up-cut and down-cut. However, in the case of up-cut processing, there was a difference in compressive residual stress of about a factor of 2 depending on whether the coating was applied or not at low processing speeds.

Finally, to investigate the change in residual stress with the depth of cut, machining was performed under the machining conditions shown in Table 7, and the results are shown in Fig. 7. In the case of the up-cut, the smaller depth of cut provided greater compressive residual stress. The difference between the case with a depth of cut of 0.2 mm and the case with a depth of cut of 2 mm was about 100 MPa at any rotational speed. Therefore, to obtain higher compressive residual stress, the smaller depth of

Fig. 6 Effect of TiN coating on the tool on residual stress for S50C

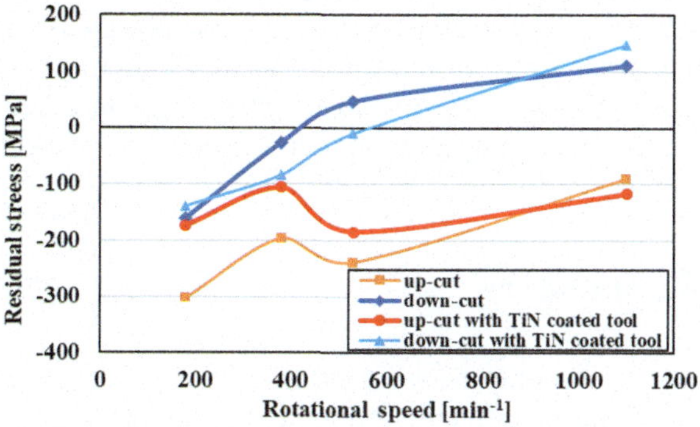

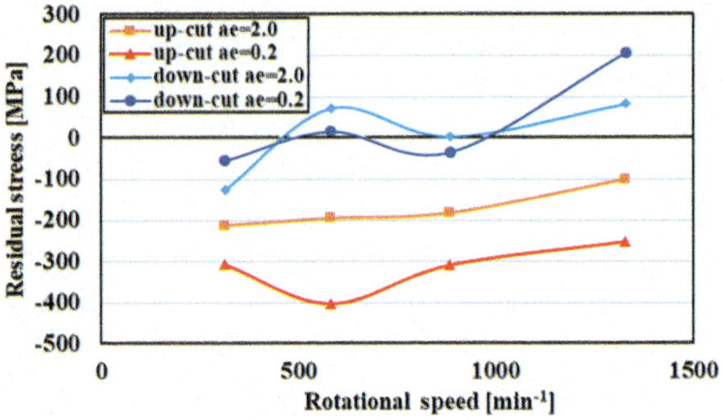

Fig. 7 Effect of depth of cut on residual stress for S50C

cut is considered effective. On the other hand, in the down-cut, almost no change in residual stress with depth of cut was observed.

These are the results of our investigation of the relationship between side processing methods and residual stress. We will try to discuss why the trend of residual stress generated differs greatly depending on the processing method. Various factors are involved in the generation of residual stress due to cutting. According to the analysis in previous studies, the stress state in the deformation zone is described as complex.

We try to consider the phenomena in terms of shear stresses generated on the machined surface. As the tool advances, the plastic deformation of the machined surface generates tensile residual stresses [4]. It is assumed that down-cut, in which cutting is performed by striking the tool edge, tends to impart tensile residual stress due to the balance of forces. On the other hand, up-cut involves cutting as if tearing, which is assumed to result in compressive residual stress due to the balance of forces.

We will also consider the phenomena from the viewpoint of indentation stress. When a large indentation load exceeding the yield stress is applied to the material surface, plastic deformation occurs from the surface to the interior, and residual stress is generated after unloading [4]. Since this occurs more strongly in the up-cut than in the down-cut, it is assumed that the up-cut imparts compressive residual stress.

In the present results, when the rotational speed was low, compressive residual stress tended to be imparted even in the down-cut (Fig. 2). In this case, the residual stress may also be influenced by indentation stress. However, in actual process, cutting is usually performed at higher rotational speeds, then it is not practical to aim for the imposition of compressive residual stress in the low-speed rotational range.

6 Summary

The relationship between the side processing method and the generated residual stress was investigated and found to be as follows.

- The up-cutting process generates compressive residual stress.

- Tensile residual stress is applied in the down-cutting process.
- In the case of multiple machining, up-cut in the final process imparts compressive residual stress.
- In the case of up-cut machining, the use of cutting fluid imparts higher compressive residual stress.
- In the up-cutting process, a smaller depth of cut imparts higher compressive residual stress.

Acknowledgements We would like to thank Pulstec Industrial Co., Ltd. for the X-ray residual stress measurement system to measure residual stress.

References

1. Pulstec Homepage, http://www.pulstec.co.jp. Accessed 25 Apr 2023
2. Yonetani S (1987) Residual stresses and countermeasures, Youkendo, Japan
3. Umeda F, Oguri S, Nakagami T, Takai K (2021) Evaluation of residual stress of machined aluminum alloy A7050 for aircraft structure. In: Proceeding of the 59th aircraft symposium, The Japan Society for Aeronautical and Space Sciences, JSASS-2021-5028, Japan
4. Kazuo K, Yamaguchi S, Adachi K, Honda S (2001) Causes and features of residual stress generated on machined surface. Reports of Industrial Technology Research Institute, pp 27–36, Osaka Research Institute of Industrial Science and Technology, Japan

Microstructure and Magnetic Properties of Non-oriented Electromagnetic Steel Sheets by Burnishing Method

Yuji Kobayashi, Yuta Saito, Masato Okada, and Ryota Watari

Abstract Burnishing is a process in which a cylindrical or hemispherical tool made of a hard material is pressed against a workpiece with a rough cut surface to obtain a smooth surface. Compared to shot peening, the process stress is easier to control and more precise. Burnishing is usually used to improve surface roughness. Depending on the processing conditions, it is possible to affect the interior of the material and obtain an effect similar to shot peening. Electromagnetic steel sheet is a material that is easy to observe changes in crystal characteristics because it is close to pure iron in terms of material composition and has large crystal grains. When the crystals change, the magnetic properties also change. Therefore, it is a material that is easy to evaluate in various aspects. In this study, the changes that occur when electromagnetic steel sheets are processed by inclined burnishing was investigated.

Keywords Burnishing · Shot peening · Electromagnetic steel · Crystal characteristics

1 Introduction

Burnishing is a process in which a cylindrical or hemispherical tool is pressed against a metal workpiece with a rough cut surface to obtain a smooth finished surface.

Burnishing is a conventional technique used to improve surface roughness. Also, depending on the processing conditions, it is possible to affect the inside of the material and achieve the same effect as shot peening.

Compared to shot peening, the burnishing process is easier to control and more precise because a machining center is used.

Y. Kobayashi (✉) · Y. Saito
Sintokogio, Ltd, 180-1, Komaki, Ohgi-Cho, Toyokawa 441-1205, Aichi, Japan
e-mail: y-kobayashi@sinto.co.jp

M. Okada · R. Watari
Faculty of Engineering, University of Fukui, Bunkyo 3-9-1, Fukui 910-8507, Japan

N. Maharjan and W. He (eds.), *Proceedings of the 3rd International Conference on Advanced Surface Enhancement (INCASE) 2023*, Lecture Notes in Mechanical Engineering, https://doi.org/10.1007/978-981-99-8643-9_21

Electromagnetic steel sheet is a thin sheet with soft magnetic properties, and its material composition is close to pure iron with large crystal grains. It is mainly used for motor cores and is an indispensable material for electric vehicles.

This study investigates the crystalline and magnetic properties that occur when electromagnetic steel sheets are processed using the burnishing process.

2 Experimental Method and Evaluation Method

The material used in the experiments was JIS 50A400. No heat treatment was applied.

For the burnishing process, a precision tabletop 3-axis robot was mounted with the motor spindle inclined 10 degrees from the normal direction.

A spring is built into the jig that fixes the workpiece to control the pressing force (F_tN). Table 1 shows the processing conditions used in this study.

An experimental machine was fabricated to investigate changes in magnetic properties due to the burnishing process. The electromagnetic steel plate that was processed is made to generate electricity by rotating it within the ground field of two magnets. Since the electromagnetic steel plate is rotated by a spindle motor, the electromotive force is constant.

Figure 1 shows image of motor core evaluation equipment.

Table 1 Condition of burnishing process

Pressing force	40, 80 N
Sliding speed	4500, 13,500 rpm
Feed speed	5.0 mm/s
Cross feed	50 μm
Oil	BP-3000

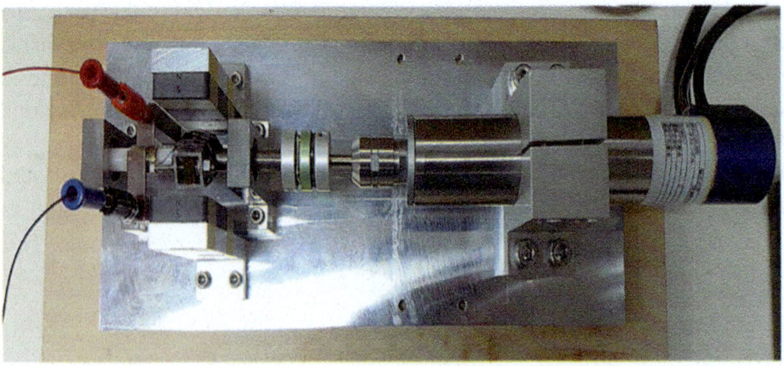

Fig. 1 Image of motor core evaluation equipment

In order to investigate the crystal properties after burnishing process, crystal orientation analysis was performed by FE-SEM-EBSD.

3 Result

The iron loss during power generation was evaluated by measuring the time integral of the electromotive force when the burnished electromagnetic steel plate was rotated by a spindle. Figure 2 shows the time integration of the electromotive force at each spindle speed. The integration time is 0.2 s at 1000 rpm and 0.05 s at 6000 rpm. At both 1000 and 6000 rpm, the electromotive force decreased compared to that of the unprocessed work. However, the range of decrease was different between 1000 and 6000 rpm, and the amount of electricity generated after burnishing at 80 N was about 75% at 1000 rpm and about 86% at 6000 rpm compared to before burnishing. Therefore, at a higher rpm, there is a smaller the decrease in power generation. The electromotive force was evaluated by the time integration of the electromotive force obtained when the motor core was rotated. Figure 2 shows the time integration of the electromotive force at each spindle speed. The integration time is 0.2 s at 1000 rpm and 0.05 s at 6000 rpm. At both 1000 and 6000 rpm, the electromotive force decreased compared to that of the unprocessed work. However, the range of decrease was different between 1000 and 6000 rpm, and the amount of electricity generated after burnishing at 80 N was about 75% at 1000 rpm and about 86% at 6000 rpm compared to before burnishing. Therefore, at a higher rpm, there is a smaller the decrease in power generation. The reason for the change in power generation with rotation speed may be due to the change in eddy current loss. Eddy current loss is a type of magnetic field loss that constitutes iron loss. It is said that the effect is larger at higher frequencies and can be improved by subdividing the magnetic domains. It is thought that the burnishing process has refined the crystal grains and subdivided the magnetic domains, resulting in a decrease in eddy current loss at high frequencies.

Next, the results of crystal evaluation by EBSD are shown. From the IPF map in Fig. 3, it was confirmed that the rolled structure by skin pass rolling on the surface layer remained after processing for both 40 N and 80 N pressing force.

Figure 4 shows the IPF and KAM maps when the pressing force is changed to 40 N and the sliding speed to 13,500 rpm. Compared to Fig. 3, crystal refinement occurs as the sliding speed increases. However, the KAM map for the case where the KAM value is higher increases to a deeper position when the pressing force is increased. Therefore, the depth of the process-affected layer is correlated to the pushing force.

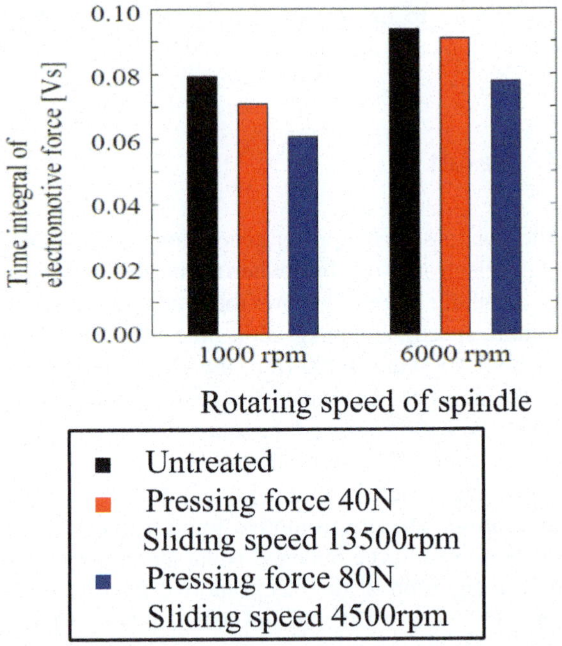

Fig. 2 Integration of the electromotive force

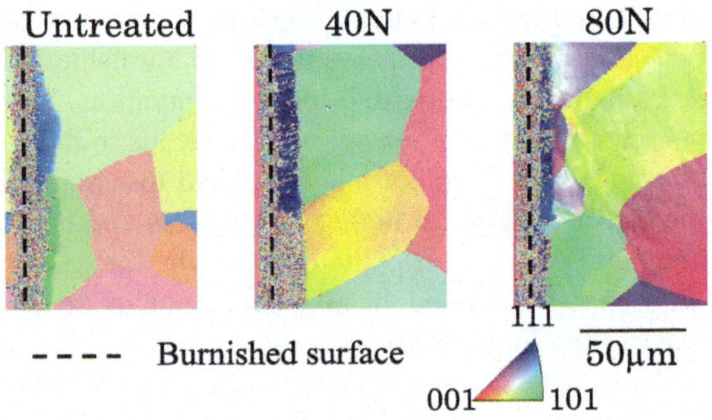

Fig. 3 IPF map of before and after burnishing process (Sliding speed = 4500 rpm)

4 Conclusion

In this study, the crystalline and magnetic properties of electromagnetic steel sheets after burnishing were investigated. The results obtained are shown below.

(1) The time-integral value of electromotive force decreased after burnishing the electromagnetic steel plate.
(2) The decrease in the time-integral value is considered to be due to a decrease in eddy current loss caused by the subdivision of magnetic domains due to the refinement of crystal grains.

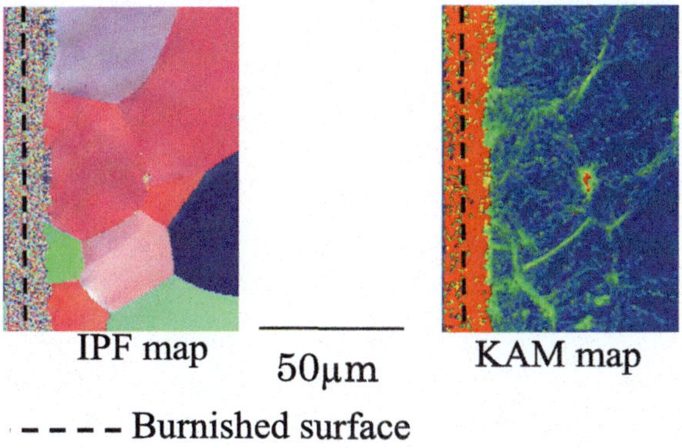

Fig. 4 IPF and KAM map of after burnishing process (Pressing force = 40 N, Sliding speed = 13500 rpm)

(3) Burnishing process on the electromagnetic steel sheet resulted in grain refinement in the surface

History of Laser Peening Without Coating (LPwC): From Invention to Future Developments

Yuji Sano

Abstract Laser peening without coating (LPwC) was invented in Japan in the 1990s. After confirming the effect of reducing the stress corrosion cracking (SCC) susceptibility of sensitized materials, LPwC has been applied to operating nuclear power reactors as a countermeasure against SCC of in-core components. Meanwhile, fiber-delivery technology for high-power laser pulses has been developed to expand the applicability of LPwC. In parallel, the effect of LPwC to improve fatigue properties has been investigated and then applied to the inside of the connecting pinholes at the root of nuclear steam turbine blades. In the last decade, various technological developments have been made to further expand the applications, such as dry LPwC, which does not use water as a confinement medium, and handheld LPwC, which uses a finger-sized microchip laser mounted directly on a robot. The author reviews the activities in LPwC over the past 30 years, focusing on the underlying physics, inventions, applications, and future developments.

Keywords Laser peening · Residual stress · Fatigue · Handheld laser · Robot arm

1 Introduction

Advances in laser science and technology are enabling novel processes and applications in industry. The development of LPwC (Laser Peening without Coating), along with monolithic microchip lasers [1], is a breakthrough in bringing high-power lasers into the field of maintenance arena [2–4]. LPwC can introduce compressive residual stresses (RSs) into metallic materials by simply exposing the surface of a water-covered component to a series of high-power nanosecond laser pulses [5–7], and has advantages over mechanical processing because it is an inertia-free process. LPwC

Y. Sano (✉)
SANKEN, Osaka University, Ibaraki 567-0047, Japan
e-mail: yuji-sano@sanken.osaka-u.ac.jp; yuji-sano@ims.ac.jp

Institute for Molecular Science, National Institutes of Natural Sciences, Okazaki 444-8585, Japan

N. Maharjan and W. He (eds.), *Proceedings of the 3rd International Conference on Advanced Surface Enhancement (INCASE) 2023*, Lecture Notes in Mechanical Engineering, https://doi.org/10.1007/978-981-99-8643-9_22

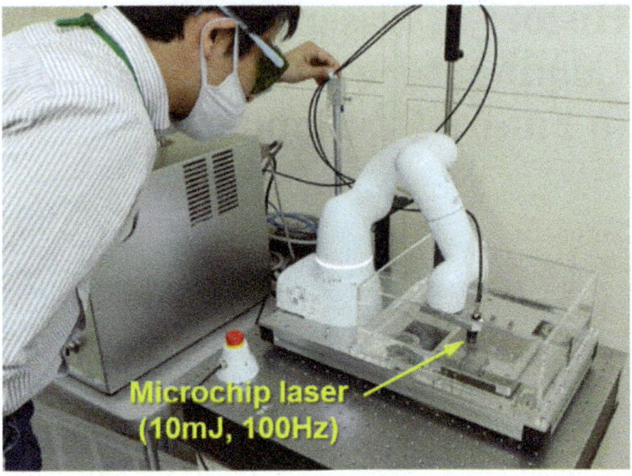

Fig. 1 Mobile LPwC system

has been applied to existing nuclear facilities since 1999 to prevent stress corrosion cracking (SCC) of nuclear internals [2].

Regarding fatigue issues, LPwC has greatly improved the fatigue strength and prolonged the fatigue life of austenitic stainless steels [8], aluminum alloys [9, 10], etc. Recently, we have confirmed the effect of improving the fatigue properties of HT780 (780 MPa grade high-strength steel) and A7075 aluminum alloy by low-energy LPwC with pulse energies down to less than 10 mJ [3, 11].

2 Development of Mobile LPwC System

To expand the application of LPwC, a mobile LPwC system has been developed with a finger-sized microchip laser mounted on a collaborative robotic arm as shown in Fig. 1 [3, 4]. The system can be stored in two suitcases and transported as airline baggage, and is ready for use within one hour of opening the packages. The maximum power consumption is less than 400 W at 100–220 V AC.

LPwC was applied to HT780 butt-welded joint samples and then the samples were subjected to fatigue testing [3]. The results showed that LPwC significantly improved the fatigue properties of the HT780 welded samples, as shown in Fig. 2.

3 Handheld Laser Development

The throughput of LPwC is roughly proportional to the laser power, i.e., pulse energy (mJ) multiplied by pulse repetition rate (Hz). Therefore, it is desirable to increase the power of microchip lasers. We have applied room-temperature bonding technology to integrate laser medium (Nd:YAG), passive Q-switch (Cr:YAG), and transparent

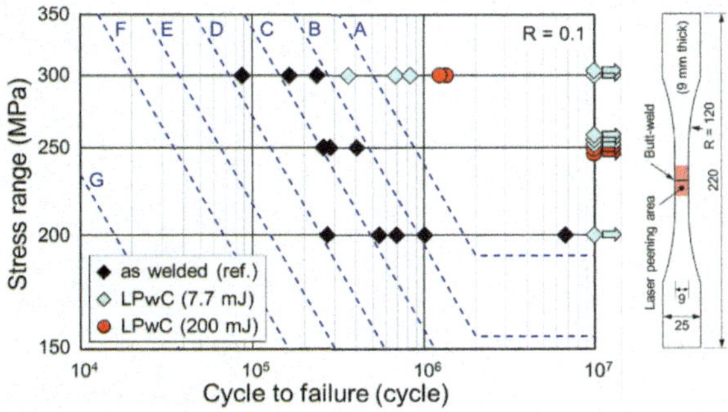

Fig. 2 Fatigue properties improvement by LPwC [3]

substrate with high thermal conductivity (sapphire) to realize high-power microchip lasers as shown in Fig. 3. LPwC experiments with the microchip laser induced compressive RSs with a depth of 0.8 mm in A7075 aluminum alloy, as shown in Fig. 4. We are confident that by further optimizing the laser irradiation conditions, 1 mm depth of compression can be achieved.

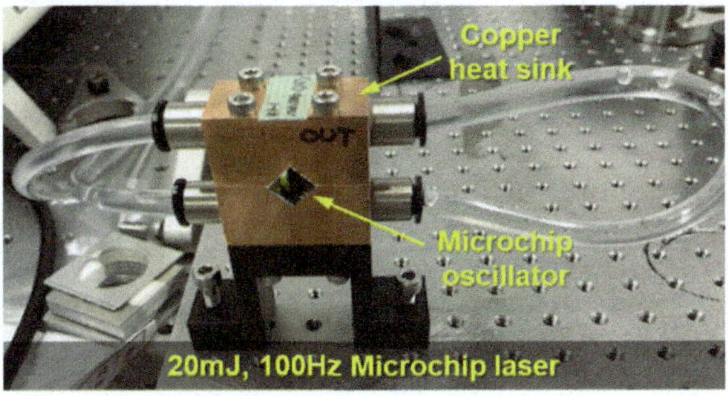

Fig. 3 Microchip laser developed at IMS

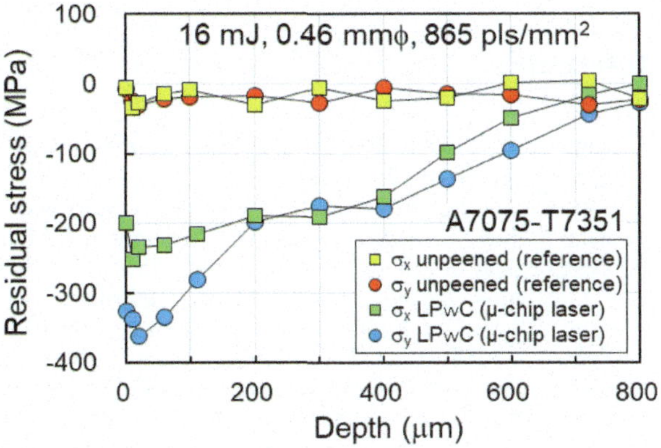

Fig. 4 Residual stresses induced by LPwC

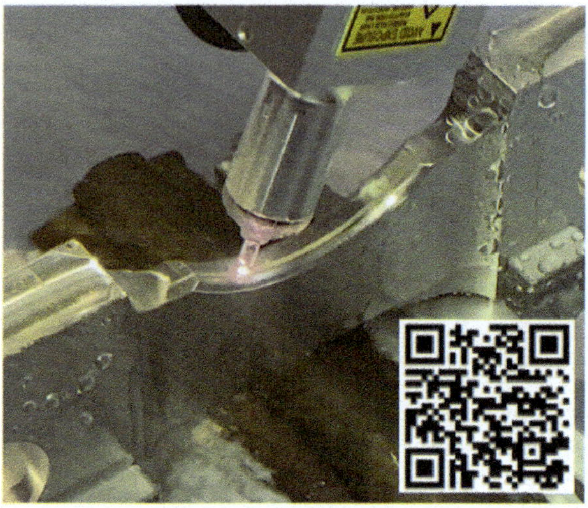

Fig. 5 LPwC of 3D object, QR code for link to video

4 Conclusions

We have 30 years of experience in the development of LPwC since the 1990s [2]. In the last decade, high-power microchip lasers have been realized. The combination of such lasers with a robotic arm has made it possible to apply LPwC to 3D objects as shown in Fig. 5 [4], facilitating its use in post-weld processing on construction sites and in the maintenance of aircraft, bridges, ships and offshore structures, etc.

Acknowledgements This work was partially supported by the JST-MIRAI Program [grant number JPMJMI17A1] and the Supporting Industry Program of the Small and Medium Enterprise Agency [grant number 20334933].

References

1. Taira T (2011) Domain-controlled laser ceramics toward giant micro-photonics. Opt Mater Express 1:1040–1050
2. Sano Y (2020) Quarter century development of laser peening without coating. Metals 10:152
3. Sano Y, Kato T, Mizuta Y, Tamaki S, Yokofujita K, Taira T, Hosokai T, Sakino Y (2022) Development of a portable laser peening device and its effect on the fatigue properties of HT780 butt-welded joints. Forces Mech 7:100080
4. Sano Y, Schneidau V (2023) Downsizing for new flexibility in laser shock peening. The Shot Peener, 6–12 (Spring 2023): https://www.shotpeener.com/library/pdf/2023014.pdf
5. Sano Y, Mukai N, Okazaki K, Obata M (1997) Residual stress improvement in metal surface by underwater laser irradiation. Nucl Instrum Methods Phys Res B 121:432–436
6. Sano Y, Akita K, Sano T (2020) A mechanism for inducing compressive residual stresses on a surface by laser peening without coating. Metals 10:816

7. Sano Y, Akita K (2022) Dependence of surface residual stress on the coefficient of thermal expansion for materials subjected to laser peening without coating. Opt Laser Technol 156:108528
8. Sano Y, Obata M, Kubo T, Mukai N, Yoda M, Masaki K, Ochi Y (2006) Retardation of crack initiation and growth in austenitic stainless steels by laser peening without protective coating. Mater Sci Eng A 417:334–340
9. Sano Y, Masaki K, Gushi T, Sano T (2012) Improvement in fatigue performance of friction stir welded A6061–T6 aluminum alloy by laser peening without coating. Mater Des 36:809–814
10. Sano Y, Masaki K, Akita K, Kajiwara K, Sano T (2020) Effect of laser peening on the mechanical properties of aluminum alloys probed by synchrotron radiation and x-ray free electron laser. Metals 10:1490
11. Sano Y, Masaki K, Mizuta Y, Tamaki S, Hosokai T, Taira T (2021) Effects of laser peening with a pulse energy of 1.7 mJ on the residual stress and fatigue properties of A7075 aluminum alloy. Metals 11:1716

Surface and Sub-surface Characterisation (In-Situ and Ex-Situ)

Corrosion Surface Morphology of 13Cr Steel in the Boiler Water Containing Chloride Ions and Formic Acid

Li-Bin Niu and Hayato Kubota

Abstract Electrochemical corrosion tests were conducted on 13Cr steel for steam turbine blade material of power plants in the simulated boiler water containing chloride ions (Cl^-) and formic acid (HCOOH). The corroded surfaces were observed and analyzed in detail. In the boiler waters containing 100 ppm Cl^-, the formation of corrosion products as passivation films on the specimen surfaces inhibited the occurrence of pitting corrosion by the additions of 25–75 ppm HCOOH.

Keywords 13Cr steel · Corrosion · Pitting corrosion · Passivation film · Boiler water · Formic acid

1 Introduction

In thermal power plants, typical corrosive chemicals such as chloride and sulfate ions (Cl^- and SO_4^{2-}) will be mixed into the water/steam cycles in the case of inadequate boiler water treatment or accidental seawater leakage [1]. In particular, Cl^- ions have the property of destroying the corrosion resistant film (passivation film) of steam turbine materials, causing pitting corrosion on the steel surfaces, and further inducing stress corrosion cracking and corrosion fatigue from the pits [2]. On the other hand, in recent years there is a tendency to add organic amines such as ethanolamine and cyclohexylamine into the boiler water for suppressing the flow-accelerated corrosion of boiler equipment materials in some thermal power plants [3]. However, these organic amines are easy to produce organic acids such as formic acid (HCOOH) by their decomposition in the high-temperature water/steam system of thermal power plants [4]. Therefore, there is concern about corrosion of plant equipment materials caused by the organic acids.

L.-B. Niu (✉)
Faculty of Engineering, Shinshu University, Wakasato 4-17-1, Nagano 380-8553, Japan
e-mail: niulibn@shinshu-u.ac.jp

H. Kubota
Former Graduate Student, Shinshu University, Nagano 380-8553, Japan

N. Maharjan and W. He (eds.), *Proceedings of the 3rd International Conference on Advanced Surface Enhancement (INCASE) 2023*, Lecture Notes in Mechanical Engineering, https://doi.org/10.1007/978-981-99-8643-9_23

In this work, corrosion surface morphology of 13Cr steel for low-pressure steam turbine blades in thermal power plants was investigated in the simulated boiler water containing Cl^- ions and HCOOH by electrochemical corrosion tests and surface analyses.

2 Methods

The 13Cr steel (0.21%C, 0.32%Si, 0.62%Mn, 13.36%Cr, 0.46%Ni, bal. Fe) was used in this work. Specimens with a size of $25 \times 15 \times 2$ mm^3 were cut from the unused steam turbine blade material. The surfaces were abraded up to 600 grit emery papers. Prior to testing, the specimens were degreased with acetone and rinsed with pure water. In addition, the specimens were insulation-treated by coating a liquid type RTV rubber on the surfaces, leaving a reaction area of 1 cm^2 exposed to the test water.

Four kinds of test water, simply named WQ1, WQ2, WQ3 and WQ4, were used in the work. As the base water, the AVT (all volatile treatment) boiler water (pH9.5, DO < 7 ppb) was firstly simulated on the base of JIS B 8223 [5] with ion-exchanged pure water. The test water WQ1 was prepared by the addition of 100 ppm Cl^- ions into the base water, while the others were prepared by the addition in combination with 100 ppm Cl^- ions and HCOOH of 25 ppm (WQ2), 50 ppm (WQ3) and 75 ppm (WQ4). The temperature of the test waters was 90 °C.

Potentiodynamic and potentiostatic polarization tests were conducted on the specimens in the four test waters, using a potentiostat with three electrodes: the specimen as the working electrode, a platinum counter electrode, and a saturated KCl-Ag/AgCl reference electrode. After the polarization tests, the specimen surfaces were observed with a digital microscope and a scanning electron microscope. The films formed on the specimen surfaces were analyzed by electron probe microanalysis (EPMA).

3 Results and Discussion

For the specimens of 13Cr steel in potentiodynamic polarization tests in the test waters WQ2—WQ4 with the addition of HCOOH, the current density in the passivation region attended to decrease compared to that in the water WQ1 with only 100 ppm Cl^- ions added. It is considered that the anodic reactions were suppressed in the waters with the HCOOH added. The potentials V'_{c100}, which have been regarded as the pitting potentials and at which the current density reached 100 μA/cm^2, for the steel in the four test waters WQ1—WQ4 were 0.33, 0.50, 0.64 and 0.67 V (versus Ag/AgCl). The V'_{c100} increased with the addition of HCOOH and showed higher values as the amount of HCOOH added in the test waters increased. This behavior suggests that the addition of 25–75 ppm HCOOH into the waters containing 100 ppm Cl^- ions suppressed the destruction of the passivation films and the initiation of pitting corrosion. It is considered that the competitive adsorptions between $HCOO^-$ ions and Cl^- ions constrained the destruction of the passivation film by the Cl^- ions. It is also presumed that $HCOONH_4$ formed in the test waters with the addition of HCOOH exhibited the inhibitory effect. Furthermore, a relatively large number of pits was observed on the specimen tested in the water WQ1, while the areas around the pits on the specimens tested in the waters WQ2—WQ4 were discolored black.

Potentiostatic polarization tests were performed on the specimens in the four test waters at 0.33 V (versus Ag/AgCl) for 20 h. The applied potential was the V'_{c100} of the 13Cr steel in the water WQ1 obtained in the above potentiodynamic polarization test. As examples, the time variations of the current density during the potentiostatic polarization tests in the test waters WQ1 and WQ3 are shown in Fig. 1. For the specimen in the water WQ1 with the addition of only 100 ppm Cl^- ions, the current density increased with time. While for the specimen in the water WQ3 with also the addition of 50 ppm HCOOH, the current density showed a remarkable decreasing tendency. After the polarization tests, the specimen surfaces were observed in detail. It was found that the number densities of pits on the specimens tested in the waters WQ2—WQ4 with the addition of HCOOH decreased by comparison to that in the water WQ1, and the number density of pits occurred in WQ4 was the least. This result suggests that the initiations of pitting corrosion were inhibited by the addition of 25–75 ppm HCOOH into the test waters, agrees with the behavior shown in the above potentiodynamic polarization tests. Figure 2 gives the EPMA results for the films and corrosion products formed on the specimens after potentiostatic polarization tests in the waters WQ1 and WQ3 at 0.33 V (versus Ag/AgCl) for 20 h. A large amount of O and Cr atoms was detected in the films formed on the specimen tested in the water WQ3. It is presumed that the anodic reactions were suppressed by the formation of adsorbents such as chromium (III) formate ($Cr(OOCH)_3$) on the specimens [6–8] in the waters WQ2—WQ4 with added HCOOH, so that the pitting corrosion was suppressed.

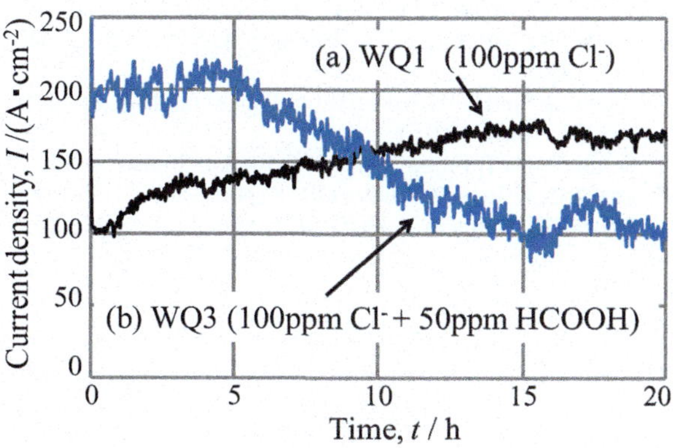

Fig. 1 Time variation of the current density during potentiostatic polarization in the waters WQ1 and WQ3 at 0.33 V (versus Ag/AgCl)

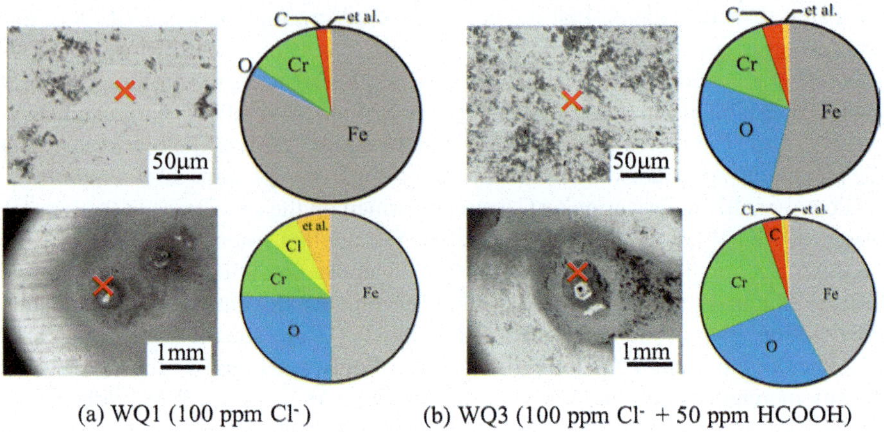

Fig. 2 EPMA results for the films and corrosion products formed on the specimens after potentiostatic polarization tests in the waters WQ1 and WQ3 at 0.33 V (versus Ag/AgCl) for 20 h

4 Conclusion

For the 13Cr steel in the simulated boiler water containing 100 ppm Cl^- ions, the addition of 25–75 ppm HCOOH into the waters suppressed the corrosion reactions as well as the destruction of the passivation films on the steel surfaces. It was suggested that the formation of Cr-rich passivation films on the steel surfaces was helped by the presence of HCOOH and it inhibited the occurrence of pitting corrosion.

References

1. Dooley B, McNaughton WP (2000) Power Plant Chem 2:709–715
2. Jonas O (1985) Mater Perform 24:9–18
3. Parker NA, Howell AG (2013) Power Plant Chem 15:212–215
4. Moed DH, Verliefde ARD, Heijman SGJ, Rietveld LC (2013) Proceedings of ICPWS16, PWS-034, London
5. Japanese Standards Association: Water conditioning for boiler feed water and boiler water; JIS B 8223, Tokyo (2015)
6. Sekine I, Ito T, Fujita T (1985) Boshoku Gijutsu 34:500–508
7. Paul RC, Kapoor P, Baidya OB, Kapoor R (2014) Z. Naturforsch B 34:160–162
8. Badea GE, Ionita D, Cret P (2014) Mater Corros 65:1103–1110

Elucidation of Fatigue Fracture Mechanism on Glass-Fiber-Reinforced-Plastics

J. Arakawa, M. Sakai, M. Hayashi, H. Akebono, A. Sugeta, J. Ohshita, H. Tanizawa, K. Shimizu, and J. Ogawa

Abstract In this study, it was elucidated that the fatigue failure mechanism of glass-fiber reinforced plastics (GFRP) with different fiber orientations, and the relationship between the fiber orientation and the failure mechanism by using a simple unidirectional fiber orientation. Fatigue tests exhibited that GFRP, whose glass fiber oriented parallel to the load direction showed high fatigue strength. Furthermore, as a result of observing the fatigue fracture, in the case of GFRP containing fibers oriented parallel to the loading direction, fatigue cracks occur at the fiber edges, propagated through between the glass fibers and the matrix. After that, the fatigue crack propagation rate

J. Arakawa (✉)
Okayama University, Okayama-Shi 3-1-1 Tsushimanaka, Kita-Ku, Okayama 700-8530, Japan
e-mail: jarakawa@okayama-u.ac.jp

M. Sakai · M. Hayashi · H. Akebono · A. Sugeta · J. Ohshita
Hiroshima University, 1-4-1 Kagamiyama, Higashi-Hiroshima, Hiroshima 739-8527, Japan
e-mail: akebono@hiroshima-u.ac.jp

M. Hayashi
e-mail: akebono@hiroshima-u.ac.jp

H. Akebono
e-mail: akebono@hiroshima-u.ac.jp

A. Sugeta
e-mail: akebono@hiroshima-u.ac.jp

J. Ohshita
e-mail: akebono@hiroshima-u.ac.jp

H. Tanizawa · K. Shimizu
Digital Monozukuri (Manufacturing) Education and Research Center, 3-10-32 Kagamiyama, Higashi-Hiroshima 739-0046, Japan
e-mail: ogawa.j@mazda.co.jp

K. Shimizu
e-mail: ogawa.j@mazda.co.jp

J. Ogawa
Mazda Motor Corporation, 3-1 Shinchi, Fuchu-Cho, Aki-Gun, Hiroshima 730-8670, Japan
e-mail: ogawa.j@mazda.co.jp

N. Maharjan and W. He (eds.), *Proceedings of the 3rd International Conference on Advanced Surface Enhancement (INCASE) 2023*, Lecture Notes in Mechanical Engineering, https://doi.org/10.1007/978-981-99-8643-9_24

gradually decreases, the fatigue crack growth stopped, and finally, GFRP material was statically fractured. Therefore, this suggests that the increase in fiber length of fibers oriented parallel to the loading direction extends the fatigue crack propagation life, leading to an increase in total fatigue life.

Keywords GFRP · Fatigue crack propagation · Matrix-fiber interface · Shear lag model

1 Introduction

In recent years, the use of glass-fiber reinforced plastics (GFRP) has been expanding. It is lightweight, has excellent corrosion resistance, and can be mass-produced even if parts are complex shape at low cost. Injection molding is especially used in the automobile industry for automobile bumpers and trunk deck boards [1–4]. In such as Injection molding process, the fibers tend to orient for specific directions in short fiber composite materials, which exhibits different fiber orientations in the skin layer and the core layer as a characteristic property. It is known that the fibers are relatively aligned in the molding flow direction in the skin layer and, aligned in perpendicular axis to the molding flow direction in the core layer. In addition, static strength and fatigue strength are greatly related to the fiber orientation in the layers such as skin or core. It has been confirmed that many fiber delamination is observed in a layer with a perpendicular fiber orientation to loading axis [5]. However, only a few studies have investigated the details of fatigue fracture, especially the initiation and growth behavior of fatigue cracks, in GFRP with skin–core-skin fiber orientation caused by injection molding. This is due to the difficulty of micro-level damage observation in specimens made by conventional injection molding, because of the complex fiber orientation in the conventional specimen. Therefore, in this study, in order to examine the effect of fiber orientations on fatigue fracture mechanism, we prepared the specimens with different fiber orientations in the skin layer and the core layer, respectively, which are similar to the condition of actual parts, and made test specimens with the fiber orientation aligned in one direction, which has no skin–core-skin layer.

2 Experimental Procedures

2.1 Type-A Specimen (Multiple Layer: Skin–core-Skin Layer)

Type-A specimen is made from injection molded plates, shown in Fig. 1, by using polypropylene (PP) as the matrix material and E-glass as the reinforcing fiber. *Type-A* specimens are prepared by cutting molded plates with the angles of 0° and 90°, respectively. In addition, as shown in Table 1, two types of E-glass with different fiber lengths were used for injection molding. The GFRP compounded with long-fiber E-glass is referred as *Type-A (long, 0°)*, and the GFRP compounded with short-fiber E-glass is referred as *Type-A (short, 0°)*. Figure 3 shows the results of examining the fiber orientation tensor distribution in the plate thickness direction using X-ray computed tomography (CT). In Fig. 3, the vertical axis indicates the plate thickness direction, with "0" representing the center of the specimen and "1" and "− 1" representing the surface layers of the specimen, and the horizontal axis indicates the fiber orientation, with "0" representing the perpendicular direction to the flow direction and "1" representing the flow direction. As shown in Fig. 3, regardless of the fiber length, the fibers tend to be oriented relatively in the direction of flow in the surface layer, while the fibers in the central layer tend to be oriented perpendicular to the direction of flow.

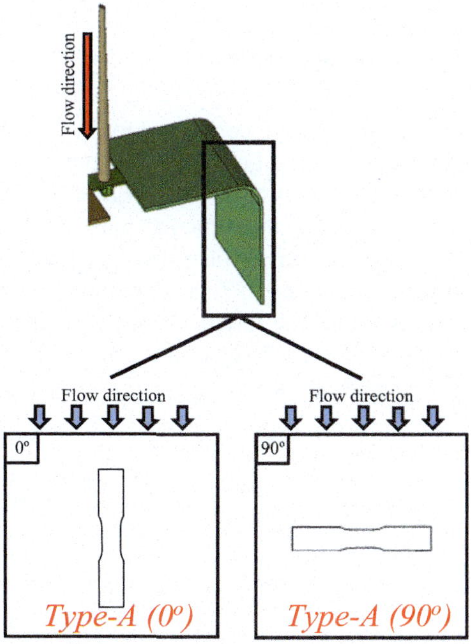

Fig. 1 Direction and location of *Type-A* specimens cut from injection molded plate

Table 1 Details of all specimens

Specimen type	Type-A(short)	Type-A(long)	Type-B
Fiber	E-glass		
Matrix	Polypropylene		
Fiber length (mm)	0.5	10	
Fiber weight ratio (%)	40		10

2.2 Type-B Specimen (Sigle Layer)

Type-B specimen, whose the fiber orientations are aligned in one direction, by using a random polypropylene (PP) copolymer to improve transparency as the matrix material, by using E-glass as the reinforcing fiber (length 10mm, weight ratio 10%). Figure 4 shows the fiber orientation tensor data in the plate thickness direction in *Type-B* obtained from X-ray computed tomography (CT). As shown in Fig. 4, although there is a little scatter in the center of the specimen, the glass fibers are generally aligned in the longitudinal direction of the specimen, and it is mentioned that *Type-B* specimen suggested a single layer. Furthermore, the specimen with fiber orientation parallel to the loading direction is called for *Type-B (Machining direction: MD)*. Then, as shown in Fig. 2b, another specimen was cut from the center of *Type-B (MD)* in the perpendicular direction to the fiber orientation, and this specimen is called for *Type-B (Traverse direction: TD)* in our study.

2.3 Fatigue Tests

The fatigue test was performed for *Type-A* specimen using hydraulic testing machine (Shimadzu corporation) as the fatigue tests at room temperature, with a frequency of $f = 5$–8 Hz and a stress ratio of $R = 0$. Then, the specimens were fixed at the same position by using the some jigs to reduce the scatter of experimental results. Furthermore, the fracture surface was observed by using a scanning electron microscope (SEM). According to *Type-B*, fatigue tests were performed at a stress amplitude $\sigma_a = 17.5$ MPa. The fatigue test was interrupted, and the initiation and propagation behavior of local fatigue cracks were tracked by observing the specimen surface with an optical microscope.

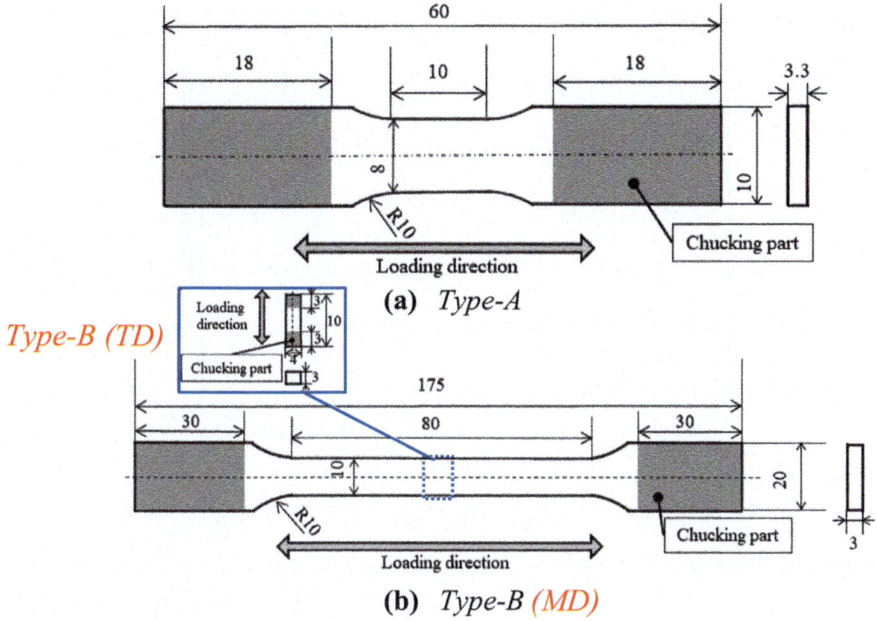

Fig. 2 Dimensions of *Type-A* and *Type-B* specimens

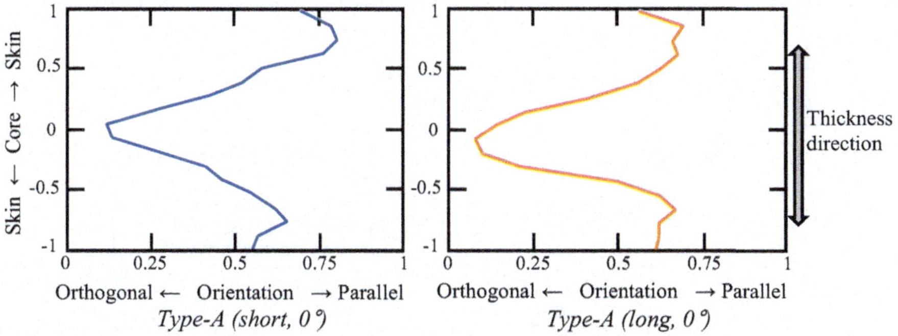

Fig. 3 Fiber orientation tensor of short and long glass fiber of *Type-A*

3 Experimental Results

3.1 Fatigue Tests Results for Type-A Specimen

Figure 5 shows the fatigue test results for *Type-A*. From the test results, it was found that *Type-A (long)* specimens with long fibers exhibited higher fatigue strength. Therefore, it was suggested that increasing the fiber length improves the fatigue

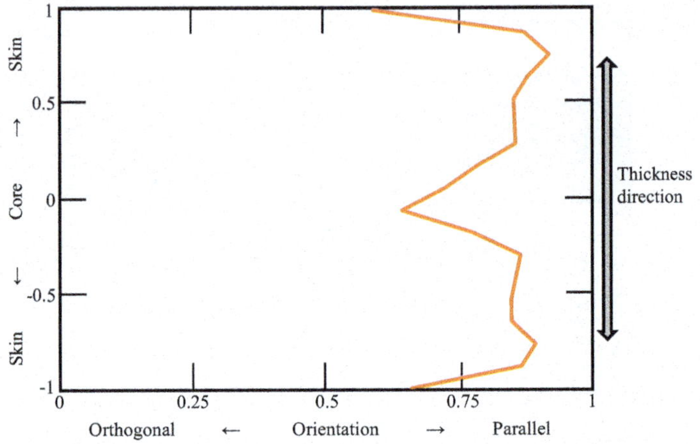

Fig. 4 Fiber orientation tensor of *Type-B*

resistance. In order to investigate the fracture mechanism of the *Type-A (long, 90°)* specimen, the fracture surface observation was performed by using SEM. The results are shown in Fig. 6. As shown in the figure, fiber fracture were observed in the core layer, while fiber delamination was confirmed in the skin layer. However, since we could not obtain useful information on the fracture origin and progression of damage, we followed up the crack initiation and propagation using *Type-B* specimens.

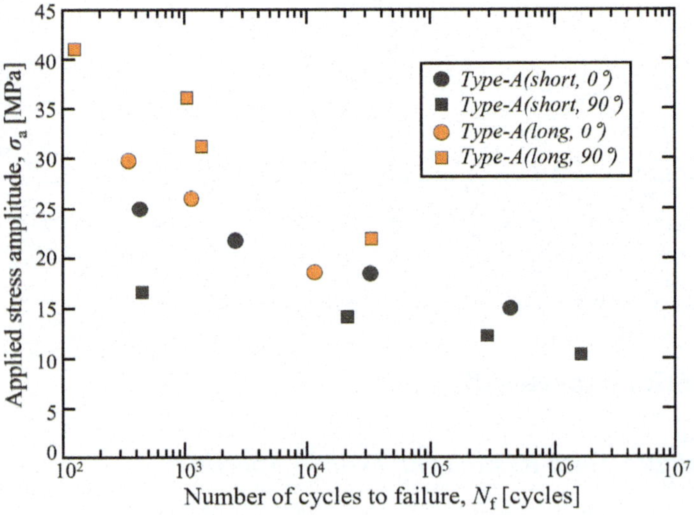

Fig. 5 Fatigue test results of *Type-A* specimen

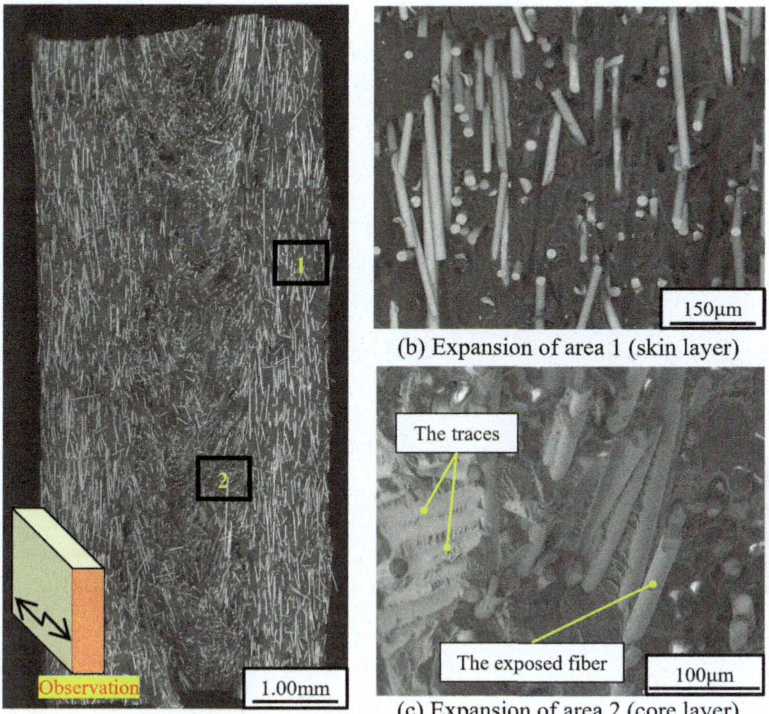

(a) Overall view
(b) Expansion of area 1 (skin layer)
(c) Expansion of area 2 (core layer)

Fig. 6 Fatigue fracture surface of *Type-A (long 90°)*

3.2 Tracking Fatigue Initiation and Propagation

In order to investigate the fatigue fracture mechanism in *Type-B (MD)*, the fatigue test was stopped as appropriate, and the area around the fiber was observed with an optical microscope. The results are shown in Fig. 7. As a result of (i) at early stage of the test, it was confirmed that cracks occur from the edge of the fiber. Then, in observation results (ii), it was confirmed that fracture progressed rapidly at the fiber-matrix interface. Finally, as a result of tracking the progress (iii), the rate of fracture growth at the fiber-matrix interface slowed down, and the crack growth generally stopped. Therefore, it is expected to extend the fatigue fracture life if the fiber length is increased. While, in *Type-B (TD)* specimens, no significant fracture occurred at the fiber-matrix interface, which is considered to depend largely on the strength of the matrix.

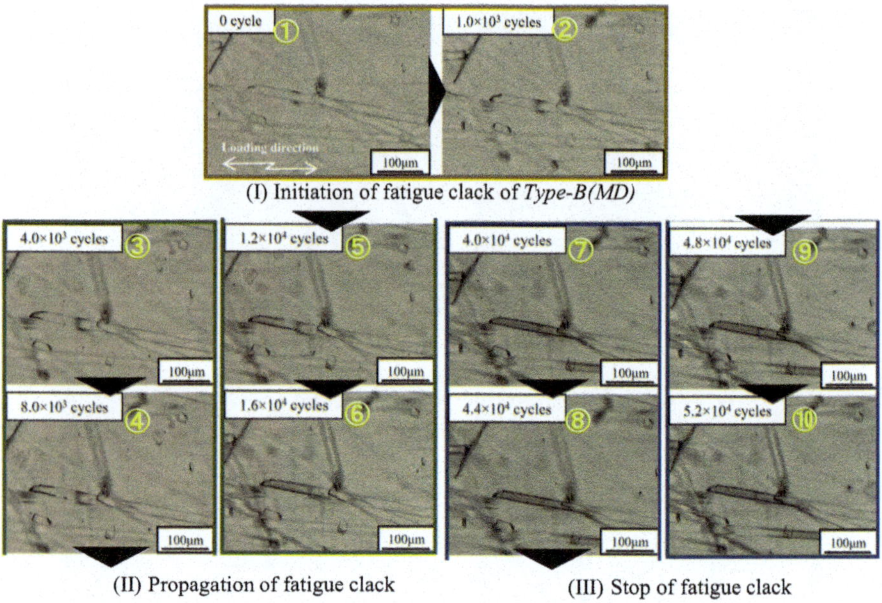

Fig. 7 Fatigue failure at the fiber-matix interface of *Type-B (MD)* ($\sigma_a = 17.5$ MPA)

3.3 Discussion Based on Shear-Lag Model

The shear stress distribution coefficient $P(x)$ was calculated based on the shear-lag theory [6] proposed by Cox et al. For the calculation, the shape factor β near the fiber was determined from Eq. (1), and $P(x)$ was calculated from Eq. (2). where, G: Transverse elastic modulus of fiber, t: Fiber diameter, E_1: Young's modulus of matrix, t_1: Thickness of matrix layer, L: Fiber length (here, *Step1*: $L = 1$ mm, *Step2*: $L = 0.5$ mm, *Step 3*: $L = 0.25$ mm). Figure 8 shows the calculation results. From this figure, it can be seen that the slope of $P(x)$ increases rapidly as the interfacial crack progresses from *Step1* to *Step3*. From this, it is considered that the shear stress gradient at the crack tip became to be higher as the interfacial crack propagated, finally it can be thought that the crack growth stopped. Therefore, in order to extend the fatigue life of GFRP, it is important to extend the fiber length and the interfacial crack to propagate as long time as possible.

$$\beta = \sqrt{G/t \cdot (2/E_1 t_1) \cdot L} \tag{1}$$

$$P(x) = \sinh(\beta \cdot x/L)/\cosh(\beta/2) \tag{2}$$

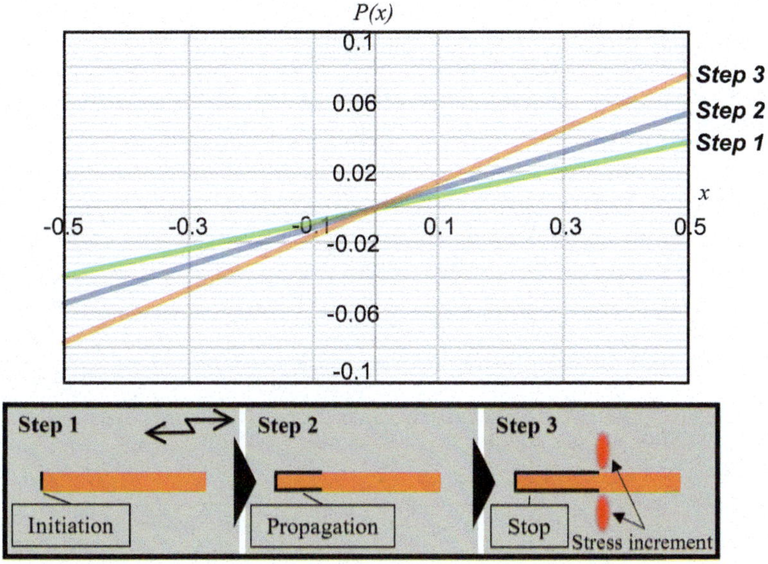

Fig. 8 Stress distribution factors $P(x)$ increasing as fatigue crack length is longer

4 Conclusions

In the case of GFRP with fiber oriented parallel to the load axis, it was found that fatigue loading causes fatigue crack initiation at the fiber edges, the crack propagated at the fiber-matrix interface and the crack growth rate gradually decreases and, finally stopped. This suggests that the increase in fiber length of fibers oriented parallel to the load axis in GFRP leads to an increase in the crack propagation life at the fiber-matrix interface, which leads to an increase in fatigue life.

References

1. Liu H, Zeng D, Li Y, Jiang L (2016) Development of RVE-embedded solid elements model for predicting effective elastic constants of discontinuous fiber reinforced composites. Mech Mater 93:109
2. Mehdipour H, Camanho PP (2019) Elasto-plastic constitutive equations for short fiber reinforced polymers. Compos Part B: Eng 165:199
3. Eftekhari M, Fatemi A (2015) Tensile, creep and fatigue behaviours of short fibre reinforced polymer composites at elevated temperatures: a literature survey. Fatigue Fract Eng Mater Struct 38(12):1395
4. Yun S, Lauke B (1998) An analytical characterization of the anisotropy of the elastic modulus of misaligned short-fiber-reinforced polymers. Compos Sci Technol 58(12):1961

5. Brunbauer J, Mösenbacher A, Guster C, Pinter G (2014) Fundamental influences on quasistatic and cyclic material behavior of short glass fiber reinforced polyamide illustrated on microscopic scale. J Appl Polym Sci 131(19):40842
6. Cox HL (1952) The elasticity and strength of paper and other fibrous materials. Br J Appl Phys 3(3):72–79

On Porosity and Its Mitigation and Closure in Additively Manufactured Alloys

Tzee Luai Meng, Na Gong, Wai Luen Chan, Rahul Karyappa, Yuefan Wei, Henry Kuo Feng Cheng, and Hongfei Liu

Abstract Porosity has long been observed in manufacturing metal alloys, which is much more frequent in additively manufactured (AM) alloys than in those manufactured by traditional methods such as cast and wrought. The common origin of porosity is strongly depending on specific manufacturing process parameters and conditions. In general, pore structures can originate from a variety of factors, including trapped gas, incomplete fusion, cooling rate, feedstock quality, processing parameters and conditions, etc. By carefully controlling the manufacturing process and conditions, the porosity in AM-alloys can be mitigated. Besides, post-AM treatment, e.g., hot isostatic pressing and/or cold working, can reduce the porosity (in both the size and density) through plastic deformation and materials flow. Here, we present a glimpse on porosity and its effect on mechanical performance of AM alloys. Mitigation and closure of porous structures will also be discussed mainly based on recent observations in aluminum- and nickel-based AM alloys, representing high-temperature high-strength and low-density applications, respectively.

Keywords Additive manufacturing · Porosity · Hot isostatic pressing · Cold-working

T. L. Meng · N. Gong · R. Karyappa · H. Liu (✉)
Institute of Materials Research and Engineering (IMRE), A*STAR (Agency for Science, Technology and Research), 2 Fusionopolis Way, Singapore 138634, Singapore
e-mail: liuhf@imre.a-star.edu.sg

W. L. Chan · Y. Wei · H. K. F. Cheng
Advanced Remanufacturing and Technology Center (ARTC), A*STAR (Agency for Science, Technology and Research), 3 CleanTech Loop, Singapore 637143, Singapore

N. Maharjan and W. He (eds.), *Proceedings of the 3rd International Conference on Advanced Surface Enhancement (INCASE) 2023*, Lecture Notes in Mechanical Engineering, https://doi.org/10.1007/978-981-99-8643-9_25

1 Introduction

Additive manufacturing (AM), which is also known as 3D printing, has seen the rapid expansion in the past decades due to its ability in rapid prototyping and cost-effective production of complex geometries [1–3]. Through building up the materials layer-by-layer [4], AM can create intricate shapes and designs that would be difficult or impossible to achieve by traditional manufacturing methods. The faster product development cycles, the reduced costs, and the increased flexibility in product design of AM made it an essential tool for a wide range of industries, ranging from aerospace and automotive to healthcare and consumer goods. Specifically, AM can offer customized products, which is particularly important in industries such as healthcare, where personalized medical devices and implants can improve patient outcomes. Furthermore, AM can be on-site processed, which is particularly important in remote or underserved areas, where access to traditional manufacturing methods may be limited.

Different from traditional manufacturing of metal alloys such as cast and wrought, where the structural defects such pores or voids can be significantly or even completely removed, post-AM process (e.g., forging and rolling) may have limited access to AM components, which provides increased requirements for quality control in AM process. The structural quality of AM materials can be affected by various factors such as temperature, humidity, and the quality of the feedstock. This can make it more challenging to ensure consistent product quality of metallic AM components.

AM of metal alloys can result in various types of structural defects, including porosity, incomplete fusion, cracking, warping, nonuniformity of elemental distribution, all related to the rapid heating and fast cooling of the molten pool (MP) during the AM process, e.g., by laser cladding (LC) and wire-arc additive manufacturing (WAAM). Such structural defects may tremendously degrade the mechanical performance and the fatigue life of AM components. In this work, we present an overview on the origin and effect of porous structures, as well as their mitigation and closure, in AM alloys. Our discussion is mainly based on recent observations in nickel- and aluminum-based AM alloys, representing high-temperature high-strength and low-density applications, respectively.

2 Origin and Effect of Porous Structures in AM Alloy

The common origin of porous structures in AM alloy can vary depending on the specific manufacturing process and conditions. However, in general, pores can result from a variety of factors such as trapped gas, cooling rate, feedstock quality, and the processing parameters. Trapped gas induced pores have been frequently observed in thin film depositions [5, 6]. During AM process, unescaped gas can become trapped in the molten metal, leading to the formation of pores when the metal solidifies. Usually, the solubility of gas molecules decreases with the temperature in metals and exhibits a sharp decrease, i.e., by order of magnitudes, when the metal is solidifying from liquid to solid near the melting point. If the segregated gas molecules during solidification are trapped at the liquid–solid interface, e.g., due to the fast deposition and/or rapid cooling, porous structures will be formed. Such a mechanism induced by hydrogen incorporations dominates the formation of porosity in aluminum alloys deposited by AM [7]. Different from the gas trapping mechanism, incomplete melting of the feedstocks, e.g., due to a mismatch between the direct-energy and the feeding rate, gives rise to incomplete fusion and thus the formation of porous structures. For the same reason, porous structures can be caused by a mismatch between the direct energy specifications and the other processing parameters. Besides, the quality of the feedstock, for example, moisture and/or contamination, can affect pore formation due to increased gas trapping and/or incomplete melting. Overall, it is important to carefully control the processing parameters and conditions to minimize the formation of porous structures in AM alloys.

At the current stage, porosity is still a common defect in AM alloys and the presence of pore structures or voids within AM alloys can essentially weaken their structural integrity and degrade their mechanical properties (i.e., the strength, ductility, and toughness) as well as fatigue performance. Besides, porosity tends to increase the susceptibility of AM alloys to cracking and failure under external stress, all because voids within the material can act as stress concentrators, leading to localized failure and reduced overall performance. Physically, the severity of the porosity impact on mechanical performance depends on a few factors, including the size, shape, and distribution of the pores, as well as the specific alloy material and AM technique [8].

Figure 1 shows that there are three critical locations where porous structures can form in AM alloys. They are the surface, subsurface, and interface of the deposited layers. Since the AM is typically a layer-by-layer process, nucleation and growth of pores may also occur at the interlayer boundaries. Once formed, the interface porosity is decisive to toughness and adhesions when external load is applied transverse to the layer orientation. The surface porosity-induced roughening tends to decrease the corrosion resistance and, meanwhile, promotes the nucleation of stress corrosion cracking (SCC) of the AM alloy. In general, all types of porosity can affect the tensile and fatigue strengths since they act as stress concentration centers, leading to localized failure and reduced overall mechanical performance.

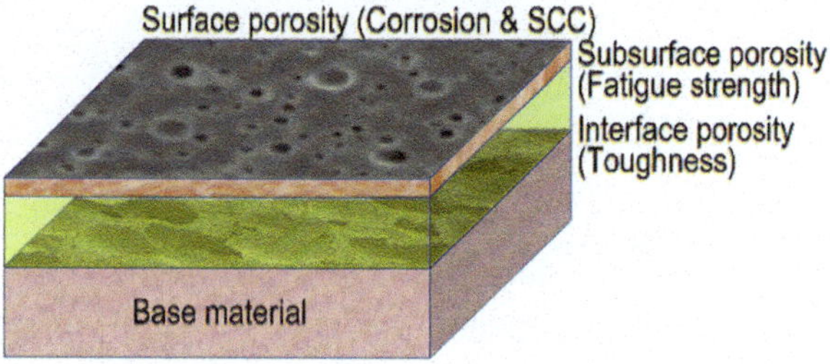

Fig. 1 Schematic diagram showing the typical effect of porous structures, including their types and locations, in AM alloys on their mechanical properties

3 Porosity Mitigation and Closure in AM Alloy

It has been mentioned above that both the feedstock quality and the processing parameters and conditions contribute to the origin of porosity in AM alloys. In this light, carefully engineering and storage of the feedstocks and optimizing the processing parameters under certain conditions can mitigate the porosity in specific AM alloys. For example, pre-heating of the wire feedstock could reduce the porosity in aluminum alloy deposited by WAAM [9]. The porosity mitigation has been attributed to removal of moisture, grease, and other hydrogen-related contaminants from the surface of the wire feedstock by the pre-heating. In this way, the hydrogen incorporated into the MP is effectively reduced.

Under certain AM process, the direct relationship between the porosity and the mechanical performance is usually correlated by introducing porosity mitigation and/or closure techniques such as interpass cold-working and/or post-AM treatments [10–15]. The general physics behind cold-working is plastic deformation and material flow under dynamic external workload while the post-AM treatments include not only cold-working but also hot isostatic pressing (HIP) and laser treatment [16–22]. Cold-working and laser-treatment are typical surface engineering techniques that can close the surface and subsurface porosity while HIP can also close the pores much deeper in the AM alloys [21–25]. For example, we have recently observed remarkable porosity closures induced by robotic hammer peening (RHP) in Inconel 718 coupons deposited by LC, especially at the near-surface regions. A typical example is shown in Fig. 2, which shows that both the size and density of the porous structures have been significantly reduced by the RHP process.

It is worth mentioning that most of the pore closure techniques developed for metal alloys manufactured by traditional method might not be suitable for AM components due to their complex configuration and the layer-by-layer process. For example, the individual blades of aerospace turbine blisks, where the limited space and specific orientation relationship between adjacent blades make many post-AM treatments inaccessible. In this regard, a lot of efforts have been made in recent years to develop AM process with reduced necessity of post-AM treatments. A typical stratage is

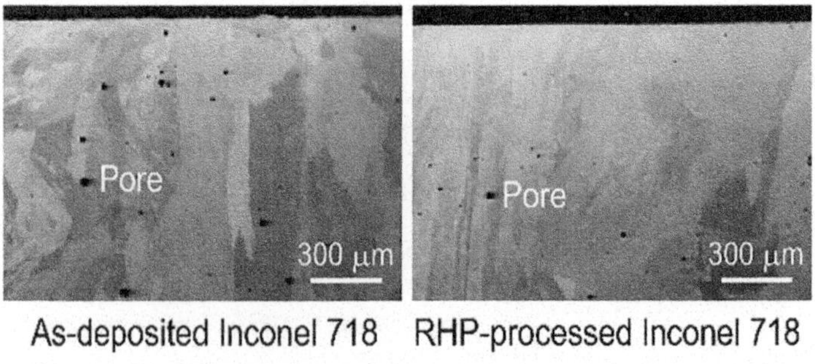

Fig. 2 Microscopic images taken from the cross-section of Inconel 718 coupons deposited by laser-cladding (LC) before (left) and after (right) robotic hammer peening (RHP)

to introduce interpass cold-working process. Considering the process compatibility, laser-shock peening (LSP), RHP, and cold-rolling have been applied by various research groups as the interpass process to enhance the component integrity of AM alloys. Our recent studies revealed that the effective thickness of the surface enhancement in Inconel 718 by RHP can be up to 1.0 mm, which is close to a single-layer deposition of AM process. This observation provides a promising perspective for monolithic integration of RHP and LC-based processes on industrial robot platforms towards high-quality AM of metallic alloys with reduced necessity for post-AM treatments.

Finally, theoretical studies, especially advanced artificial intelligence (AI) methods, have recently shown great help in building the feedstock-process-performance relationship for AM of metallic alloys. Together with advanced in-situ process-monitoring and data-collection techniques, e.g., nondestructively visualizing the porosity by X-ray computed tomography (XCT), the optimization of AM process for prototyping specific alloys can be significantly accelerated.

4 Conclusion and Outlook

Along with the rapid developments in precisely control and digitalizable automation, AM, such as LC and WAAM, has attracted increasing research interest in recent years, not only in academic communities but also in industrial applications. Although AM has many advantages, especially the ability to enable rapid prototyping and cost-effective production of customizable components with complex geometries, there are some challenges in achieving high-quality metal alloys by AM as compared with traditional methods. Porosity is a common defect in AM alloys and its origin is associated with both the quality of the feedstock and the AM processing parameters and conditions. By careful engineering the feedstock and optimizing the proceeding parameters and conditions, the porosity can be greatly mitigated in AM alloys. Post-AM treatments can also be employed to close the porous structures and enhance the mechanical performance of metallic alloys, typically through introducing plastic deformation and material flow at either room temperature (i.e., cold-working) or elevated temperatures (e.g., HIP and laser-treatment). Compared with post-AM treatments, in-situ integrity enhancement of AM alloys has attracted more interest in recent years. Interpass cold-working process, including peening and rolling, has been extensively studied and shown promising impact in AM process of metallic alloys. It is believed that theoretical modeling and advanced AI techniques, together with recent advances in in-situ process-monitoring and data-collection techniques, will greatly accelerate the optimization of AM process for designing and prototyping specific metal alloys with reduced porosity and thus enhanced mechanical properties.

References

1. Gong N, Meng TL, Cao J et al (2023) Laser-cladding of high entropy alloy coatings: an overview. Mater Technol 38(1):2151696
2. Lim WYS, Cao J, Suwardi A et al (2022) Recent advances in laser-cladding of metal alloys for protective coating and additive manufacturing. J Adhes Sci Technol 36(23–24):2482–2504
3. Liu H, Tan CKI, Meng TL et al (2022) Direct deposition of low-cost carbon fiber reinforced stainless steel composites by twin-wire arc spray. J Mater Process Technol 301:117440
4. Liu H, Tan CKI, Wei Y et al (2020) Laser-cladding and interface evolutions of inconel 625 alloy on low alloy steel substrate upon heat and chemical treatments. Surf Coat Technol 404:126607
5. Gong N, Seng HL, Jin H et al (2023) Interfacial hydrogen-related species in low-vacuum-deposited coatings on extrinsic substrates: Insights from experimental and theoretical studies. Mater Today Commun 34:105050
6. Liu H, Guo S, Yang RB et al (2017) Giant blistering of nanometer-thick Al_2O_3/ZnO films grown by atomic layer deposition: mechanism and potential applications. ACS Appl Mater Interfaces 9(31):26201–26209
7. Bai J, Ding HL, Gu JL et al. (2017) Porosity evolution in additively manufactured aluminium alloy during high temperature exposure. In: IOP conference series: materials science and engineering, vol 167(1). pp 012045
8. Al-Maharma AY, Patil SP, Markert B (2020) Effects of porosity on the mechanical properties of additively manufactured components: a critical review. Mater Res Expr 7(12):122001
9. Fu R, Tang S, Lu J et al (2021) Hot-wire arc additive manufacturing of aluminum alloy with reduced porosity and high deposition rate. Mater Des 199:109370
10. Liu H, Meng TL, Cao J et al. (2022) In: Advanced surface engineering and protective coating. Springer, Singapore, pp 138–141
11. Xie C, Wu S, Yu Y, et al. (2021) Defect-correlated fatigue resistance of additively manufactured Al-Mg4.5Mn alloy with in situ micro-rolling. J Mater Process Technol 291:117039
12. Gu J, Yang S, Gao M et al (2020) Micropore evolution in additively manufactured aluminum alloys under heat treatment and inter-layer rolling. Mater Des 186:108288
13. Gu J, Ding J, Williams SW et al (2016) The effect of inter-layer cold working and post-deposition heat treatment on porosity in additively manufactured aluminum alloys. J Mater Process Technol 230:26–34
14. Kalentics N, de Seijas MOV, Griffiths S et al (2020) 3D laser shock peening—a new method for improving fatigue properties of selective laser melted parts. Addit Manuf 33:101112
15. Liu H, Tan CKI, Dong X, Meng TL, Cao J, Wei Y (2022) Laser-cladding and robotic hammer peening of stainless steel 431 on low alloy steel 4140 for surface enhancement and corrosion protections. J Adhes Sci Technol 36(21):2313–2327
16. Gong N, Luai Meng T, Kiang Ivan Tan C et al. (2023) high energy laser-shock induced phase transformation and micro-spallation on surface of stainless steels: the effect of stacking fault energy and deformation mechanisms. Appl Surf Sci 613:156013
17. Liu H, Meng TL, Cao J et al (2022) Comparisons on localized surface modifications of stainless steels induced by laser shock peening and robotic hammer peening. Proc CIRP 108:118–122
18. Liu H, Wei Y, Tan CKI et al (2020) XRD and EBSD studies of severe shot peening induced martensite transformation and grain refinements in austenitic stainless steel. Mater Charact 168:110574
19. Liu H, Ivan Tan CK, Wei Y et al. (2020) Robotic hammer peening-induced martensite in austenitic steels: spatial distributions of plastic deformation and phase transformation. Proc CIRP 87:297–301
20. Liu H, Tan CKI, Cheng WS et al (2020) Effects of robotic hammer peening on structural properties of Ni-based single-crystal superalloy: dislocation slip traces and crystallographic reorientations. Metall and Mater Trans A 51(6):3180–3193
21. du Plessis A, Macdonald E (2020) Hot isostatic pressing in metal additive manufacturing: X-ray tomography reveals details of pore closure. Addit Manuf 34:101191

22. McLean N, Bermingham MJ, Colegrove P et al (2022) Effect of hot isostatic pressing and heat treatments on porosity of wire arc additive manufactured Al 2319. J Mater Process Technol 310:117769
23. Gong N, Wei Y, Meng TL et al (2023) Laser-treatment-induced surface integrity modifications of stainless steel. Mater Res Expr 10(3):034002
24. Gong N, Maharjan N, Liu H et al (2022) Laser-induced in-plane curving of ripples on biomedical stainless steel and their relationship to biological functions. Mater Technol 37(14):3089–3099
25. Liu H, Wei Y, Ivan Tan CK, et al. (2021) Laser-treatment-induced morphology and structure modifications of stainless steel: element segregations and phase evolutions. Mater Chem Phys 266:124570

Effect of Plasma Electrolytic Oxidation on the Fatigue Strength of Al Alloys: Mechanistic Understanding from Surface and Sub-Surface Characterizations

Xiaohu Huanga, Jaime Jia Min Pang, and Yee Ng

Abstract Plasma electrolytic oxidation (PEO) is a promising technology to improve the durability of aluminum alloys. However, herein we find that PEO process reduces the fatigue strength of the aluminum substrate, and the reduction varies with the PEO coating process. Detailed surface and sub-surface characterizations are conducted to understand the mechanism. Our study provides in-depth understanding on the effect of PEO coating on the properties of the base material, thus has important implications on their applications.

Keywords Plasma electrolytic oxidation · Coating · Al Alloys · Fatigue · Surface characterizations

1 Introduction

Light-weight metals are key materials in advanced manufacturing including aerospace, automobile, consumer electronics, and etc [1]. However, due to their active chemical properties and low hardness, light-weight metals, such as aluminum (Al) alloys, are susceptible to both corrosion and erosion. Surface and sub-surface plays a crucial role in determining the durability of materials, as both corrosion and erosion starts from the surface. Therefore, coating is a general strategy for surface protection and enhancement. Various coating technologies have been developed to protect light-weight metals, including physical vapor deposition, chemical vapor deposition, sol–gel, spray, anodization, etc [1]. Most of these technologies suffers from issues such as high costs, poor adhesion, poor stability. Among the various coating technologies, plasma electrolytic oxidation (PEO) is an appealing technology to enhance the corrosion and erosion resistance via a conversion coating process [1, 2].

X. Huanga (✉) · J. J. M. Pang · Y. Ng
Institute of Materials Research and Engineering (IMRE), Agency for Science, Technology and Research (A*STAR), 2 Fusionopolis Way, Innovis #08-03, Singapore 138634, Republic of Singapore
e-mail: xhhuang.nano@gmail.com

N. Maharjan and W. He (eds.), *Proceedings of the 3rd International Conference on Advanced Surface Enhancement (INCASE) 2023*, Lecture Notes in Mechanical Engineering, https://doi.org/10.1007/978-981-99-8643-9_26

However, there is not enough study on the fatigue properties of the PEO-coated Al alloys, which is very important for applications in the aerospace industry. Here we present our study on the fatigue properties of PEO-coated Al alloys, and try to shed some light on the mechanism through detailed surface and interface characterizations.

2 Methods

Al 7075-T6 alloys were machined into flat tensile coupons. PEO coating process were conducted under constant voltage mode with different process voltages, as reported in our earlier report [2]. We characterized the PEO coatings with x-ray diffraction (XRD), scanning electron microscopy (SEM) with energy dispersive spectroscopy (EDX), and x-ray residual stress. Tensile and fatigue strength of the PEO-coated Al coupons were tested according to ASTM E8 and ASTM E466, respectively. For each PEO process condition, three samples were grown with the same parameters and used for the tensile and fatigue tests.

3 Results and Discussion

XRD and EDX results show that the PEO coating is mainly composed of Al_2O_3, in consistent with the conversion coating nature. During the coating process, the surface aluminum is oxidized and transformed into Al_2O_3. SEM results show the presence of round-shaped craters on the surface of the PEO coating, which is a characteristic feature of PEO coatings. The morphology of the PEO coating varies slightly with the PEO process voltage. PEO coating provides good protection against corrosion [2], however the mechanical strength of the Al base material is weakened by the PEO coating process. For instance, the Young's modulus of Al coupon is reduced by 4% after the PEO coating process was conducted at 520 V. The fatigue strength of the coated Al coupons is deteriorated a lot after the PEO coating process. For example, compared to the bare Al coupon, Al coupon with PEO coating grown at 520 V exhibits a 45% reduction in fatigue life. The reduction in fatigue life increases with increasing the PEO process voltage, which is contrary to the effect of PEO process voltage on the corrosion protection [2]. Therefore, we need to balance the corrosion protection and fatigue (Fig. 1). The reduction in fatigue strength will greatly compromise the structural integrity of the Al alloys, thus it must be addressed properly for some applications.

To understand the mechanism behind the PEO-induced fatigue strength reduction of Al, we conducted cross-section SEM characterizations to look at the interface between the PEO coating and Al base (Fig. 1). As the thickness of the PEO coating is far thinner than that of the Al coupon, it is unlikely that the reduced fatigue strength is caused by the variation in the thickness of the base material. However, compared to the mirror flat surface of the uncoated Al coupon, we observed micro-meter scale

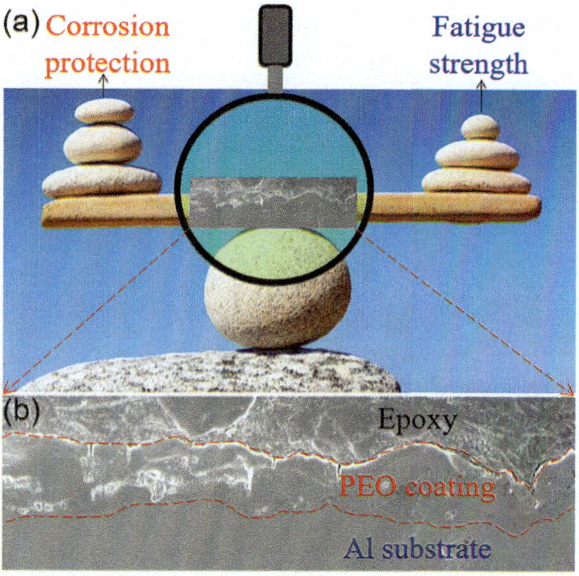

Fig. 1 **a** A schematic illustration on the balance of corrosion protection and fatigue strength of PEO-coated Al alloys. **b** A cross-section SEM image of the PEO-coated Al alloy

roughness was induced on the Al side at the interface of PEO-coating/Al. Besides, we also conducted x-ray residual stress analysis of the Al substrate before and after PEO coating, and detected tensile residual stress on the surface of the PEO-coated Al samples. Based on these surface and interface characterizations, we propose the reduced fatigue strength of PEO-coated Al is mainly caused by the tensile residual stress as well as the increased roughness on the surface of Al. This mechanistic understanding can serve as an important guideline to develop more advanced PEO process to minimize their adverse effects on the fatigue strength of light-weight metals.

4 Conclusion

Inclusion, we have grown PEO coating on Al alloys, and found the PEO coating has a detrimental effect on the fatigue strength of the Al alloys. Surface and interface characterizations provide clues on the mechanism of the fatigue strength weakening. The results provide a deeper understanding on the effect of PEO coating on the base materials, thus help to provide direction towards more advanced PEO process.

Acknowledgements The authors acknowledge the support from A*STAR Career Development Fund (SC25/21-803617) and Singapore Aerospace Programme Cycle 14 project (SC25/20-8R1709). XH wants to acknowledge Dr Zhang Zheng's help with the x-ray residual stress measurements.

References

1. Famiyeh L, Huang XH (2019) Plasma electrolytic oxidation coatings on aluminum alloys: microstructures, properties, and applications. Mod Concepts Mater Sci 2:000526
2. Liu J, Huang XH, Ren Y, Wong LM, Liu HF, Wang SJ (2022) Galvanic corrosion protection of Al-alloy in contact with carbon fiber reinforced polymer through plasma electrolytic oxidation treatment. Sci Rep 12:1–10

Effect of Heat Treatment on the Residual Stress and Mechanical Properties of an Additively Manufactured High Strength Aluminum Alloy

Nur Syafiqah Johan, Muhammad Syafiq Bin Azrin, Mohammed Faizal, and Sankaranarayanan Seetharaman

Abstract Aluminum (Al) alloys with low density are extensively used in ground transport and aerospace applications to attract fuel energy savings. While the additive manufacturing of Al alloys is not new, precipitation hardened alloys pose severe challenges due to their susceptibility to high residual stress and cracking. This study aims to report on the influence of heat treatment in minimizing the surface residual stress and improving the room temperature tensile response of a precipitation hardened high strength Al alloy (Al2139). The changes to mechanical properties will be presented by invoking the process-microstructure-property relationships.

Keywords Additive manufacturing · Selective laser melting · Al alloys · Precipitates · Residual stresses · Mechanical properties · Microstructure

1 Introduction

High strength aluminum (Al) alloys with low density are extensively used in ground transport and aerospace applications to attract fuel energy savings [1]. Traditional manufacturing of High strength Al are often inefficient and energy intensive, as a result additive manufacturing (AM) is considered. Laser powder bed fusion (LPBF) of Al alloys faces unique obstacles such as surface oxidation sensitivity and hot tearing susceptibility during rapid solidification [2]. By tackling these challenges, properties will be enhanced by residual stress relaxation and heterogeneous grain

---

N. S. Johan · M. S. B. Azrin · M. Faizal · S. Seetharaman (✉)
Advanced Remanufacturing and Technology Centre (ARTC), Agency for Science, Technology and Research (A*STAR), 3 Cleantech Loop, #01/01 CleanTech Two, Singapore 637143, Republic of Singapore
e-mail: sankar_seetharaman@artc.a-star.edu.sg

N. S. Johan
e-mail: nur_syafiqah_johan@artc.a-star.edu.sg

N. Maharjan and W. He (eds.), *Proceedings of the 3rd International Conference on Advanced Surface Enhancement (INCASE) 2023*, Lecture Notes in Mechanical Engineering, https://doi.org/10.1007/978-981-99-8643-9_27

nucleation. This study aims to report on the LPBF processing of EOS Al2139 and the effects of heat treatment on the properties of printed parts.

2 Methodology

An EOS M290 Direct Metal Laser Sintering (DMLS) machine based on LPBF technology was used for the printing of test coupons. The printed coupons were subjected to T4 heat treatment which involved solution annealing at 490 °C for 45 min before water quenching and natural ageing for 3 days at room temperature. The LPBF processed Al2139 coupons were tested and examined for density, microstructure features and mechanical tests. Surface residual stresses measurements were also obtained. Lastly, the results were also analyzed in terms of dimensional variation using CMM measurements. For comparison purpose, the residual stress and dimensional variation were also modelled using MSC Simufact software.

3 Results and Discussion

Figure 1a shows the Archimedes density measurements of the as-printed (AP) and heat-treated (HT) Al2139 coupons. AP revealed more than 99% density which was not significantly affected after heat treatment. The optical microscopy results also confirmed the fabrication of dense materials with porosity less than 0.5% (Fig. 1b). The microstructure of Al2139 coupons is shown in AP and HT conditions. While the AP coupon had a fine grain structure with severely mis-indexed regions (Fig. 1c), HT coupon shows a relatively larger grain size with lesser mis-indexed regions (Fig. 1d). The XRD spectrum also reveals Cu-rich grain boundary intermetallic Al4Cu9 (Fig. 1e) in the AP state.

The results consisting of microhardness and tensile measurements are listed in Table 1. The microhardness and tensile strength properties improved after HT and the achieved yield strength value have surpassed the traditional cast and wrought alloys. LPBF alloys generally have high strength because of rapid cooling, resulting in fine microstructures. The strength of Al-Cu alloys has significantly increased after HT, primarily due to precipitation hardening effects [3]. It is also key that the improvement in yield and tensile strengths occur without affecting the ductility.

Figure 2 displays the results of residual stress, dimensional measurements, and simulations for the cantilever coupons (Fig. 2a-d). While the results of dimensional check by CMM revealed deviation below (±) 1 mm on the free end of the cantilever prior to wire cutting, XRD measurements highlighted a drop-in mean residual stress from ~ 120 MPa in AP to < 20 MPa in HT (Fig. 2e). Simulation results confirmed successful printing of Al2139, with measured residual stress below its yield strength.

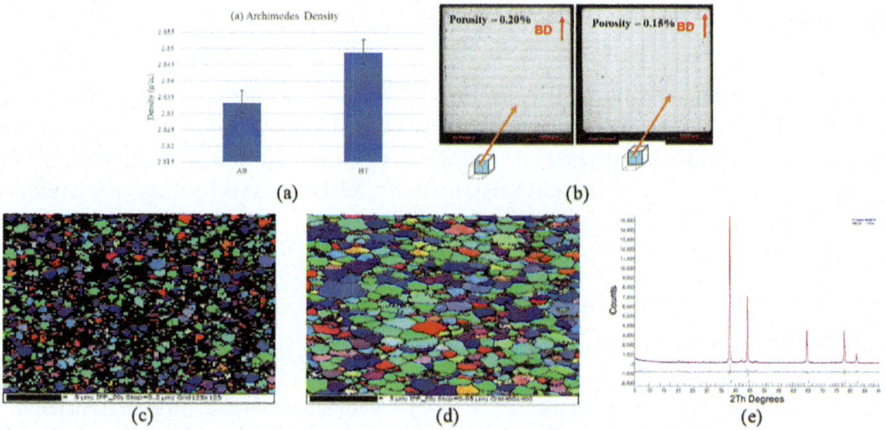

Fig. 1 Bar chart showing the density results (1a), 2D micrographs of AP and HT coupons (1b) and generated micrographs showing grain features in AP (1c), HT (1d) and XRD spectra of AP sample (1e)

Table 1 Microhardness and tensile properties of LPBF processed Al2139

Alloy	Processing method	Condition	Microhardness (Hv)	0.2 YS (MPa)	UTS (MPa)	Ductility (%)
Al2139—current study	LPBF	As-printed	145 ± 2	405 ± 2	445 ± 2	8.9 ± 0.9
		Heat treated (T4-HT)	174 ± 4	485 ± 4	572 ± 7	9.3 ± 0.3
Al2139 (reference)	Cast	T6	-	320	350	8–10
	Wrought	T6	-	440	490	15

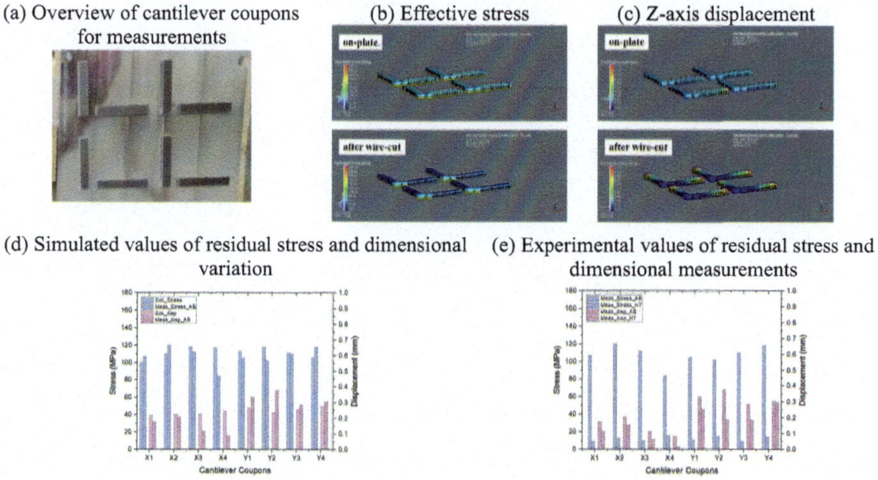

Fig.2 Results of residual stress and dimensional measurements

4 Summary

This project evaluates the LPBF printing feasibility and mechanical properties of Al2139. Test coupons were printed to assess mechanical properties and residual stress. Mechanical testing showed a superior response similar to high strength Al alloys processed by AM and traditional methods. XRD analysis revealed a significant reduction in residual stress after heat treatment.

References

1. Georgantzia E, Gkantou M, Kamaris GS (2021) Aluminium alloys as structural material: a review of research. Eng Struct 227:111372
2. Altıparmak SC, Yardley VA, Shi Z, Jianguo L (2021) Challenges in additive manufacturing of high-strength aluminium alloys and current developments in hybrid additive manufacturing. Int J Lightweight Mater Manuf 4(2):246–261
3. Mair P, Kaserer L, Braun J, Weinberger L, Letofsky-Papst I, Leichtfried G (2021) Microstructure and mechanical properties of a TiB2-modified Al–Cu alloy processed by laser powder-bed fusion. Mater Sci Eng, A 799:140209

Blanked Surface Characteristics of Amorphous Alloys with Local Microstructure Modification by Ultrashort Pulsed Laser

Chieko Kuji

Abstract Amorphous alloys with no crystal structure, particularly Fe-based ones, exhibit excellent soft magnetic properties and are expected to be applied to motor cores because of their high magnetic flux density. However, they are difficult to machine because of their mechanical properties, such as high strength and toughness, owing to their unique structure. These mechanical properties of amorphous alloys change due to thermal microstructural changes, that is, structural relaxation and crystallization; however, the excellent magnetic properties also degrade. Therefore, this study proposes a new blanking method that improves the machinability of amorphous alloys without degrading their magnetic properties by thermally changing the microstructure of the local area where blanking is performed. This study investigated the effectiveness of ultrashort pulsed lasers, which have little thermal effect, on the local heat treatment of amorphous alloys. Furthermore, blanking tests were performed on the locally heat-treated amorphous alloys, and the blanking resistance values and characteristics of the blanked surfaces were evaluated.

Keywords Amorphous alloy · Ultrashort pulsed laser · Local heat treatment · Blanking · Surface characteristic

1 Introduction

Amorphous alloys exhibit a unique structure without crystals, which is achieved by rapid quenching from the molten state to suppress the formation of a long-range ordered structure [1]. In particular, Fe-based amorphous alloys exhibit excellent material properties, such as high soft magnetic properties, strength, and toughness. Because of their high magnetic flux density and low iron loss, they are expected to be applied in motor cores with high energy-saving effects [2]. However, their

C. Kuji (✉)
Tohoku University, Sendai 980-8579, Japan
e-mail: c-kuji@tohoku.ac.jp

N. Maharjan and W. He (eds.), *Proceedings of the 3rd International Conference on Advanced Surface Enhancement (INCASE) 2023*, Lecture Notes in Mechanical Engineering, https://doi.org/10.1007/978-981-99-8643-9_28

tough mechanical properties render them difficult to machine. All of these material properties are caused by the amorphous structure.

Therefore, this study develops a new blanking method that locally transforms the amorphous microstructure, which is the cause of various material properties, and improves only the machinability [3–5]. Because amorphous alloys are metastable, they transform their structure through structural relaxation [6] and crystallization [7], in which atoms are shifted to stable positions by heat. In this study, the microstructure was gradually changed by adjusting the heat treatment temperature [3], and the mechanical properties in a few-micrometer region drastically changed depending on the microstructure [4]. Furthermore, by using ultrashort pulsed lasers, which have a low thermal effect, for heat treatment applications, it was possible to thermally transform only the machining area while suppressing thermal diffusion, and the machining resistance was successfully reduced [5].

In this paper, the effect of microstructure on blanked surface characteristics is discussed by comparing the blanked cross sections of samples heat-treated with ultrashort pulsed lasers and conventional methods.

2 Methods

Iron-based amorphous alloy foil strip $Fe_{77}B_{16}Cr_2Si_5$ (Metglas 2605S-3A, PROTERIAL, Ltd.) with approximately 20-μm thickness, whose microstructure changes due to heat treatment had been investigated in the past [3], was used for the evaluation. As the microstructure of amorphous alloys changes drastically with a difference of only a few degrees in the heating temperature [3], normal heat treatment was performed using a thermogravimetric differential thermal analyzer (TG-DTA: TG8120, Rigaku Corp.), which enables precise temperature control. Heat treatment was performed in a reducing atmosphere (0.998% H_2-Ar gas) to prevent oxidation. The samples were heated from 300 K at 20 K/min to 713 and 873 K for structural relaxation and full crystallization, respectively. Subsequently, it was cooled to 300 K at 20 K/min without holding the samples isothermally at the highest temperature. An ultrashort pulsed laser (LPF-2, Lastec Co., Ltd.) with a laser beam diameter ($1/e^2$ width) of 65.6 μm was used for local heating. The laser irradiation conditions were an oscillation central wavelength of 1,030 nm, power of 1 W, pulse width of 260 fs, and pulse repetition frequencies of 50 and 200 kHz. Laser irradiation at 50 kHz causes structural relaxation, whereas irradiation at 200 kHz causes crystallization [5]. The laser was irradiated at equal intervals with a 0.1 mm spacing after 20,000 laser pulses per spot, and local heat treatment was performed in a dotted pattern along the contour of the tool geometry (see Fig. 1a). Moreover, four pinholes were created to fix the sample to the blanking machine. The as-received, normally heated, and locally heated samples were blanked under the same conditions. The machine shown in Fig. 1b was used for the blanking test. The clearance between the punch and die was set to 2 μm. A rectangular punch with a 4-mm width and 12-mm length was used. To accurately blank out the heated position of the locally heated specimen, the punch was fixed to

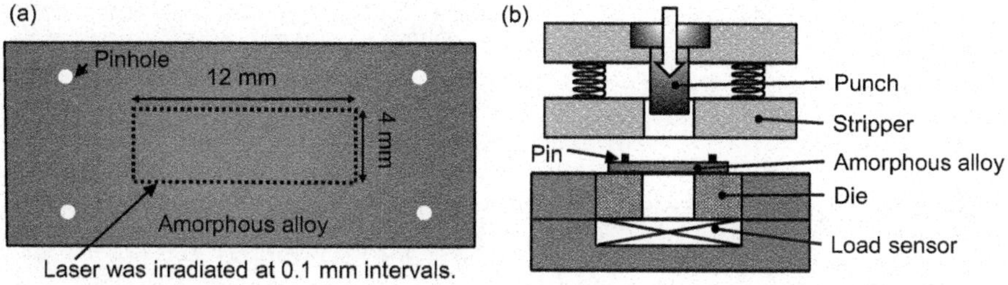

Fig. 1 Schematic diagram of **a** the laser irradiated specimen and **b** the blanking machine

four pins installed on the blanking machine. The blanking speed was set to 20 mm/s, and the blanked fracture surface was observed using a scanning microscope (SEM, JXA-8530F, JEOL).

3 Results and Discussion

Fig. 2 shows the SEM images of the blanked fracture surfaces of the as-received and normally heat-treated samples. Fig. 2a, b, and c refer to the as-received sample, structurally relaxed sample heated to 713 K, and (c) fully crystallized sample heated to 873 K. In all the cases, the tool was fed from the top to the bottom of the image. Generally, the blanking surface of a metallic material consists of four elements: (i) a shear droop caused by material deformation, (ii) a sheared face created by friction between the tool and material, (iii) a broken face created by crack growth, and (iv) a burr, which is a machining defect at the bottom of the material. The as-received specimen in Fig. 2a has high toughness, and faces (i) and (ii) were generated by shear slip and tool friction. In Fig. 2b, (iii) appeared all over the surface due to embrittlement, and in Fig. 2c, (iv) was caused by the recovery of the plastic deformation ability due to crystallization. In the unobserved area, large cracks and wavy surfaces were observed in all cases, and it is difficult to conclude that the surfaces were well-machined.

Figure 3 shows the SEM images of the blanked surfaces of the samples locally heated by an ultrashort pulsed laser. Figure 3a shows a sample of structural relaxation when the pulse repetition frequency is 50 kHz, and Fig. 3b shows a crystallized sample when the pulse repetition frequency is 200 kHz. In both cases, the laser irradiation initially causes ablation, followed by local heating due to latent heat, resulting in a hole-like cut line at the laser-irradiated point (arrow mark in Fig. 3). Figure 3 shows that (ii) and (iii) appear above and below the fracture surface, respectively, regardless of the microstructure. Furthermore, neither (i) nor (iv), which can lead to machining defects, were observed. Even in the unshown region, an ideal machined surface without cracks or waviness was obtained. As shown in Fig. 2, the specimens with normal heating exhibited significant changes in mechanical properties [4] and

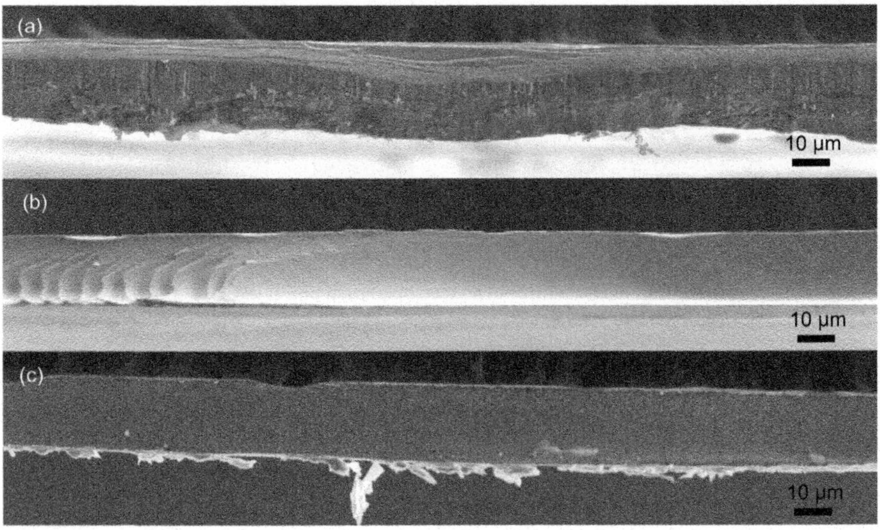

Fig. 2 Blanked surface characteristics of **a** the as-received specimen, **b** the specimen annealed up to 713 K, and **c** the specimen annealed up to 873 K, as observed by SEM

fracture surface shape [8], depending on the microstructure of the specimens. This is attributed to the original undulation of the specimen, which preferentially deforms the convex part of the specimen, resulting in an unevenly blanked surface. In contrast, when ultrashort pulsed laser local heating was used, the small holes formed during the heat treatment caused a notch effect, and deformation proceeded from the stress concentration points, resulting in almost the same fracture surface shape, regardless of the microstructure.

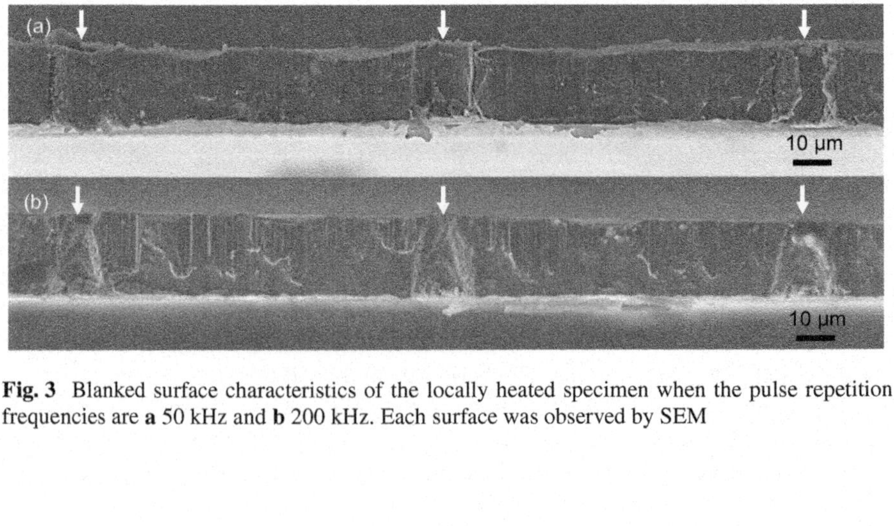

Fig. 3 Blanked surface characteristics of the locally heated specimen when the pulse repetition frequencies are **a** 50 kHz and **b** 200 kHz. Each surface was observed by SEM

4 Conclusions

In this study, the blanked surface characteristics of Fe-based amorphous alloys subjected to different heat treatment methods were evaluated using SEM images. For the normal heat treatment method, the blanked surface characteristics significantly changed depending on the change in the microstructure, resulting in the appearance of an uneven blanked surface. However, for local heating using an ultrashort pulsed laser, the notch effect caused by the small holes formed during the heat treatment resulted in the appearance of a uniform and ideal blanked surface, even for specimens with different microstructures.

Acknowledgements The author would like to express her deepest gratitude to Professor Emeritus Tsunemoto Kuriyagawa and his colleagues at Tohoku University, Japan for their invaluable guidance in the conduct of this research. Special thanks to the Materials Processing Technology Department of the Iwate Industrial Research Institute, Japan for the use of their scanning electron microscope. Part of this work was supported by JSPS KAKENHI, grant number 23K13228, and The Amada Foundation, AF-2022238-C2.

References

1. Chen HS, Miller CE (1970) A rapid quenching technique for the preparation of thin uniform films of amorphous solids. Rev Sci Instrum 41:1237–1238
2. Enomoto Y (2019) Radial gap type motor using amorphous metal for iron core. IEEJ Transact Industr Appl 139(6):606–611
3. Kuji C, Takenaka K, Mizutani M, Shimada K, Kuriyagawa T, Konno TJ (2021) Crystallization behavior and machining properties of annealed Fe–Si–B–Cr amorphous alloys. J Mater Sci 56(29):16697–16711
4. Kuji C, Mizutani M, Shimada K, Suzuki K, Miura H, Kuriyagawa T (2022) Evaluation of fracture properties of annealed Fe–Si–B–Cr amorphous alloys using micro-tensile tests and blanking machinability. Mater Sci Eng A 848:143483
5. Kuji C, Mizutani M, Takenaka K, Shimada K, Konno TJ, Soyama H, Kuriyagawa T (2023) Effect of local heating by ultrashort pulsed laser on Fe–Si–B–Cr amorphous alloys and its influence on blanking machinability. Precis Eng 81:135–144
6. Antonione C, Baricco M, Riontino G (1988) Structural relaxation kinetics in FeSiB amorphous-alloys. J Mater Sci 23(6):2225–2229
7. Zaluska A, Matyja H (1983) Crystallization characteristics of amorphous Fe-Si-B alloys. J Mater Sci 18(7):2163–2172
8. Kuji C, Mizutani M, Takenaka K, Shimada K, Konno TJ, Yokobori AT, Kuriyagawa T (2023) Relationship between blanking performance and microstructure of annealed Fe–Si–B–Cr amorphous alloy sheets. Precis Eng 82:33–43

Exploring the Effects of Prolonged Vibratory Polishing on Shot Peened Ti-64

Abhay Gopinath, Luei Kah Han, Wan Yin Chi, Chan Wai Luen, and A. Senthil Kumar

Abstract Shot peening is a common surface treatment technique used to improve the fatigue life of aerospace materials such as Titanium-64. However, shot peening increases the surface roughness of the part and hence, an additional vibratory polishing process is usually employed as a post-treatment step to reduce the surface roughness. Unfortunately, vibratory polishing is a material removal process, and its effect on the residual stress and cold work in the part is not well understood. In this paper, we explore the impact of prolonged vibratory polishing on shot peened coupons. The study focuses on the residual stress, cold work, and microstructure of the test coupons.

Keywords Vibratory polishing · Residual stress · Surface roughness · Microstructure · Ti-64

1 Introduction

Vibratory polishing (VP) is a surface treatment technique that is commonly applied to aerospace parts that have been previously shot peened (SP). The aim is to obtain a particular surface finishing required to achieve a certain aerodynamic performance [1–3]. SP improves the fatigue life of aerospace parts through plastic deformation and the introduction of cold work at the surface, but at the same time increases the surface roughness of the part [4, 5]. VP improves the finishing of the surface through

A. Gopinath (✉)
Rolls-Royce Singapore Pte Ltd, 1 Seletar Aerospace Crescent, Singapore, Singapore
e-mail: e0444185@u.nus.edu

L. K. Han · W. Y. Chi · C. W. Luen
Advanced Remanufacturing and Technology Centre, 3 Cleantech Loop, #01/01 CleanTech Two, Singapore 637143, Singapore

A. Gopinath · A. Senthil Kumar
Department of Mechanical Engineering, National University of Singapore, 21 Lower Kent Ridge Rd, Singapore, Singapore

N. Maharjan and W. He (eds.), *Proceedings of the 3rd International Conference on Advanced Surface Enhancement (INCASE) 2023*, Lecture Notes in Mechanical Engineering, https://doi.org/10.1007/978-981-99-8643-9_29

Fig. 1 VP setup for blades (Courtesy Walther Trowal)

a material removal process by reducing the distances between the peaks and troughs of the surface, hence reducing the surface roughness which is described as Ra [6, 7]. Koltsov had earlier discussed the effect of grinding on shot peening in a previous work [8]. Prolonged VP is generally not prescribed due to the material removal nature of the process, as prolonged VP may negate the advantages gained from SP. In this paper, the authors explored the effect of prolonged VP on SP Ti-64 coupons, particularly looking into the residual stress, microstructure, and surface roughness of the coupons. This will enable design and manufacturing engineers to reconsider the effect of VP on the aerospace parts and relax tolerances which may be beneficial in terms of cost and sustainability (Fig. 1).

2 Experimental and Research Methodology

SP was carried out with a Progressive robotic machine operating at an intensity of 0.15 ± 0.3 mmN. VP was carried out with a vibratory trough from Walther Trowal (TMV175/85), running at various process times of 2, 12 and 20 h. The vibratory motion of the trough is achieved by rotating the unbalanced shaft of the trough with a motor, which causes the media to flow in a specific direction in the trough. Test coupons were mounted on to a fixture that is a representative of an aerospace component, follow which the entire setup is lowered into the trough for polishing. The combination of the polishing media hitting on the coupon surface, together with the coupon also hitting the media (due to vibration) results in a necessary pressing action that gives the polishing effect and the creation of a smoother surface. Residual stress was measured from the center of the Ti-64 coupons using an XSTRESS Robotic X-Ray Diffraction (XRD) measurement system from Stresstech Oy, Finland. The residual stresses were measured at various depths, with the material removal process carried out via electropolishing. In terms of cold work, Full Width Half Max (FWHM) returns the amount of cold work within the part. The higher the FWHM, the higher the cold work. A BS EN 15305:2008 test standard was used for the XRD measurements. The material properties of the Ti-64 coupons used in this work are as follows: (i) Young's Modulus of 115000 MPa, and (ii) Poisson ratio of 0.32. Surface roughness was measured with a Talysurf PGI 840 contact stylus, with a sampling length of 12 mm in the longitudinal direction and 4 mm in the transverse direction (Fig. 2). The experimental settings of the VP setup is as shown in Table 1.

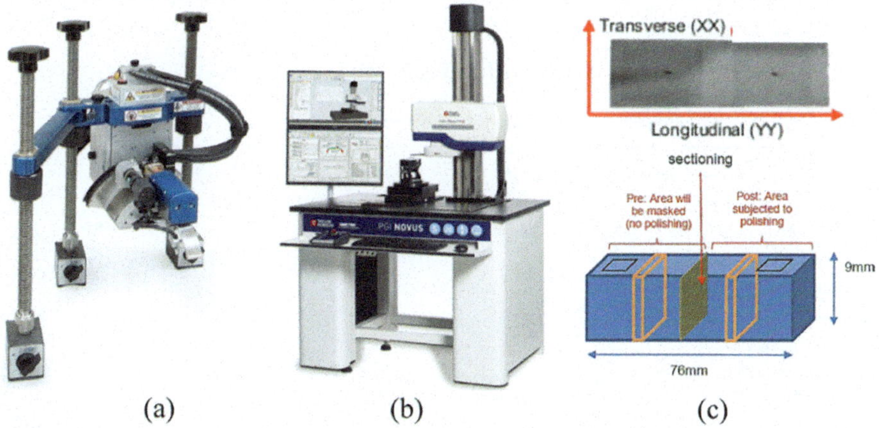

Fig. 2 Instruments **a** XSTRESS XRD machine, **b** Talysurf PGI 840 for surface roughness and **c** specimen sectioning used for characterization

Table 1 Experimental settings for trials

Machine settings	Value
Type	TMV 175
Motor RPM	1500-clockwise and anticlockwise rotation for uniformity
Amount of water, litre/hr	200
Compound type	KS66
Compound ratio	2
Media type	Walther trowel V2030 (polishing media)-Mass 835 kg
Aerospace material	Ti-64
Process times	120 min, 720 min, 1200 min

3 Results and Discussions

3.1 *Residual Stress and Cold Work*

Figure 3 shows that the residual stresses of shot peened (SP) coupons subjected to various duration of Vibropolishing (VP). Coupons that were only SP gave a residual stress reading of around 700 MPa compressive. It has been found that there is no relaxation of the stresses even after 2, 12 and 20 h of VP. The variation in residual stress is within the measurement error. Depth-dependent XRD is a destructive method and hence, different coupons may have a slight difference in the RS measured. Similar observations for cold work were made, where the differences are not significant in terms of reduction of cold work in the coupons after VP. Even after 20 h VP, cold work is still present for a depth of around 90–100 microns (Fig. 4).

From the IPF plots of EBSD scanning, we first observed that there is no preferred grain growth orientation. In addition, from the KAM map, a layer of high dislocation

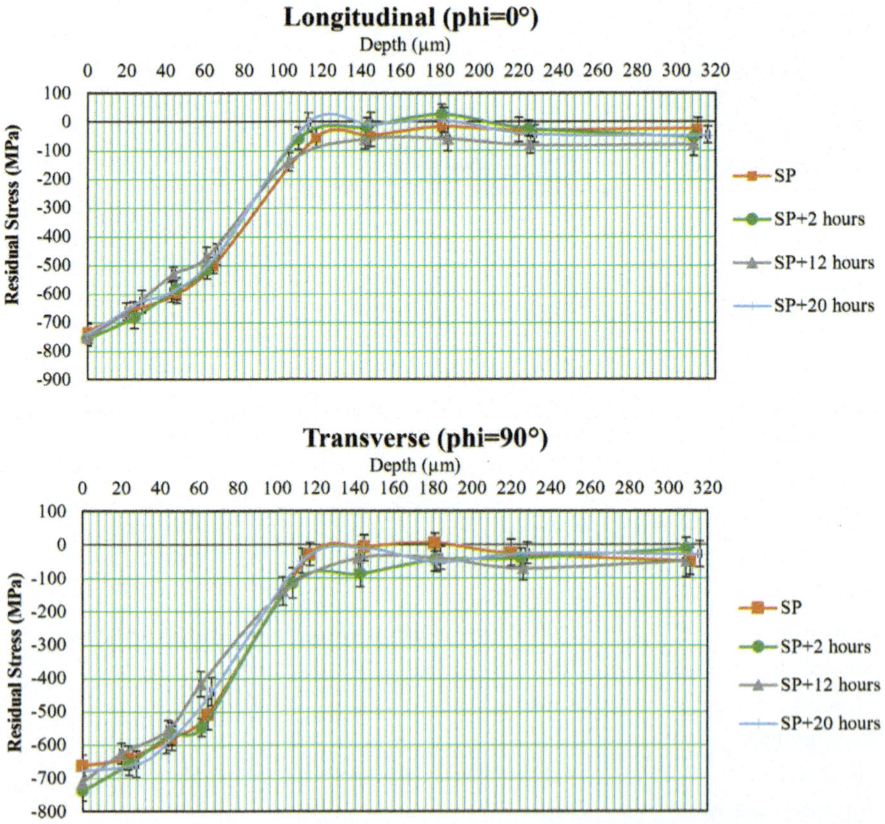

Fig. 3 Residual stress profile of different hours of Vibropolishing using V2030 media on SP

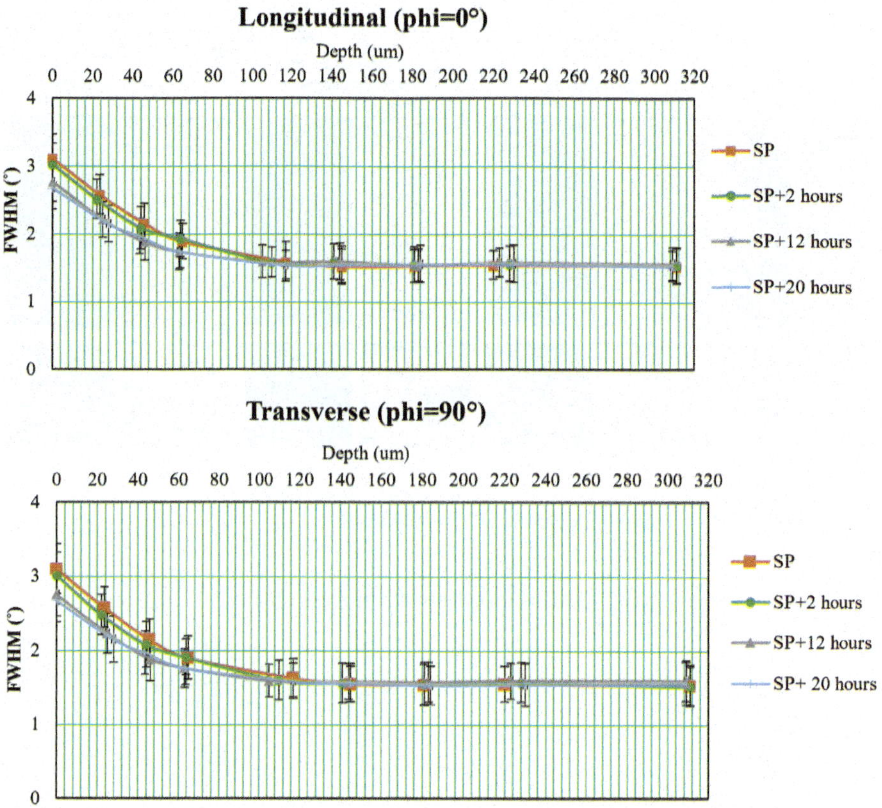

Fig. 4 FWHM profiles of different hours of Vibropolishing using V2030 media on SP

density is seen in the SP coupon, indicating that there is significant cold work in the surface due to the peening effect from SP. However, this dislocation layer is observed to decrease in thickness or be removed after certain a duration of VP, confirming the material removal nature of the VP process. The grain orientation of coupons subjected to VP of 2 and 12 h are similar, as well as the dislocation density as seen from KAM. The dark space can either be due to internal defects (voids) from the material itself, or from grain size being too small to be resolved under the EBSD. This hypothesis will require a Transmission Electron Microscope (TEM) to verify. There is a region with preferred grain growth orientation (Ti cubic 111, Ti hex 0110) at around 10 μm below the surface. The coupon that was subjected to 24 h of VP seems to have slightly larger grain size as seen in Fig. 5. Apart from the removal of the dislocation layer after VP for 24 h, a region of preferred grain orientation (Ti cubic 111, Ti hex 0110) can be seen around 400 μm below the surface. However, this may be due to the coupon's inherent microstructure makeup, or it was formed from the raw material manufacturing route. By comparing different treatment times, we observed that as the treatment time increases, the space between the grains decreases. It could be due

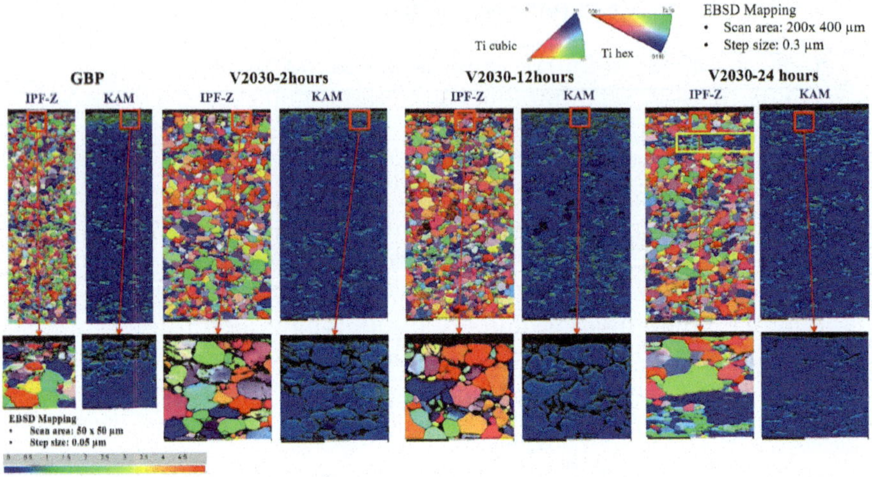

Fig. 5 EBSD map for different durations of Vibropolishing

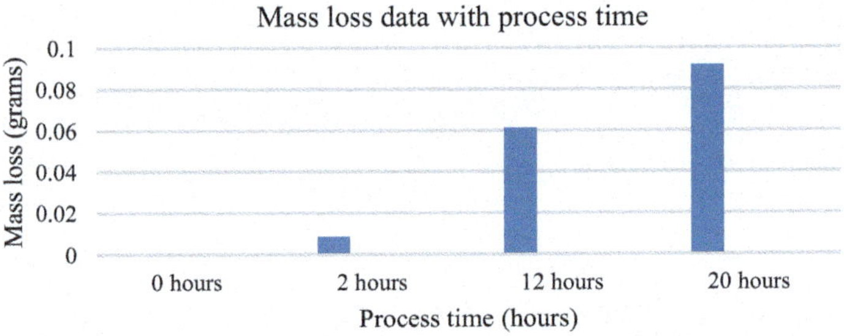

Fig. 6 Mass loss profiles after vibropolishing process

to either the VP process reducing the porosity between the grains, or the VP process removing material from the surface, and the grain structure observed after 24 h could be from the deeper portions of the original material. Since VP is a material removal process as shown in Fig. 6, the latter could be more prominent.

4 Conclusions and Outlook

In this work, a high-level study is done to understand the effect of prolonged Vibropolishing (using Walther Trowal V2030 media) on shot peened Ti 64 coupons. There is a limited loss of residual stress from the surface or the bulk, even after 20 h of polishing. For future work, the amount of material removal at sharp corners, Residual Stress and coldwork could be studied using cutting media (example: Walther trowel

SP media). Also, a study of the rounding of edges could be incorporated if it's a real-life aerospace part study.

Acknowledgements The authors would like to thank National University of Singapore (NUS), Advanced Remanufacturing Technology Centre (ARTC) and Rolls-Royce for providing equipment, materials, measurement resources and facilities. The authors would also like to thank Haris Fadzullah Bin Taufik, Ahmad Alif Bin Hamid, Ahmad Suhaimi Bin Ahmad Redzuan from ARTC for their experimental efforts. The authors also thank Dr. Thomas Haubold and Dr. Goetz Feldmann from Rolls-Royce Deutschland for technical guidance.

References

1. Hennig W, Feldmann G, Haubold T (2014) Shot peening method for aerofoil treatment of blisk assemblies. In: Procedia CIRP, Elsevier B.V., pp 355–358. https://doi.org/10.1016/j.procir.2014.04.060.
2. Davidson DA (2003) Developments in mass finishing technology. Met Finish 101:49–56. https://doi.org/10.1016/S0026-0576(03)90203-0
3. Canals L, Badreddine J, McGillivray B, Miao HY, Levesque M (2019) Effect of vibratory peening on the sub-surface layer of aerospace materials Ti-6Al-4V and E-16NiCrMo13. J Mater Process Technol 264:91–106. https://doi.org/10.1016/j.jmatprotec.2018.08.023
4. Gibson GY, Perkins KM, Gray S, Leggett AJ (2016) Influence of shot peening on high-temperature corrosion and corrosion-fatigue of nickel based superalloy 720Li. https://doi.org/10.1080/09603409.2016.1161945
5. Bogachev I, Knowles KM, Gibson GJ (2020) Electron backscattered diffraction analysis of cold work in a shot peened single crystal nickel superalloy. Materialia 14:100860. https://doi.org/10.1016/J.MTLA.2020.100860
6. Dwivedi YD, Wahab A, Pallay AD, Shesham A (2022) Effect of surface roughness on aerodynamic performance of the wing with NACA 4412 airfoil at Reynolds number 1.7 × 105. Mater Today Proc 56:468–476. https://doi.org/10.1016/j.matpr.2022.02.153
7. Van Rooij RPJOM, Timmer WA (2003) Roughness sensitivity considerations for thick rotor blade airfoils. J Sol Energy Eng Trans ASME 125:468–478. https://doi.org/10.1115/1.1624614
8. Koltsov VP, Vinh LT, Rakitskaya VB (2020) To determination of removable material during grinding with flap wheels after shot peen forming. In: IOP conference series materials science of engineering, Institute of Physics Publishing. https://doi.org/10.1088/1757-899X/709/2/022023

Simulation and Modelling of Surface Integrity

Advancing Plating Efficiency with Simulation Technology

Klaus Schmid

Abstract Powerful tools for improving the layer thickness distribution in electroplating are available through simulation techniques for calculating the electric field. To get the best out of these simulation technologies, a good knowledge of the possibilities of simulation tools, the needed requirements as well as typical, proven workflows are important. This paper uses an industrial case study to cover, present and discuss the above-mentioned points as well as the used approach used and the achieved results. Both the individual results achieved and the improved understanding of the process noticeably improve the efficiency in electroplating.

Keywords Advanced electroplating · Plating efficiency · Simulation technology · Current density distribution · Layer thickness distribution

1 Introduction

1.1 Plating Efficiency as Key Element for Increased Sustainability in Electroplating

Sustainability in its true sense is difficult to achieve in electroplating. The metals used for coating purposes come from non-renewable sources and when coated components are being recycled, the layer material is rather difficult to retrieve. Key element for the most sustainable use of electroplating processes is therefore undoubtedly the highest possible increase in efficiency. Especially the minimization of the amount of coating metal required to achieve desired properties represents a promising approach.

Typically, a certain minimum layer thickness must be applied to components in order to maintain the necessary functional properties over the service life of the

K. Schmid (✉)
Fraunhofer Institute for Manufacturing, Engineering and Automation IPA, Nobelstrasse 12, 70569 Stuttgart, Germany
e-mail: Klaus.schmid@ipa.fraunhofer.de

N. Maharjan and W. He (eds.), *Proceedings of the 3rd International Conference on Advanced Surface Enhancement (INCASE) 2023*, Lecture Notes in Mechanical Engineering, https://doi.org/10.1007/978-981-99-8643-9_30

products. However, according to Faraday's law, the influence of the electric field leads to significant differences in the layer thickness. This means that in most cases, in order to achieve the necessary minimum layer thickness, valuable metals have to be applied to a higher extent than actually necessary with a more uniform layer thickness distribution.

Certainly, many electroplating companies achieve good coating thickness distributions, but their expertise is usually based on a good knowledge of electroplating fundamentals and a lot of experience. The development of new knowledge as well as the adaptation and enlargement of existing knowledge to new conditions such as new products, new plants or new processes is therefore mainly empirical and therefore time-consuming. The use of simulation techniques promises both a reduction of the empirical effort as well as the generation of further experience, as simulations can be varied far more flexibly and extensively than test coatings.

1.2 Simulation Technology for Electroplating

It can be assumed that there are many possible applications for simulation techniques in electroplating. In some cases, the systems or models are still being developed, e.g. for predictive electrolyte management [1], or for support in electrolyte testing or development [2]. However, this does not apply to the layer thickness distribution. With simulation techniques for calculating the electric field, powerful tools are available for improving the layer thickness distribution [3], which have already proven their practical suitability [4].

Some approaches already exist to use simulation techniques for the steering of plating systems in order to achieve a better layer distribution by moving the components relative to the anode [5, 6], some models are set up in widespread modelling kernels [7] or applications are developed on available finite element software [8]. In addition to software and solutions from ELSYCA [3, 9], COMSOL Multiphysics [10, 11] provide well applicable simulation tools.

It must be assumed that the use of simulation techniques means a significant effort and implies certain requirements for approach and execution in order to achieve relevant, practicable results with an acceptable effort. The development of appropriate operational methods is a worthwhile object for research and development. In the following, the use of simulation technology for the sustainable improvement of the layer thickness distribution is presented based on a job-plater case study.

2 Using Simulation Technology for Improving Plating Efficiency

2.1 Starting Points and Requirements

In principle, the choice of the software itself is of secondary importance. In the selected example, the two-dimensional software tool Elsy2D® was used, mainly because the components to be calculated were easy to model in 2D and the computing power requirements for the software are rather low. All necessary calculations were carried out on site with a usual office laptop.

In order to carry out a simulation successfully, the geometric data of the component in use, the plating tank and the anode technology are required in any case. Depending on the application, various methods can be used; from recording the data on site and re-modelling the plating situation to importing existing CAD data as a starting point for the modelling. In the process of gathering the geometric data, it is also necessary to determine the actually available optimization prospects. For example, not every plant allows the anode distances to be varied and components cannot be positioned in any desired position, etc.. Only an exact adjustment ensures the achievement of satisfactory results.

Furthermore, layer thickness measurements of the relevant component are important for adjusting the simulation quality. Here it is especially important to discuss the exact position of the measuring points as well as the chosen measuring method in advance in order to avoid misunderstandings and additional effort.

The qualitative and quantitative target setting is very important, too. If the goal is to improve the existing layer thickness distribution without having to calculate the exact layer thicknesses, a calculation of the so-called primary current density distribution is sufficient and no further information is required apart from the geometric data. The primary current density distribution neglects the equalising effect of the overpotential [12]; the layer thickness distribution is therefore practically overdrawn. Pure optimization is possible without any problems, but if exact layer thicknesses are to be calculated, additional electrochemical data of the plating process must be taken into account. Depending on the availability of the data, this can result in significant effort without the optimization results being able to improve significantly in terms of quality. Here, a corresponding cost-benefit analysis is appropriate.

2.2 Typical Workflow

The typical, proven process of a simulation project is divided into these individual points:

- Problem analysis and target setting
- Data gathering

- Setting up the calculation geometry
- Determination of the reporting method
- Calculation of the initial state
- Structured implementation of calculation studies
- Determination of optimization data
- Documentation, presentation and discussion

The individual points are described in the following case study.

3 Case Study—Effect of Current Thief in Hard Chromium Plating

3.1 Introducing the Part and Its Plating

The selected case study concerns the hard chrome plating of a dressing roll for steel strip production. It should be noted, that the roll surface is sandblasted before the plating in order to achieve a sufficient grip of said plated roll due to the resulting roughness. The plating of the roll is carried out hanging in the longitudinal direction in a deep bath, according to the state of the art, the concentric arrangement of the anodes ensure a good plating thickness distribution in the circumferential direction.

Using an upper and lower current thief controls the layer thickness distribution in the longitudinal direction and was already controlled but not optimised in the initial state. The objective was to improve the layer thickness distribution as much as possible and to determine general findings about the optimal positioning of current thieves, which should also be transferable to other parts. Additionally, it was also investigated whether the size of the current thieves could be reduced; layer metal deposited on the current thieves does not contribute to the desired plating of the part and is both a waste in material and energy.

3.2 Starting Point—Initial Situation

Based on the data collected on site, a simulation geometry was set up as a two-dimensional rotationally symmetrical model with the y-axis as the axis of rotation. The design corresponds to a typical deep bath for hard chrome with anodes arranged all around and reaching into the depth of the bath and current thieves installed in direct part extension, see Fig. 1.

The calculation of the primary current density distribution was used to carry out the study. This is only dependent on the geometry and allows the current density used for the calculations to be normalised to an average value of 100 A/m^2, which means that the calculated current density values can directly be interpreted as percentage of

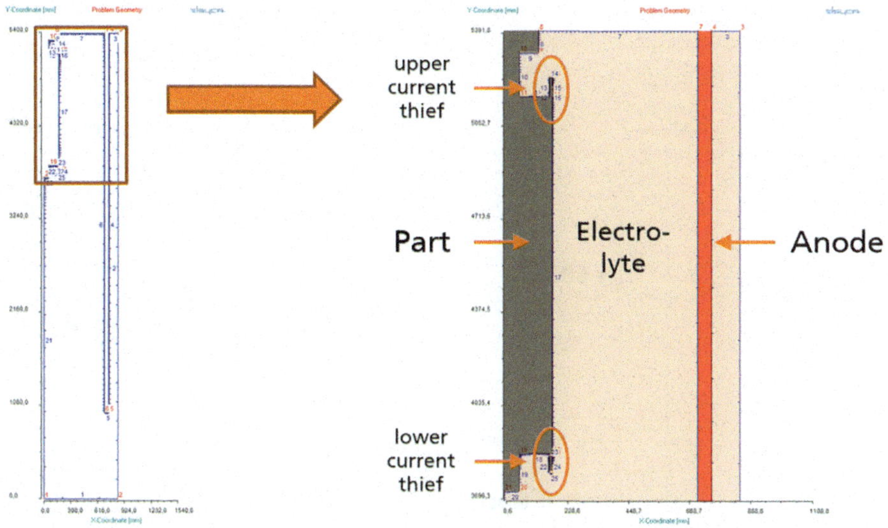

Fig. 1 Initial situation with regard to the entire plating tank as a rotationally symmetrical simulation (left) with the relevant area in detail (right)

the average current density. This made it easier to compare and explain the results, so that a strongly graphically oriented approach could be chosen for the reporting in order to achieve a simple and clear presentation of the calculated effects. The current density was therefore usually displayed graphically along the developed length along the part, see Fig. 2.

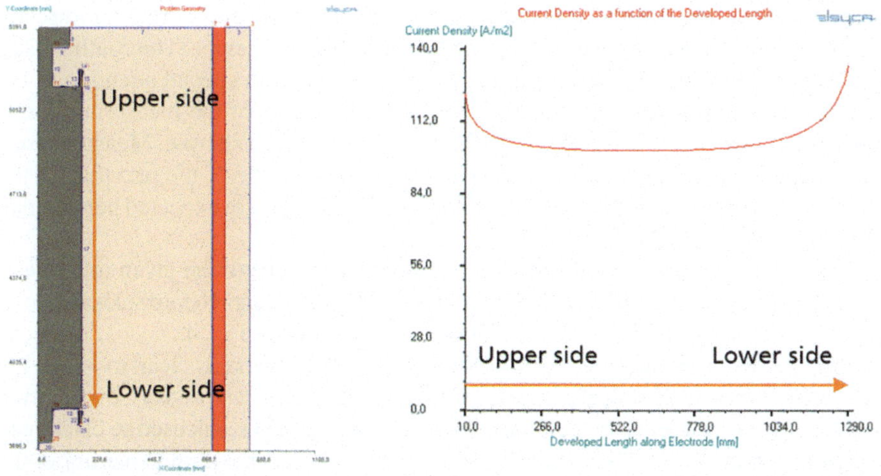

Fig. 2 Representation of the current density normalised to an average of 100 A/m^2 along the developed part length

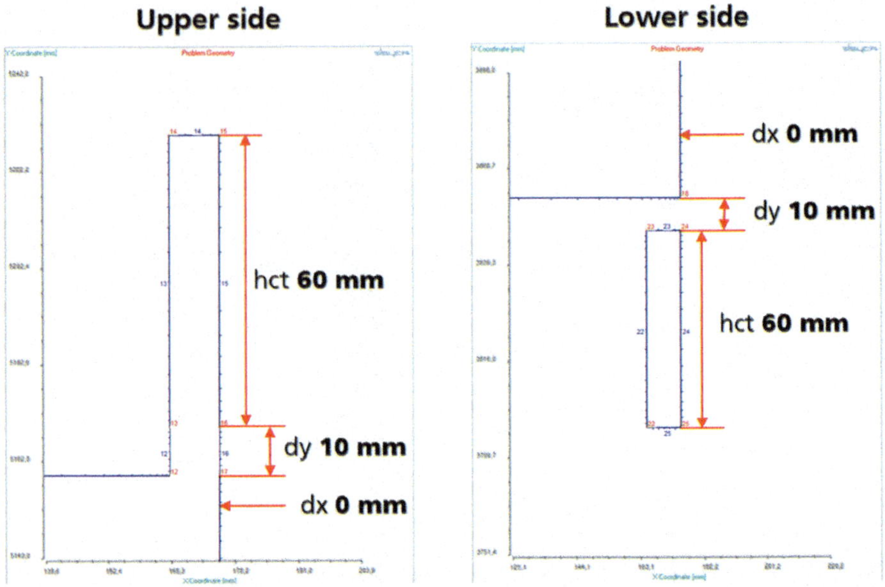

Fig. 3 Upper and lower current thief of the initial geometry. The upper current thief is attached directly to the end of the part; the lower one is positioned at a distance. hct: **h**eight of **c**urrent **t**hief, dy: **d**istance in direction **y** (vertical axis), dx: **d**istance in direction **x** (diameter)

Dimensioning and positioning of the current thieves in the initially setup are shown in Fig. 3:

To check the simulation results, measurements was carried out on site. Since chromium was a non-magnetic layer on parts made of magnetisable steel, the magnetic induction method was used. The measurements were carried out directly after coating with the company´s own hand-held measuring device. The coating itself was carried out with parameters typical for production, with special attention to the exact positioning of part and current thieves. The position of the measuring points was agreed in advance and marked on the component with a sharpie. Measurements were taken at closer intervals at the upper and the lower side of the part than in the middle, since the greatest differences in coating thickness were expected here, which was confirmed by the measurements.

There were strong variations in the measured values due to the given roughness, which could not be completely eliminated even by averaging. Nevertheless, a good picture of the layer thickness distribution was obtained, see Fig. 4.

The already mentioned equalising effect of the overpotential is clearly visible, the measured values represent the course of the calculated current density distribution well, but the layer thickness itself only shows about half the variance as the calculated primary current density. The primary current density correspondingly overdraws the real layer thickness distribution; this effect is to be expected and irrelevant for the intended optimization.

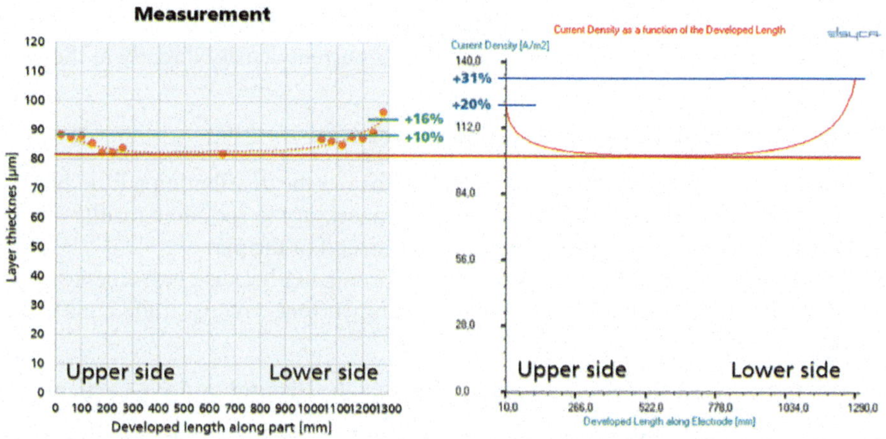

Fig. 4 Measured layer thicknesses (left) versus calculated primary current density distribution (right). The measured values are the average of 2–3 measured values. The variation of the measured values results from the given roughness of the sandblasted substrate

3.3 Optimization of the Current Density Distribution

To carry out the optimization, certain improvement approaches that could be realised in the plant were identified in advance, both qualitatively and quantitatively. Table 1 provides an overview of the selected optimization approaches:

The main findings were summarized and can be generalized as follows:

Moving current thieves in direction x—horizontal

- If moved towards the anode (+ x), the current thief works way more effective
- Resetting the current thief from the surface of the parts (- x) is ineffective

Table 1 Approaches and parameter ranges of the studies carried out

Study	Approach	Moving direction/relevant size	Range [mm]
1	Moving current thieves in one direction	Horizontal—x	−10 to +10
2	Moving current thieves in one direction	Horizontal—x	+5 to +20
3	Moving current thieves in one direction	Vertical—y	−5 to +10
4	Moving current thieves in two directions	Horizontal—x	0 to +20
		Vertical—y	0 to +20
5	Reducing size of current thieves	Height of current thieves	+20 to +60
6	Additional movement lower current thief	Horizontal—x	+5 to +20
		Vertical—y	0 to +10

- Moving + x in a range of 10 to 20 mm gives the best results
- Moving more than 20 mm can cause a lower current density directly at the ends of the part

Moving current thieves in direction y—vertical

- Moving away from the part ends (+ y) flattens out the distribution a little bit
- Moving + y in a range of at least 10 mm is necessary to see some results
- Doing so, the current density increases on the end of the part
- Combining + x and + y seems to be very promising because movement in + y allows more movement in + x without losing current directly at the ends of the part

Moving current thieves in two directions x—horizontal and y—vertical

- Gives the best results so far; without any further measures, the current density distribution can be lowered to around 50% in comparison to the initial situation

Reducing size of current thief

- The smaller the current thieves, the less current they attract and the more current is on the part
- The possible reduction in size is limited because a smaller current thief is less effective than a bigger one, so hct values below 20 mm effects the current density distributing in a negative way
- That effect is even worse at the lower side of the part
- Additional movement of the lower current thief can counteract that

3.4 *Final Result*

Despite the already good initial situation, a significant improvement of the current density distribution could be achieved. Compared to the originally up to approximately 31% higher current density at the lower end of the part, a reduction of this value to approximately 11% could be achieved. Based on the measured values, a reduction of the layer thickness distribution to approximately 5 % can be concluded (see Fig. 5). At the same time, the percentage of current loss on the current thieves could be slightly reduced from approximately 21% to approximately 19%. The percentage of current on the part increased to 81% accordingly.

The best results were achieved by a combination of different measures. The optimal values finally determined for the given part dimensions are shown in Fig. 6.

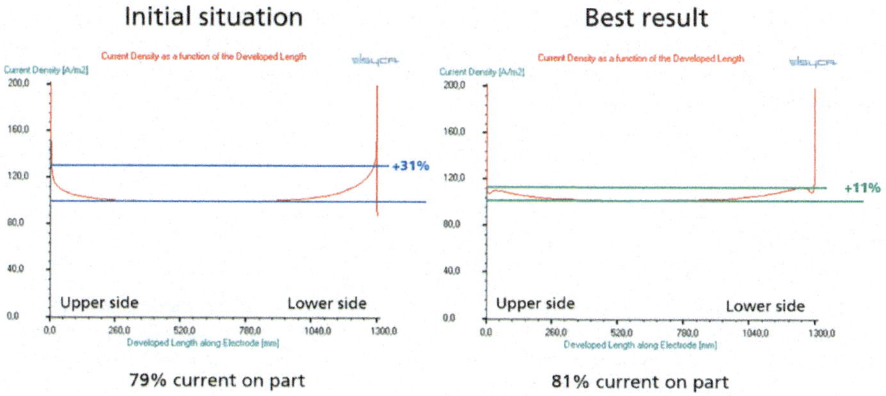

Fig. 5 Best achieved current density distribution compared to the initial state

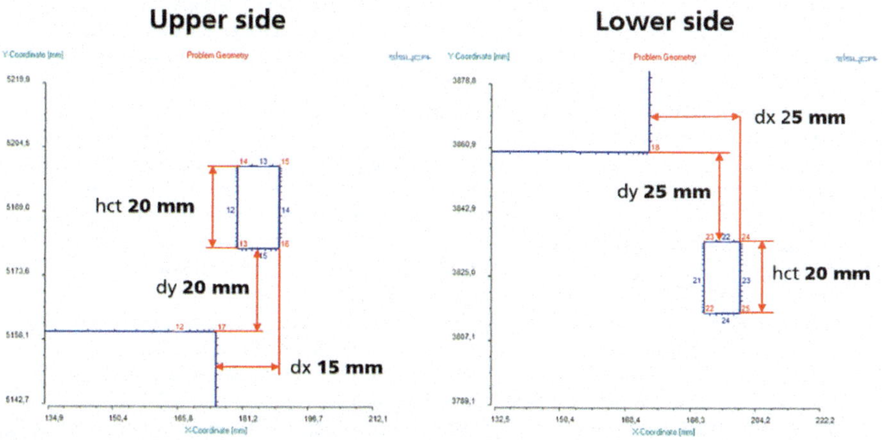

Fig. 6 Upper and lower current thief of the best geometry. Both current thieves are positioned at a distance. hct: **h**eight of **c**urrent **t**hief, dy: **d**istance in direction **y** (vertical axis), dx: **d**istance in direction **x** (diameter)

4 Summary

By using current density simulation with a proven procedure, it was possible to improve the plating situation with regard to the example part. In addition, it was possible to determine basic effects of current thieves that can be transferred from the company to other parts without further simulation calculations. The further improvement of the layer thickness distribution allows a more efficient use of layer metal and electrical energy and therefore contributes to a more sustainable use of electroplating.

Acknowledgements The author received financial support from the German Federal Ministry of Education and Research (BMBF) for the project "ECOPLATE: Station IPA-Chula "Energieeffizienz elektrochemischer Beschichtungsprozesse und -anlagen" (Grant ID: 01DP17044). The author would also like to thank the project partners for their support and especially Prof. Dr. Yuttanant Boonyongmaneerat, Deputy Director of the Metallurgy and Materials Science Research Institute MMRI of Chulalongkorn University Bangkok, Thailand.

References

1. Leiden A, Kölle S, Thiede S, Schmid K, Metzner M, Herrmann C (2020) Model-based analysis, control and dosing of electroplating electrolytes. Int J Adv Manufact Technol 111:1751–1766
2. Chotirach M, Rattanawaleedirojn P, Boonyongmaneerat Y, Chanajaree R, Schmid K, Metzner M, Rodthongkum N (2022) Systematic investigation of brighteners effects on alkaline non-cyanide zinc electroplating using HPLC and molecular modeling. Mater Chem Phy 277
3. Van Den Bosche B, Rose A, Phillips J (2009) Maximizing decorative electroplating productivity by optimizing the rack design for a family of ten different door handles. Plat Surf Finish 46–54
4. Schmid K, Schwanzer P (2018) Improving plating efficiency with simulation technology. In: National association for surface finishing: national association for surface finishing manufacturing & technology trade show & conference 2016: 6–8 June 2016, Las Vegas, Nevada, USA. Washington DC, USA, pp 412–425
5. Hernández P, Socas A, Benítez A, Diaz N, Marrero D, Ortega F, Monzón M (2013) Computer aided electroforming. Elecform3DTM. Procedia Eng 63:532–539
6. Hernández P, Socas A, Benítez A, Ortega F, Diaz N, Marrero D, Monzón M (2014) Boundary element method applied to electroforming process. Mater Sci Forum 797:125–132
7. Druesne F, Afzali M (2003) Electroplating simulation and design tool. In: Proceedings of the institution of mechanical engineers, vol 217, pp 705–707
8. Hezhong P, Zhang J, Zhang G, Huang P (2012) Research to the uniformity of Ni-Co alloy electroforming. Adv Mater Res 479–481:497–503
9. ELSYCA Homepage. Retrieved from https://www.elsyca.com. Accessed on May 05, 2023
10. COMSOL Homepage, https://www.comsol.com. Accessed on May 05, 2023
11. Tamara LO, Lipkin MS, Lukovkin MV (2021) Effect of Current Density Distribution in the Zinc Plating Process. Solid State Phenom 316:839–844
12. Keyword "current distribution", in: Gold Book of the International Union of Pure and Applied Chemistry IUPAC, https://goldbook.iupac.org/terms/view/C01456, last accessed 2023/05/05

Numerical Simulation of Particle Transport from Dispenser to Target in Micro-Shot Peening Applications

Sundaravadivelu Kannan, Te Ba, Stephen Wan, and Chang Wei Kang

Abstract Three dimensional numerical simulations of micro shot peening process are carried out for different operating and nozzle surface conditions. The influence of two different turbulence model variants namely the SKE model and AKN model are employed to investigate their capability to accurately predict the shot peening process. Nozzle surface is assumed to be both smooth and rough and their influence on the solid discrete particle dynamics is investigated. Particle velocity and splat sizes resulting from the shot peening process are numerically simulated and compared against the measurements. It is inferred from simulations that the AKN turbulence model predicts the surface deposition characteristics both qualitatively and quantitatively closer to the measurements than that of the SKE model. Nozzle surface roughness is found to influence the particle velocity, particle rebounding angles and therefore resulting in wider particle dispersion leading to larger particle splats on the target substrate.

Keywords Shot peening · Turbulence · RANS model · CFD

1 Introduction

A relatively new variant of shot peening, micro-shot peening or fine particle peening features micro-shots in the size range of 0.02 to 0.2 mm diameter, propelled to speeds up to 200 m/s [1, 2]. The micro-shots themselves are made of ceramics, which are harder than the cast iron shots commonly applied in conventional shot peening. The present study stemmed from a need to develop a reasonably accurate prediction of particle density (coverage) and impact velocity distribution on a target surface for the purpose of exploring various designs of micro shot peening nozzles. One concern is the effect of the surface roughness of the nozzle wall, on which these small shots would impact before leaving the nozzle. This is because micro-shots would not

S. Kannan (✉) · T. Ba · S. Wan · C. W. Kang
Institute of High Performance Computing (IHPC), Agency for Science, Technology and Research (A*STAR), 1 Fusionopolis Way, Connexis No. 16–16, Singapore 138632, Singapore
e-mail: kannan_sundaravadivelu@ihpc.a-star.edu.sg

N. Maharjan and W. He (eds.), *Proceedings of the 3rd International Conference on Advanced Surface Enhancement (INCASE) 2023*, Lecture Notes in Mechanical Engineering, https://doi.org/10.1007/978-981-99-8643-9_31

encounter a flat nozzle surface, but a terrain populated with geometrically significant and irregular obstacles. For example, through their experiments with spherical glass beads ranging from 125 to 150 microns in diameter, accelerated through a stainless steel nozzle, Shipway and Hutchings [3] demonstrated that the rougher the surface of the nozzle bore, the larger the dispersion of the glass particles issuing from the nozzle; and that surface roughness has also the effect of decreasing the magnitudes of, while increasing their spread, of particle exit velocities. Shipway and Hutchings [3] reasoned that a greater surface roughness effectively increases particle rebound angles, thus particles would tend to exit the nozzle at greater angles and so increase the particle dispersion. Larger particle rebound angles would also mean that, within the nozzles, there would be a greater number of impacts leading to a greater loss of particle kinetic energies. Subsequently, to account for the effect of surface roughness, Sommerfeld and Huber [4] adjusted the impact angle by an inclination angle, γ_w relative to a theoretically smooth wall, assuming that the angle, γ_w has a Gaussian distribution with a zero mean and a standard deviation Another concern is that, apparently, through our own computational fluid dynamics (CFD) runs in the past, we found that the standard Reynolds averaged Navier–Stokes (RANS) turbulence models are unable to predict the correct magnitude of particle dispersion forces and hence cannot reproduce experimentally observed results of impact particle density distribution. For this reason, the present study attempts to carried out a comparative study of a standard two equation RANS turbulence model and its modified version to correctly predict the particle dynamics in terms of impact velocity and particle splat size distributions on the target surfaces. The CFD setup for the present study was first calibrated against the experimental results from Shipway and Hutchings [3]. Particle wall interaction was captured by a surface roughness model [4] and a particle rebound model developed from previous numerical studies is employed. Simulations were then carried for various nozzle surface roughness including the case of a theoretically smooth wall and at different operating pressure conditions ranging from 0.1 MPa to 1 MPa, usually employed in micro-shot peening applications. Numerical results were the analyzed in terms of particle impact velocity distribution and particle splat size distributions on the given target surface.

2 Numerical Model

The design and geometrical dimensions of the micro-shot peening nozzle used in this study is that of the Shipway and Hutchings [3] gas-blast erosion test rig and is shown in Fig. 1. Air is pumped into the nozzle through an inlet at a specified driving pressure ranging from 0.1 Mpa to 1 MPa. The nozzle is assumed to be made up of 304 stainless steel tube with 6.35 mm outer diameter and 4.72 mm inner diameter. The nozzle is designed to be a parallel walled tube of 308 mm in length and a further longer tube length does not significantly affect the measured particle exit velocity [3]. Internal surface of the nozzle is assumed to be rough with the mean roughness ranging from Ra = 0.25 μm to Ra = 0.94 μm. For the sake of clarity, we here

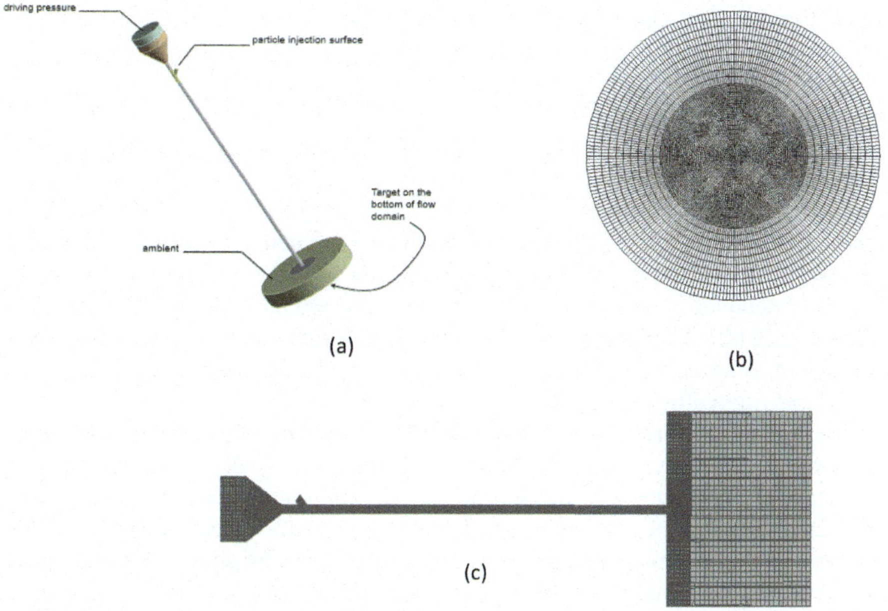

Fig. 1 **a** Schematic of the micro-shot peening nozzle and ambient flow domain, **b** mesh distribution on the target surface and **c** mesh distribution on the centre-plane of the flow domain

refer to the nozzle with Ra = 0.25 μm as smooth nozzle and with Ra = 0.94 μm as a roughest nozzle. Discrete ballotini particles are assumed to be suspended into the nozzle through a particle injector at the rate of 0.25 g/min giving a maximum particle concentration of ~ 3 mg/l of air at 0.095 MPa driving pressure. Ballotini particles as reported in [3] is assumed to be spherical in shape, with the mean size of 137.5 microns and 2600 kg/m^3 being its mass density [3].

Assuming the air flow to be compressible and turbulent, the classical Reynolds Averaged Navier–Stokes (RANS) equations in the Eulerian frame coupled with rhe discrete phase model (DPM) are employed to simulate the mean fluid flow and the particle dynamics. Two variants of the *k*-ε turbulent model viz., the standard *k*-ε turbulent model [5, 6], the Abe, Kondoh & Nagano (AKN)- *k*-ε turbulent model [5] are used here to calculate the turbulent energy and its dissipation whose standard forms are given below.

The Abe, Kondoh, Nagano (AKN)- *k*-ε turbulent model

$$\frac{\partial \rho k}{\partial t} + \frac{\partial}{\partial x_j}\left[\rho u_i k - \left(\mu + \frac{\mu_t}{\sigma_k}\right)\frac{\partial k}{\partial x_i}\right] = \rho P_K - \rho \varepsilon, \tag{1}$$

$$\frac{\partial \rho \varepsilon}{\partial t} + \frac{\partial}{\partial x_j}\left[\rho u_i \varepsilon - \left(\mu + \frac{\mu_t}{\sigma_\varepsilon}\right)\frac{\partial \varepsilon}{\partial x_i}\right] = C_{\varepsilon 1}\frac{\varepsilon}{k}\rho P_K - C_{\varepsilon 2} f_\varepsilon \rho \frac{\varepsilon^2}{k}, \tag{2}$$

where the turbulent viscosity is defined by

$$\mu_t = C_\mu f_u k T \tag{3}$$

and the damping functions are given in [7]. The standard k-ε turbulence model proposed by Launder and Spalding can be recovered from the AKN k-ε model by simply replacing the damping function with $f_m = f_e = 1$ in Eqs. (2) and (3). In this analysis we do not use the standard k-e model as such instead we employ using the enhanced wall treatment option as explained in [7] in order to reproduce the near wall behavior more correctly. In simple words we have attempted to make the AKN model free from being dependent on experiments like tuning the realizability coefficient. Instead, the turbulence production at the walls, especially the impinging walls, was limited by limiting the strain rates as below.

The boundary conditions for this problem are no slip velocity conditions at the solid walls, zero flux condition for the turbulent energy and its dissipation at the walls. At the flow ambient, outside the nozzle, the gauge pressure is set to zero and zero normal gradient condition is applied to all other flow variables. At flow inlet to peening nozzle gauge pressure is fixed and varied from 0.1 bar to 1 bar for different flow simulations. As far as the particle boundary conditions are concerned particles are assumed to enter through the particle inlet at a pre-specified flow rate with a slightly high pressure compared to the nozzle inlet air pressure in order to avoid particle escaping from the particle inlet. At the target surface located at 20 mm from the nozzle exit we assume particles get trapped once they hit the target surface. Particles are assumed to escape along the ambient flow boundaries. In micro shot peening applications, the particle size is fixed, assumed to be spherical in shape as per the experimental conditions. The Navier Stokes equations of motion coupled with equations of the turbulent energy and its dissipation expressed in Eqs. (1) to (3) are solved using the commercial CFD code Fluent [7] by employing a suitable pressure based compressible flow solver. Spatial derivatives of the flow variables are approximated using second order upwind scheme built within the SIMPLE finite volume algorithm. The AKN k-ε turbulence model is currently not available in Fluent and hence need to be coded through the user defined scalar transport theory option available in the Fluent. Hence, the commercial CFD code Fluent used only as a flow solver while the AKN k-ε turbulence model is programmed in C++ and implemented with the help of user defined scalar transport equations and coupled with the Fluent momentum equations through the turbulent viscosity, μ_t. The computational mesh spread over the entire flow domain is made up of fully structured mesh consisting of 1.8 million three dimensional hexahedral cells. Mesh inside the nozzle is more uniform but with staggering near the walls to capture the wall effects as required by the blending functions. Outside the nozzle the mesh is arranged in such a way that cells get denser near the target surface to capture the flow impingement effects particularly the stagnation regions. The mean flow is assumed to steady and hence the transient effects are ignored to compare against the measurements. Numerical iterations are assumed to be converged when the cell weighted residuals fall below 10^{-7}. Once the gas flow is converged discrete particles are injected at the particulate inlet and solved using the Eulerian approach [6]. In doing the particle analysis we have considered only the one-way coupling approach assuming the particle loading is too small to alter the gas flow characteristics.

3 Results and Discussion

Three-dimension numerical simulations of the micro-shot peening process are carried out based on an experimental study reported in the literature [3]. Turbulent flow simulations are carried out using two different models viz., the standard *k*-ε model with enhanced wall treatment and the AKN *k*-ε model as explained in the previous section. Particle mean velocities obtained from the standard *k*-ε model and the AKN *k*-ε model are shown in Fig. 2. It is clear from the figure that the SKE model over predicts particle mean velocity.

Not only at low inlet pressures but also under high inlet pressure conditions AKN model is found to yield velocity magnitudes in good agreement against the reported measurements thus validating the use of AKN model for simulating micro shot peening processes. Also reported in the literature is that particle velocities are sensitive to the nozzle surface roughness [3]. Most of the nozzles used in shot peening process are populated with geometrically significant surface irregularities either by its manufacturing process or caused due to erosion by its repeated use. Therefore, a popular and widely used surface roughness model [7] to mimic the particle wall reflection effects is employed here along with the discrete phase model to simulate the particle mean velocity. Particle size is assumed to be uniform here, 137 μm, that represents the mean size of the particle distribution used in the experiments [3]. The results of the simulations are shown in Fig. 3a and b. Particle mean velocities are drawn against different surface roughness from 0.25 to 0.95 μm as employed in the measurements and for $P_{in} = 0.1$ bar. It is very clear once again that the AKN model accurately predicts the particle velocity for different roughness values and are in good agreement with the measurements than compared to its the SKE model. Higher inlet pressure conditions are not investigated for the sake of simplicity and also due to unavailable of their measurements. It is inferred from Fig. 3 that the

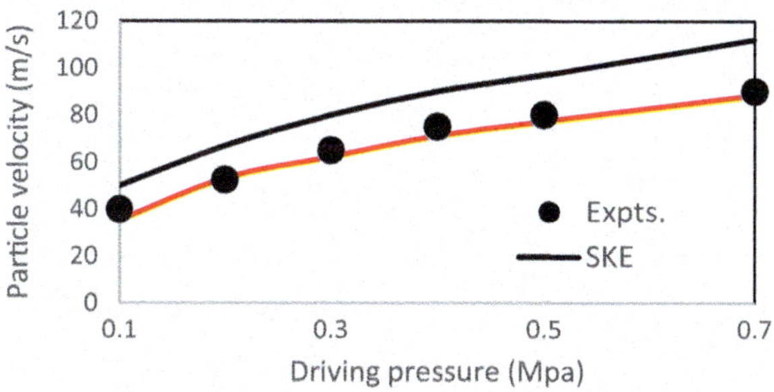

Fig. 2 Particle mean velocity at the target for different inlet pressure conditions

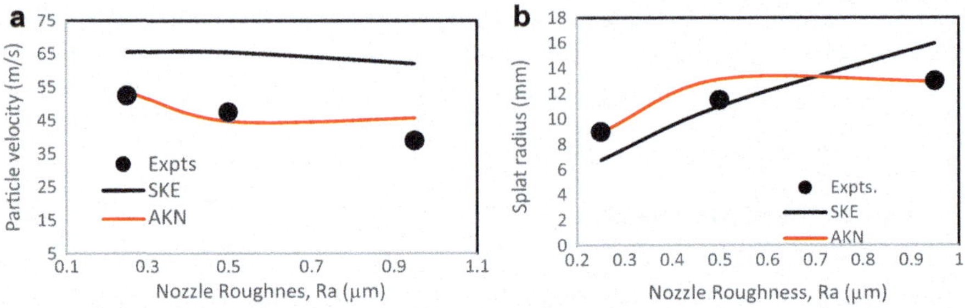

Fig. 3 **a** Particle mean velocity at the target for different nozzle roughness conditions. **b** Particle splat size on the target surface for different nozzle roughness conditions

particle velocity reduces in magnitude with increasing surface roughness. This may be attributed to the fact that the surface roughness retards the rebound velocity by influencing the rebound angles largely and therefore resulting in wider particle splats on the target surface. Hence in order to understand the influence of surface roughness on the particle splat sizes, the particle impact distributions on the target surface were analyzed for different surface roughness conditions, Fig. 3. Interestingly, the SKE model predicts a correct splat size for a specific surface roughness, $Ra = 0.5$, but deviating from the measurements for other values of *Ra*. In fact, the trend of the particle splat radius predicted by the SKE model for various *Ra* is found to increase linearly, which is different from that of the measurements. Whereas the AKN model accurately predicts not only the splat radius for various of surface roughness but also the asymptotic trend that exist at higher surface roughness values, *Ra*. Scatter plots of particle impact distribution on the target surface for inlet pressure of 0.2 bar and $Ra = 0.25$ μm are predicted using both the SKE and AKN models are shown in Fig. 4. Inlet nozzle pressure, $P_{\text{in}} = 0.2$ bar, is chosen rather than $P_{\text{in}} = 0.1$ bar here just to compare against the reported particle splat size in the measurements [3]. Scatter plots depicted in Fig. 4 clearly reveal that the SKE model results in a smaller splat size, 12 mm, whereas the AKN model results in a relatively larger splat size, 18 mm. As discussed earlier this is because of inaccurate predictions of both the velocity and turbulent energy by the SKE model. Whereas with higher level of turbulent energy predicted by the AKN model the particle dispersion becomes wider than that of the SKE model resulting in a larger splat size.

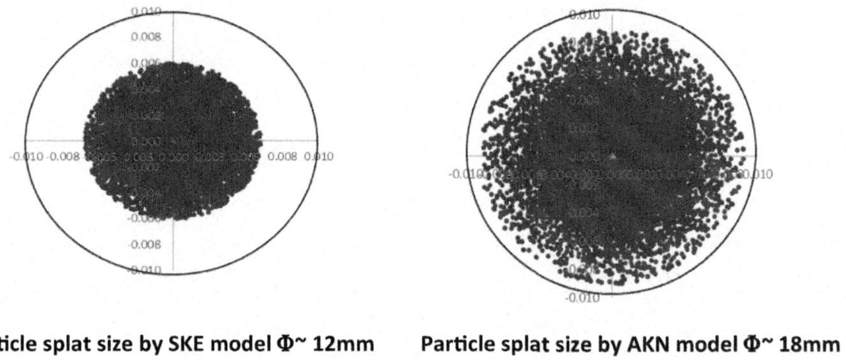

Fig. 4 Scatter plots of the particles on the target surface by the SKE and AKN turbulent models

4 Conclusion

Turbulent simulations of the micro shot peening process are carried out for various nozzle inlet pressure and nozzle roughness conditions. Two different RANS modes are employed to investigate their capability to accurately predict the shot peening process. It is inferred from simulations that the AKN model predicts both qualitatively and quantitatively the particle dynamics closer to the measurements when compared to the SKE model. Nozzle surface roughness is found to influence the particle dynamics resulting in wider particle splats on the substrates.

References

1. Chia-Ying K, Tai-Cheng C, Leu-Wen T (2023) Effects of micro-shot peening on the stress corrosion cracking of austenitic stainless steel welds. Metals 69:1–15
2. Boyu Z, Peitang W, Huaiju L, Huan Y, Mario G (2022) Effect of fine particle peening on surface integrity of flexspline in harmonic drive. Surf Coat Technol 433:128133
3. Shipway PH, Hutchings IM (1993) Influence of nozzle roughness on conditions in a glass-blast erosion rig. Wear 162:148–158
4. Sommerfeld M, Huber N (1999) Experimental analysis and modeling of particle-wall collisions. Int J Multiph Flow 25:1457–1489
5. Abe K, Kondoh T, Nagano Y (1994) A new turbulence model for predicting fluid flow and heat transfer in separating and reattaching flows-I. Flow field calculations. Int J Heat Mass Transfer 37(1):138–151
6. Launder BE, Spalding DB (1972) Lectures in mathematical models of turbulence. Academic Press, London, England
7. ANSYS Fluent 6.3 (2006) Users guide. Fluent Inc., New York, NY, USA

Numerical Study of Stream Finishing of a Fairway Golf Club

Stephen Wan, Shengwei Ma, Cary Turangan, Keni Chih-Hua Wu, Jun Ming Tan, Wei Shin Cheng, and Kai Liang Tan

Abstract The present study extends a previously calibrated rheology and wall slip model to computational fluid dynamics (CFD) simulations of material removal distribution over the surface of a fairway golf club undergoing the stream finishing process. Simulations were undertaken for the cases of the golf club immersed in a stream finishing drum, fixed for two different orientations and with the drum rotated either clockwise or counterclockwise. Owing to the complex media flow over the golf club resulting from the interaction between the media and the three-dimensional surface of the golf club, predicted values of wall pressure and slip velocity distributions span over a wide range of magnitudes, implying that achieving uniform material removal would be a particularly interesting challenge. Lastly, insights gained from the present study will inform a careful design of the physical experiments, which are meant to further improve the stream finishing material models.

S. Wan (✉) · S. Ma · C. Turangan · K. C.-H. Wu
Agency for Science Technology and Research (A*STAR), Institite of High Performance Computing (IHPC), 1 Fusionopolis Way, No. 16–16 Connexis, Singapore 138632, Republic of Singapore
e-mail: wansym@ihpc.a-star.edu.sg

S. Ma
e-mail: masw@ihpc.a-star.edu.sg

C. Turangan
e-mail: cary@ihpc.a-star.edu.sg

K. C.-H. Wu
e-mail: keni-wu@ihpc.a-star.edu.sg

J. M. Tan · W. S. Cheng · K. L. Tan
Advanced Remanufacturing and Technology Centre (ARTC), Agency for Science Technology and Research (A*STAR), 3 Cleantech Loop, No. 01/01, CleanTech 2, Singapore, Republic of Singapore
e-mail: Tan_Jun_Ming@artc.a-star.edu.sg

W. S. Cheng
e-mail: chengws@artc.a-star.edu.sg

K. L. Tan
e-mail: tan_kai_liang@artc.a-star.edu.sg

N. Maharjan and W. He (eds.), *Proceedings of the 3rd International Conference on Advanced Surface Enhancement (INCASE) 2023*, Lecture Notes in Mechanical Engineering, https://doi.org/10.1007/978-981-99-8643-9_32

Keywords Stream finishing · Computational fluid dynamics · Three dimensional surface · Material removal prediction

1 Introduction

The present study extends a previously calibrated rheology and wall slip model [1] to computational fluid dynamics (CFD) simulations of material removal distribution over the surface of a fairway golf club undergoing the stream finishing process.

Stream finishing may be considered as the reverse of the more commonly known drag finishing process, in which components are dragged through media in a stationary drum. On the other hand, in stream finishing, components are processed by flowing media over them in a rotating drum.

Whereas there have been several experimental investigations reported in the open literature [2–6], there appears to be a paucity of numerical analyses related to the stream finishing process. Apart from the continuum-based approach reported by Wan et al. [1] and Curangan et al. [7] it appears that the only other attempt at numerical prediction is that by Zanger et al [4], who employed the discrete element method.

2 Methods

2.1 Calibrated Models

As outlined in [1], the rheology of the media, assumed to be flowing as a continuum, is described by an inelastic non-Newtonian fluid, the viscosity, η, being given by:

$$\eta = frac\mu(I)P|\gamma| \tag{1}$$

in which, P is the isotropic pressure; $|\gamma| = \sqrt{\frac{\gamma_{ij}\gamma_{ij}}{2}}$ is the second invariant of the strain–rate tensor; and $\mu(I)$ is a friction coefficient given by:

$$\mu(I) = \mu_1 + \frac{\mu_2 - \mu_1}{I_0/I + 1} \tag{2}$$

where $I = |\gamma|d/\sqrt{P/\rho_p}$ is the inertial number; d and ρ_p are the media particle size and density, respectively; with μ_1, μ_2 and I_0 being the calibrating parameters.

For the media used in the calibration test, represented by a diameter, $d = 4$ mm and density, $\rho_p = 2083$ kg m^{-3}, the values of the calibrating parameters were found to be: $\mu_1 = 0.3$, $\mu_2 = 0.65$ and $I_0 = 0.3$.

The wall slip model, which calculates the velocity, u_w of the media in contact with the surface of the component being processed, is given by:

$$u_w = u_s \left(\frac{\tau_w}{\tau_n}\right)^m \exp\left(-\frac{BP}{\tau_n}\right) \tag{3}$$

in which, τ_w is the wall shear stress; P is wall pressure; τ_n is a scaling wall shear—stress; and u_s, m and B are the calibrating model parameters. The wall slip model was calibrated with these values: $u_s = 0.3\ \text{ms}^{-1}$, $m = 1$ and $B = 0.12$.

2.2 *Numerical Simulation Setup*

Details of the simulation setup can be found in [1] and are outlined here for ease of reference.

Numerical simulations were carried out using the computational fluid dynamics (CFD) software package, ANSYS CFX version 17.2, with the viscosity function, given by Eqs. (1) and (2), coded in accordingly. For the present study, simulations were carried out with the golf club fixed at two different orientations, depicted as orientation 1 and orientation 2 in Fig. 1; and with the media bowl rotating clockwise or counterclockwise. Hence, four cases were investigated, as listed in Table 1.

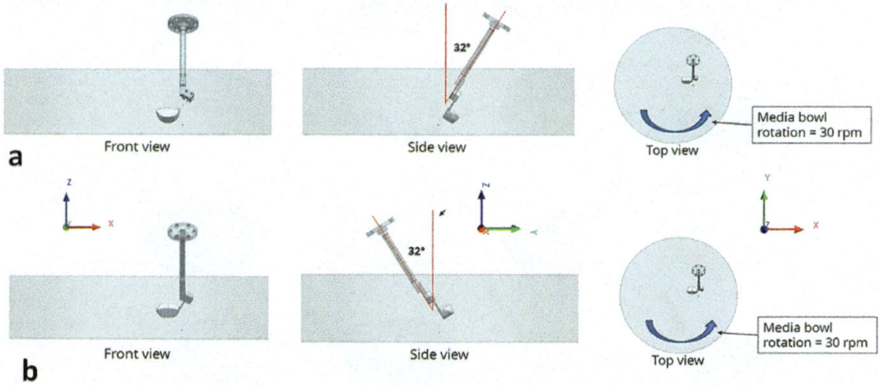

Fig. 1 Golf club fixed at **a** orientation 1 and **b** orientation 2

Table 1 List of simulation cases

Simulation case number	Golf club orientation	Media bowl rotation
1	1	Clockwise
2	1	Counterclockwise
3	2	Clockwise
4	2	Counterclockwise

The wall slip boundary condition, as described by Eq. (3) was imposed on the surfaces on the golf club and the fixture; the non-slip condition, on the media bowl rotating walls and bottom; and the pressure opening condition on the media surface.

3 Results and Discussion

The distributions of predicted media pressure P and wall slip velocity u_w over the surface of the golf club for the four cases under consideration are graphically shown in Figs. 2 and 3, respectively. We observe that, on account of the complex three-dimensional surface of the golf club, both media pressure P and wall slip velocity u_w vary greatly over the surface of the golf club. In fact, values of media pressure P span over four orders of magnitude.

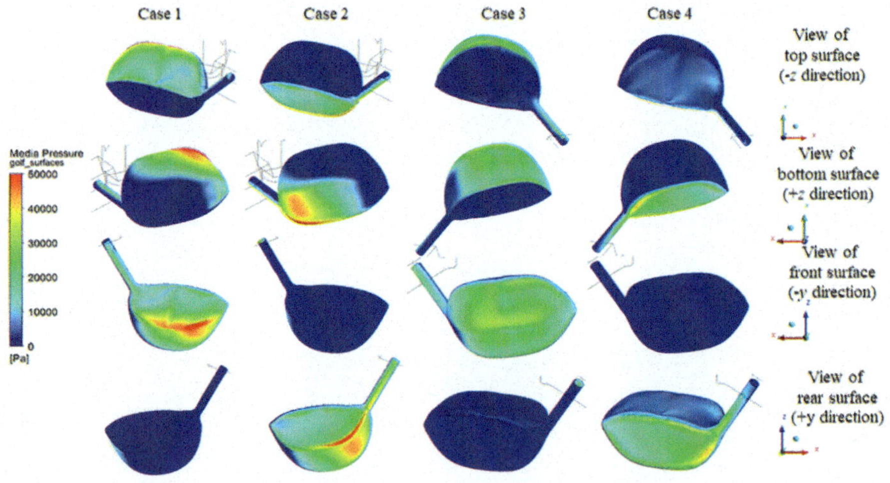

Fig. 2 Distribution of predicted media pressure P over the surface of the golf club

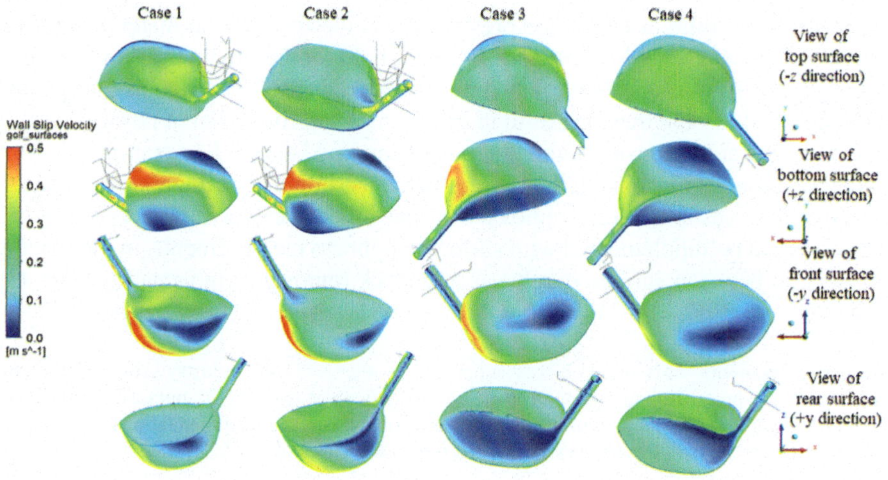

Fig. 3 Distribution of predicted wall slip velocity u_w over the surface of the golf club

4 Conclusion

Material removal by stream finishing is fundamentally due to abrasive wear and hence is a function of normal load (media pressure, P) and sliding velocity (wall slip velocity, u_w). For example, Wan et al. [8], showed that for a variety of media finishing processes (vibratory finishing, abrasive flow machining and magneto—rheological finishing), the stock removal, h and the surface roughness, R_a distribution evolution may be described by:

$$R_a = (R_i - R_\infty)e^{-\frac{k_T P u_w t}{H}} + R_\infty \tag{4}$$

$$h = a_R(R_i - R_\infty)\left(1 - e^{-\frac{k_T P u_w t}{H}}\right) + \frac{k_S P u_w t}{H} \tag{5}$$

in which, R_i is the initial surface roughess; k_T is a transient wear coefficient; H is the hardness of the worn surface; R_∞ is the limiting surface roughness; k_S is a steady—state wear coefficient; and a_R is essentially a geometrical factor linking the surface roughness with the height of the surface asperity.

Figures 2 and 3 indicate that both the media pressure P and the wall slip velocity u_w are highly non-uniform; and together with Eqs. (4) and (5), we see that achieving uniform material removal over the surface of the golf club would be a particularly interesting challenge. Hence, at this stage, it is difficult to see which particular gold club orientation would yield optimal finishing results. Rather, the prognosis is that high removal rates at certain locations would be unavoidable and so, it may be necessary to resort to a material compensation strategy, such as that described in [9].

At the same time, it cannot be presumed that the form of the material removal models encapsulated by Eqs. (4) and (5) would hold for the stream finishing process, since it is much more aggressive than those examined in [8]. However, insights gained from this study would then be useful in informing a careful design of physical experiments aimed at further improving the models outlined above. For example, coordinate measurement machine (CMM) measurements should be sufficiently accurate to reveal differences in the stock removal over the golf surface for various orientations. These should be supplemented with surface roughness measurements, in view of the connection between surface roughness and stock removal as indicated by Eqs. (4) and (5).

Acknowledgements This research was supported by the A*STAR Industry Alignment Fund-Advanced Post-processing and Non-Destructive Evaluation for Additive Manufacturing (IAF-PP, Grant no. A20F9a0045). Any opinions, findings and conclusions or recommendations expressed in this material are those of the authors and do not reflect the views of the A*STAR.

References

1. Wan S, Ma S, Turangan C, Wu CK, Itoh S, Cheng WS, Tan KL, Ho J (2022) Calibration of a continuum-based granular flowdynamics model for the numerical simulation of the stream finishing process. Proceedings of the 2nd international conference on advanced surface enhancement (INCASE 2021), pp 199–202
2. Schulze V, Gibmeier J, Kacaras A (2017) Qualification of the stream finishing process for surface modification. CIRP Ann Manuf Tech 66:523–526
3. Kacaras A, Gibmeier J, Zanger F, Schulze V (2018) Influence of rotational speed on surface states after stream finishing. Procedia CIRP 71:221–226
4. Zanger F, Kacaras A, Neuenfeldt P, Schulze V (2019) Optimization of the stream finishing process for mechanical surface treatment by numerical and experimental process analysis. CIRP Ann Manuf Tech 68:373–376
5. Budzisz W, Marciniec A (2022) The new gear finishing method research for highly loaded gears. Aerospace 9(3):131
6. Jamal M, Morgan MN (2017) Design process control for improved surface finish of metal additive manufactured parts of complex build geometry. Inventions 2(4)
7. Turangan CK, WanYee Ming S, Itoh S, Ho J (2020) Development of rheology and computational flow model for robotized external finishing on additively manufactured components. In: Itoh S, Shukla S (eds) INCASE 2019. LNME. Springer, Singapore, pp 234–243
8. Wan S, Liu Y, Woon KS, Tnay GL (2014) A material removal and surface roughness evolution model for loose abrasive polishing of free form surfaces. Int J Abrasive Tech 6(4):269
9. Kum CW, Wu CH, Wan S, Kang CW (2020) Prediction and compensation of material removal for abrasive flow machining of additively manufactured metal components. J Mat Process Tech 282:116704

Shear Viscosity Effect on High-Speed Deformation of Copper

Sabeur Msolli, Yuefan Wei, and Zhiqian Zhang

Abstract The paper presents many study cases on the effect of shear viscosity on the deformation and jet formation posterior to the impact of copper powder onto copper substrates. Thanks to a physics-based plasticity model dedicated to high strain rates processes, we performed a batch of numerical finite element simulations in order to understand the way copper particle flatten and expulse material outside the zone of contact. We considered a continuous function of viscosity as a first shot followed by other cases where temperature thresholds are considered to enable the action of shear viscosity. In the present paper we are not aiming to compare numerical results to experimental data rather than depicting the influence and the importance of physical parameters such as shear viscosity on the development of jetting in impacted copper.

Keywords Shear viscosity · Material jet · Copper · Impact

S. Msolli (✉)
Université de Bourgogne Franche-Comté—UTBM, Laboratoire Interdisciplinaire Carnot de Bourgogne, UMR 6303 CNRS, 90100 Belfort, France
e-mail: sabeur.msolli@utbm.fr

Y. Wei · Z. Zhang
A*STAR Research Entities, Singapore Institute of Manufacturing Technology, 3 Cleantech Loop, #01/01 CleanTech Two, Singapore 637143, Singapore
e-mail: wei_yuefan@artc.a-star.edu.sg

Z. Zhang
e-mail: zhangz@ihpc.a-star.edu.sg

Z. Zhang
Institute of High-Performance Computing, A*STAR Research Entities, 1 Fusionopolis Way, #16–16 Connexis, Singapore 138632, Singapore

N. Maharjan and W. He (eds.), *Proceedings of the 3rd International Conference on Advanced Surface Enhancement (INCASE) 2023*, Lecture Notes in Mechanical Engineering, https://doi.org/10.1007/978-981-99-8643-9_33

1 Introduction

The high bonding strength of assembled metals is crucial for an enhanced lifetime such as in cold sprayed assemblies. The bonding strength is directly related to the impact properties [1, 2]. It has been also demonstrated that the bonding strength of metals is dependent on the quality of the surfaces involved during the impact and the bonding mechanism is monitored by a solid-state process. The surface roughness and the surface preparation before the impact have been found to play a major role in guarantying a strong interface bonding [3, 4]. Regarding many pieces of research, reaching a critical velocity is key for achieving a strong bonding between the substrate and the deposition material [5–7]. The elucidation of the bonding mechanisms is long-standing and different fundamental theories emerged to explain the way bonding occurs at the coating substrate interface. A significant body of research presented the adiabatic shear instability phenomena as the main mechanism for bonding [1, 8–10]. Under severe plastic deformation during the impact at critical velocity, the materials can reach the shear localization leading to material softening. Material softening at the particle substrate interface promotes plastic flow instabilities. At this stage, the temperature is high and the softened materials at the particle substrate interface act as a fluid media with different shear viscosities. The dynamic of fluids of different properties in contact with each other can be explained by the Kelvin-Helmholtz Instability (KHI). Accordingly, the consequence of this difference is the formation of irregular structures such as rollups and vortices, leading to an increase of contact surface and mechanical hooking or interlocking. There is also the formation of the interfacial jet produced at the particle/substrate interface which improves the surface cleaning and leads to intimate contact between the particle and the substrate. In the subsequent sections, we show the influence of shear viscosity on the formation of material jet.

2 Methods

The cases considered are based on one continuous evolution of shear viscosity but with different enabling thresholds. The shear viscosity of copper at a temperature near melting temperature (≈ 5T/6), beyond (1600 K), and in the relevant range of strain rate proper to the present application (1/s to 10^{12}/s) is determined from molecular dynamics using a many-body embedded-atoms model potential and reported in [11]. A resultant relationship is obtained:

$$\mu_v(\text{Pa.s}) = \mu_1 T^2 - \mu_2 T + \mu_3 \tag{1}$$

The parameters μ_1, μ_2 and μ_3 are 3.79×10^{-9} Pa.s.T^{-2}, 1.38×10^{-5} Pa.s.T^{-1}, and 1.53×10^{-2} Pa.s, respectively. The following diagram (Fig. 1) shows how the

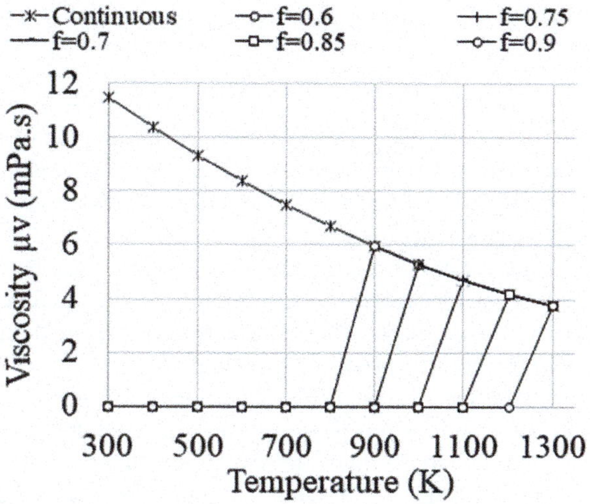

Fig. 1 Evolution of the shear viscosity from room to melting temperature with other cases of study at different ratios *f* corresponding to different thresholds of temperature (900 K, 1000 K, 1100 K, 1200 K and 1300 K)

shear viscosity evolves with respect to temperature. It decreases drastically but quasi-linearly from 11 mPa.s at room temperature to almost 4 mPa.s at melting temperature. As such, the cohesion stress is devoted to drop leading to a material more favorable to a fluid flow. In all cases, even though shear viscosity is high at room temperature, the corresponding stress still substantially lower than that the one generated by the material hardening unless the material experiences important heating imposed by plastic deformation. Therefore, more cases of study can also be addressed here on the effect of shear viscosity on the deformation. In addition to a continuous shear viscosity evolution from room temperature to melting temperature, one can see the effect of the viscosity at different thresholds of temperature depending on the ratio $f = T/T_{melt}$. Therefore, typical values studied hereafter are $f = 0.6$, and $f = 0.7$.

3 Results and Discussions

The findings concern the impact at ultrahigh velocity of a single copper particle of 60 μm on to a semi-infinite copper substrate. In the present, only copper particle is shown for sake of clarity. Again, the results cannot be interpreted in term of the observable deformation but more based on formation of material jet and the precursor mechanisms that trigger it.

3.1 FEM Simulations with Hydrostatic Pressure Effect (EOS)

The following Fig. 2 show different cartographs of the deformed particle and the associated reference in term of shear viscosity value. Three scenarios (continuous viscosity, $f = 0.6$ and $f = 0.7$) are compared for two study cases: thermal dissipation and adiabatic simulations.

Figure 2 does not show a noticeable difference between the different stress values and the deformation shapes because of the smaller temperatures reached due to thermal dissipation. The deformation profile at the particle rim is inclined which demonstrate a prominent hardening effect.

In the case of adiabatic simulations as shown in Fig. 3, the particle severely deforms, and its rim is almost vertical. The particle experiences a clear jetting of material due to the higher temperature at the particle substrate interface. It is already evidenced from experiments that jetting is obtained at ultrahigh impact velocity of copper particle suggesting that material viscosity plays a prominent role even at temperatures above 0.7 T_{melt}.

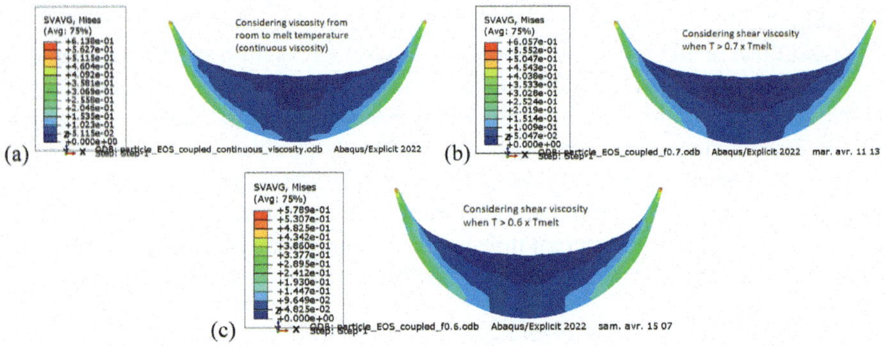

Fig. 2 Deformation of the copper particle at ultrahigh strain rate in the case of coupled thermomechanical simulations. **a** continuous shear viscosity, **b** shear viscosity activated at $T > 0.7 \times T_{melt}$ ($f = 0.7$) and **c** $T > 0.6 \times T_{melt}$ ($f = 0.6$)

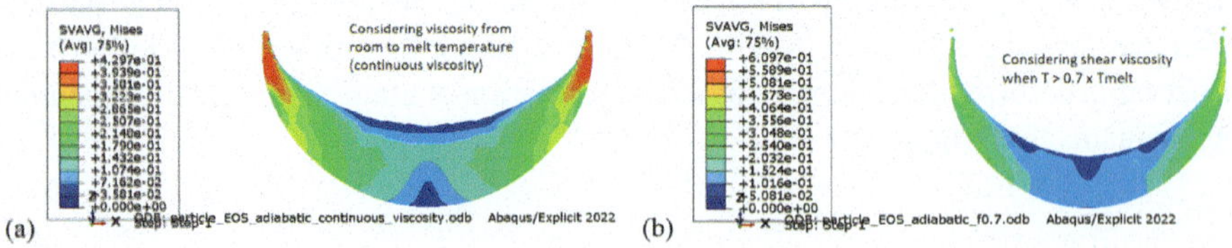

Fig. 3 Deformation of the copper particle at ultrahigh strain rate in the case of adiabatic simulations. **a** continuous shear viscosity, and **b** shear viscosity activated at $T > 0.7 \times T_{melt}$ ($f = 0.7$)

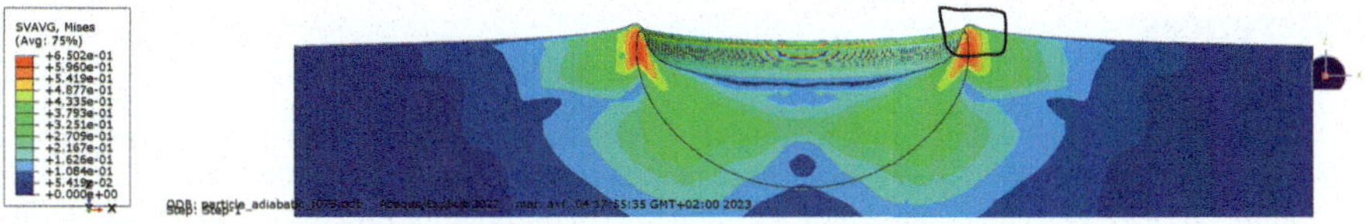

Fig. 4 Trapping of the particle inside the substrate due to a hooking at the border of the particle substrate interface

3.2 FEM Simulations Without Hydrostatic Pressure Effect (Isotropic Elasticity)

Another batch of simulations are performed without accounting for hydrostatic pressure dependency (EOS) acting at very high strain rates. Hence, simple isotropic elasticity is assumed for the reversible regime of deformation. The final equilibrium state is shown in Fig. 4. The figure shows that the particle deformation is less intense at the center part compared to the previous simulations. This can be referred to the absence of hydrostatic pressure responsible of extensive squeezing of material outside the contact region. The simulations at different shear viscosity thresholds does not show significant effect on the deformation. Nevertheless, as demonstrated in Fig. 4, a hooking phenomenon occurs at the borders of the particle substrate interface leading to the trapping of the particle inside the substrate. This appears as the possible bonding mechanism of copper particle to copper substrate at ultrahigh speed.

4 Conclusion

The shear viscosity plays a crucial role on the formation of material jet. When the material exhibits extensive softening leading to severe plastic deformation, high heating intensity, it behaves more as a fluid, and generates rupture of the surrounding rim of the particle which is itself ejected at high speed outside the surface of contact. The higher is the shear viscosity, the more is the material quantity rejected. It is also observed that some hooking phenomena occurs at the border of the particle substrate from where the material is expulsed. The relevant phenomena seem to be related to what has been indicated in some recent references on the importance of hydrostatic pressure on the particle substrate bonding and will be studied in-depth in the coming frameworks.

Acknowledgements The authors acknowledge the UTBM (Université de Technologie de Belfort-MontBéliard) through the funding sources, CNRS Tremplin 2022 and "ACCUEIL DE NOUVELLE ÉQUIPE DE RECHERCHE" (ANER) 2023 jointly accorded to Dr. Sabeur Msolli. Note that ICB Laboratory is supported by the EUR-EIPHI Graduate School (Grant No. 17-EURE-0002).

References

1. Assadi H, Gärtner F, Stoltenhoff T, Kreye H (2003) Bonding mechanism in cold gas spraying. Acta Mater 51(15):4379–4394. https://doi.org/10.1016/S1359-6454(03)00274-X
2. Wang X et al (2015) Characterization and modeling of the bonding process in cold spray additive manufacturing. Addit Manuf 8:149–162. https://doi.org/10.1016/j.addma.2015.03.006
3. Kumar S, Bae G, Lee C (2016) Influence of substrate roughness on bonding mechanism in cold spray. Surf Coat Technol 304:592–605. https://doi.org/10.1016/j.surfcoat.2016.07.082
4. Sun ZC, Wu HL, Cao J, Yin ZK (2018) Modeling of continuous dynamic recrystallization of Al–Zn–Cu–Mg alloy during hot deformation based on the internal-state-variable (ISV) method. Int J Plast 106:73–87. https://doi.org/10.1016/j.ijplas.2018.03.002
5. Fardan A, Berndt CC, Ahmed R (2021) Numerical modelling of particle impact and residual stresses in cold sprayed coatings: a review. Surf Coat Technol 409:126835. https://doi.org/10.1016/j.surfcoat.2021.126835
6. Ghelichi R, Bagherifard S, Guagliano M, Verani M (2011) Numerical simulation of cold spray coating. Surf Coat Technol 205(23):5294–5301. https://doi.org/10.1016/j.surfcoat.2011.05.038
7. Henao J et al (2017) Deposition mechanisms of metallic glass particles by cold gas spraying. Acta Mater 125:327–339. https://doi.org/10.1016/j.actamat.2016.12.007
8. Grujicic M (2007) 9—particle/substrate interaction in the cold-spray bonding process. In: Woodhead publishing series in metals and surface engineering, V. K. B. T.-T. C. S. M. D. P. Champagne, Ed. Woodhead Publishing, pp 148–177. https://doi.org/10.1533/9781845693787.2.148
9. Grujicic M, Zhao CL, DeRosset WS, Helfritch D (2004) Adiabatic shear instability-based mechanism for particles/substrate bonding in the cold-gas dynamic-spray process. Mater Des 25(8):681–688. https://doi.org/10.1016/j.matdes.2004.03.008
10. Schmidt T et al (2009) From particle acceleration to impact and bonding in cold spraying. J Therm Spray Technol 18(5):794. https://doi.org/10.1007/s11666-009-9357-7
11. Desgranges C, Delhommelle J (2008) Shear viscosity of liquid copper at experimentally accessible shear rates: application of the transient-time correlation function formalism. J Chem Phys 128(8):84506. https://doi.org/10.1063/1.2829872

Airflow Analysis Inside and Outside the Nozzle in Shot Peening Process

Koichiro Nambu, Yusuke Saeki, and Masahiro Okumiya

Abstract Surface modification treatments using particle impact, such as shot peening and fine particle peening, are used in many parts such as automotive gears. However, the surface modification effect is affected by many factors such as the material of the projectile, particle size, and nozzle diameter. In the case of the air injection type, the airflow inside and outside the nozzle has a particularly large effect. However, few studies have examined the effects of airflow inside and outside the nozzle. In this study, the airflow inside and outside the nozzle was analyzed using the finite element method. As a result, it was found that the airflow velocity increases as the nozzle diameter increases. It was also clarified that the factors causing this increase were the lengthening of the potential core section and the deceleration caused by the wall resistance.

Keywords Shot peening · Fine particle peening · FEM · Airflow

1 Introduction

Shot peening (SP) and fine particle peening (FPP) are cold forming processes that improve fatigue strength. They are used in a wide range of industries, including automotive and aerospace. SP and FPP are processes in which the shot impinges on the specimen. Work hardening increases the hardness of the specimen surface and improves the fatigue strength of the component.

One method of flying these particles is the air injection method. In this method, shots are ejected from a nozzle by a stream of air. As the particles are pushed out by the air stream, the air stream affects the velocity of the particles. Outside the nozzle,

K. Nambu (✉)
Osaka Sangyo University, 3–1–1, Nakagaito, Daito, Osaka, Japan
e-mail: Aknambu.mech@ge.osaka-sandai.ac.jp

Y. Saeki · M. Okumiya
Toyota Technological Institute, 2–12–1, Hisakata, Tenpaku, Nagoya, Aichi, Japan

N. Maharjan and W. He (eds.), *Proceedings of the 3rd International Conference on Advanced Surface Enhancement (INCASE) 2023*, Lecture Notes in Mechanical Engineering, https://doi.org/10.1007/978-981-99-8643-9_34

the spread of the airflow causes a spread in the trajectory of the shot. However, few studies have investigated the relationship between airflow and particle velocity.

Nambu et al. showed that the velocity of the shot increases as the nozzle inlet pressure increases and that there is a linear relationship between the two. They also proposed the optimum distance between the nozzle outlet and the sample [1]. Maeda et al. showed that the part of the nozzle where the shot is most accelerated is the parallel part of the nozzle, and that the optimum length of the parallel part of the nozzle must be considered to obtain a high-speed shot [2]. Ogawa et al. showed that in direct pressure shot peening, the shot velocity is proportional to the 0.57th power of the nozzle inlet pressure [3]. Ota et al. showed that the depth of compressive residual stress due to shot peening is proportional to the vertical impact velocity component when the shot diameter and material are the same [4]. Ota et al. also showed the relationship between measured shot velocity, angle of incidence and measured residual stress distribution [5]. As mentioned above, studies on particle velocity can be found, but there are no studies on airflow inside or outside the nozzle.

Therefore, in this study, we decided to analyze the airflow inside and outside the nozzle using finite element analysis. The effect of nozzle diameter on air velocity was also evaluated.

2 Analysis Conditions

A three-dimensional finite element method (FEM) steady-state analysis was used to analyze the airflow inside and outside a shot peening nozzle. Ansys Fluent was used for the FEM analysis of the airflow. The analytical model used in the airflow analysis was created using CAD software. Figure 1 shows a cross-sectional view of the analytical model. The nozzle model simulates the nozzle used in the direct pressure type blasting equipment (P-DSU III model) manufactured by Fuji Manufacturing Co. The dimensions of the substrate are 50 mm wide, 50 mm long, and 5 mm thick. The center axis of the nozzle is the y-axis, and $y = 0$ is the nozzle exit. The air flow was air (ideal gas), and the properties of the air are shown in Table 1. The k-ω SST model was used for the turbulence viscosity model, which correctly captures turbulence near the nozzle wall and turbulence away from the wall.

The three-dimensional mesh was created using Ansys Fluent's polyhedral meshing method with a maximum mesh length of 0.001 mm. The minimum and maximum surface mesh lengths were 0.05 mm and 0.77 mm, respectively. The diameter of the nozzle inlet connected to the hose was 20 mm. The nozzle outlet diameter and the distance between the nozzle outlet and the substrate were varied, while the nozzle inlet pressure was fixed at 2000 hPa. The nozzle exit diameter was varied from 3 to 10 mm (in 1 mm increments), and the distance between the nozzle exit and the substrate was investigated at 100, and 150 mm. Atmospheric pressure was set at 1013 hPa. The effect of gravity was neglected in this analysis.

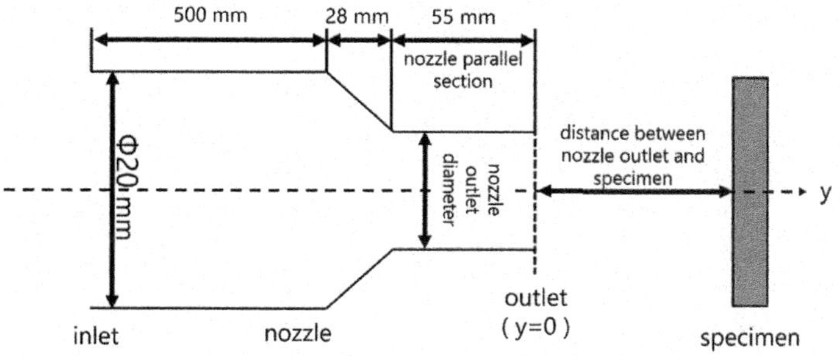

Fig. 1 Nozzle and specimen dimensions

Table 1 Physical properties of air

Cp (specific heat) [J/kg K]	Thermal Conductivity [W/m K]	Viscosity [Pa s]	Molecular Weight [kg/kmol]
1006.43 (constant)	0.0242 (constant)	1.7894e-5 (constant)	28.966 (constant)

3 Analysis Result and Discussion

3.1 Air Velocity Analysis at the Center of the Nozzle

Figures 2 and 3 show the results of air velocity analysis at the nozzle center for distance between nozzle outlet and specimens of 100 mm and 150 mm, respectively. The figures show that the airflow rapidly accelerates to a speed of about 250 m/s just before the nozzle parallel ($y \leq -55$), accelerates slowly at the nozzle parallel ($-55 \leq y \leq 0$), and reaches its maximum speed (almost the speed of sound) near the nozzle exit ($y = 0$). After the nozzle exit, the airflow decelerates slowly, and then suddenly decelerates just before the substrate. The airflow velocity in front of the substrate is considered to decrease rapidly not only due to air resistance but also due to the bounce of the airflow from the substrate. As the nozzle outlet diameter increases, the airflow velocity increases, especially outside the nozzle ($y \geq 0$). In general, one would expect air velocity to increase as the nozzle outlet diameter decreases. In this study, however, the opposite is true.

The pressure ratio of the nozzle outlet pressure to the nozzle inlet pressure was 1013/2000 = 0.5065, which is smaller than the critical pressure ratio (ratio of critical pressure to nozzle inlet stagnation point pressure: 0.528), and the nozzle outlet was clogged for all nozzle diameters. Nozzles with nozzle exit diameters of 10 mm or less are considered to be clogged. Therefore, at nozzle parallel ($-55 \leq y \leq 0$), the air velocity distributions are nearly equal. Further upstream from the nozzle parallel ($y < -55$), the air velocity increases slightly as the nozzle exit diameter increases. This may be because the airflow is blocked inside the nozzle ($y \leq 0$) as the nozzle exit diameter becomes smaller. Figure 4 shows the pressure distribution of the airflow

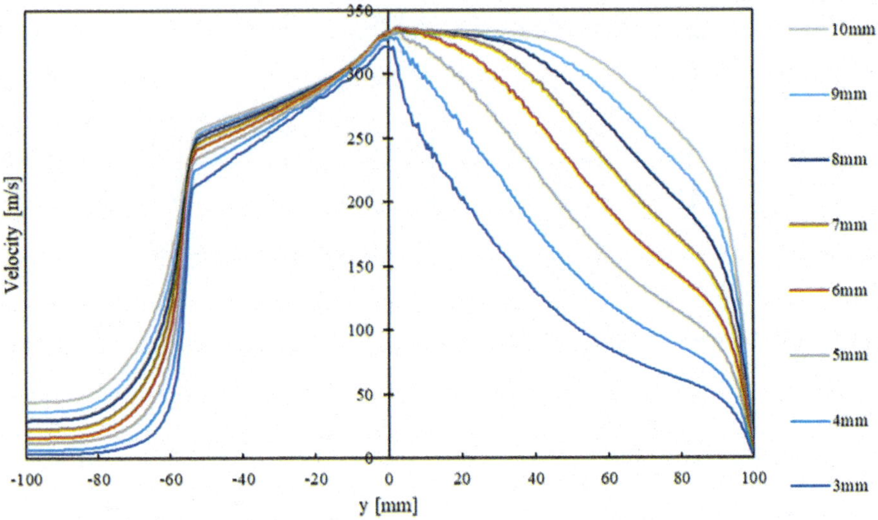

Fig. 2 Air velocity analysis results at the center of the nozzle (The nozzle at a distance between nozzle outlet and specimen of 100 mm)

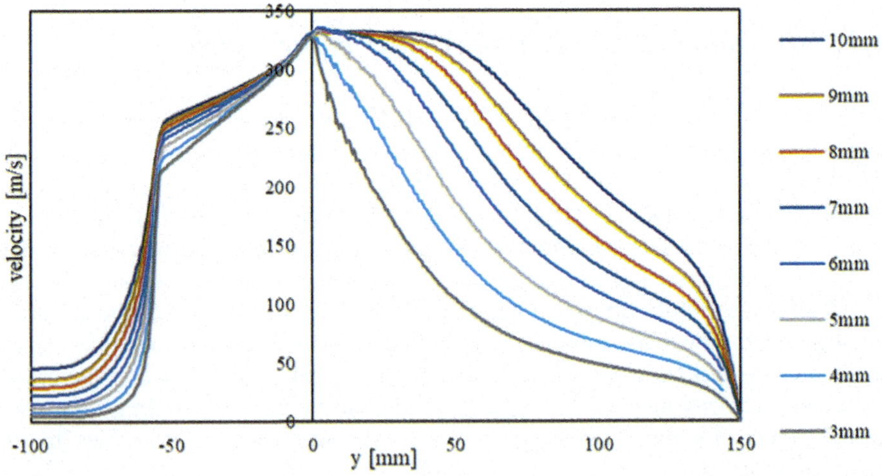

Fig. 3 Air velocity analysis results at the center of the nozzle (The nozzle at a distance between nozzle outlet and specimen of 150 mm)

on the y-axis. The figure shows that the pressure of the airflow inside the nozzle increases as the nozzle exit diameter decreases. This is thought to be due to the fact that as the nozzle exit diameter decreases, the pressure of the airflow inside the nozzle increases because the airflow is clogged at the nozzle parallels. The airflow velocity

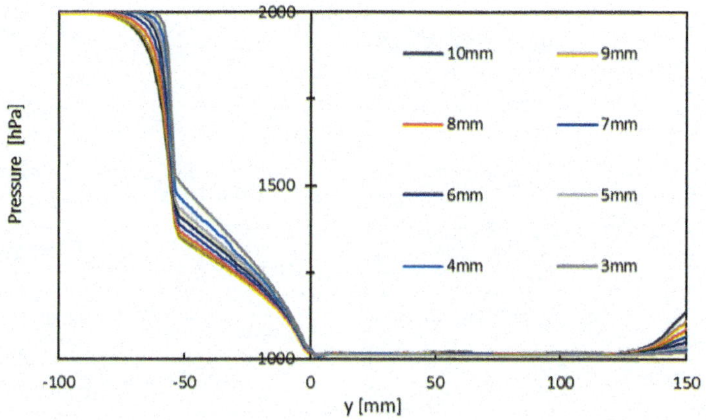

Fig. 4 Airflow pressure distribution at the center (The nozzle at a distance between nozzle outlet and specimen of 150 mm)

inside the nozzle is thought to decrease with decreasing nozzle exit diameter due to the effect of airflow clogging inside the nozzle.

The slower decrease in air velocity after the nozzle exit as the nozzle exit diameter increases is thought to be due to the flow velocity at the nozzle exit. As shown in Figs. 2 and 3, the flow velocity at the nozzle exit is almost the same for all nozzle exit diameters. Since the flow velocity at the nozzle exit is proportional to the nozzle exit cross-sectional area, the flow velocity at the nozzle exit increases as the nozzle exit diameter increases. As the flow rate at the nozzle outlet increases, the area of high velocity airflow increases. Since fast-flowing fluid is resisted by slower-flowing fluid, an increase in the area of fast-flowing airflow may allow the airflow near the nozzle's central axis (Y-axis) to maintain its fast velocity farther from the nozzle outlet.

3.2 *Results of Airflow Analysis at Nozzle Outlet*

The velocity vectors of the airflow after the nozzle exit were examined. Figure 5 shows the velocity vector of the airflow after the nozzle exit at a distance of 150 mm from the nozzle exit to the substrate. The black rectangle at the right end of each figure is the substrate, and the black rectangle surrounding the velocity vector and the substrate is the area of the airflow after the nozzle exit that was analyzed. Figure 5 shows that as the nozzle exit diameter increases, airflow with higher velocity spreads farther from the nozzle exit. Looking at the nozzle exit area, the airflow velocity is almost the same for all nozzle exit diameters, but the area of airflow with higher velocity increases as the nozzle exit diameter increases. Therefore, the flow rate at the nozzle exit increases as the nozzle exit diameter increases. The increase in the

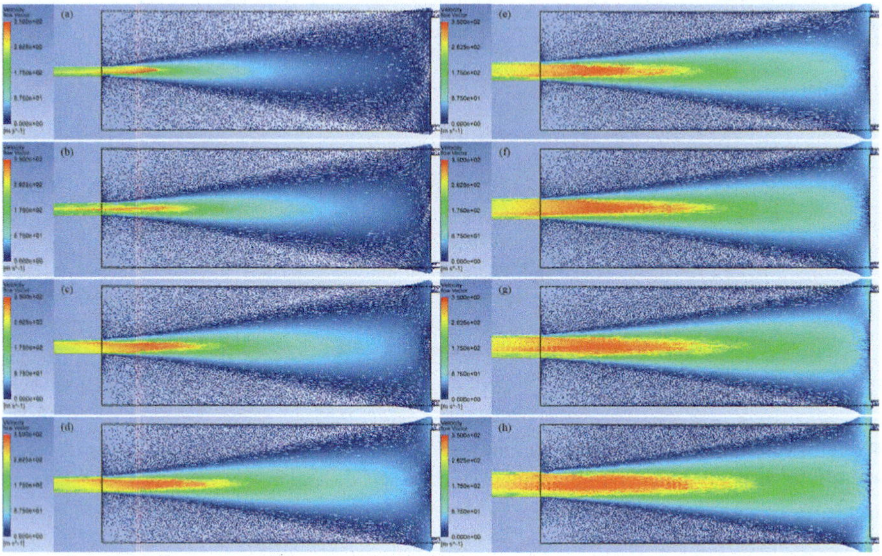

Fig. 5 Velocity vectors of airflow after the nozzle exit at a distance between nozzle outlet and specimen of 150 mm, **a** 3 mm, **b** 4 mm, **c** 5 mm, **d** 6 mm, **e** 7 mm, **f** 8 mm, **g** 9 mm, **h** 10 mm

flow rate at the nozzle exit is thought to be due to maintaining a higher velocity of airflow near the nozzle center axis (y-axis), a result that supports the discussion in section 3.1.

4 Conclusions

The airflow analysis inside and outside the nozzle resulted in the following conclusions.

a. The airflow accelerates rapidly at the nozzle tip, accelerates slowly at the nozzle parallel, and reaches its maximum velocity near the nozzle outlet.
b. As the nozzle exit diameter increases, the airflow can maintain a high velocity up to a point farther from the nozzle exit.
c. As the nozzle exit diameter increases, the flow rate at the nozzle exit increases, which increases the area of high airflow velocity and prevents the velocity from decreasing.

References

1. Nambu K et al (2010) Numerical calculation of particles velocity under the compressible fluid in fine particle bombardments. Trans JSME(ver.C) 67(660):306–312 (in Japanese)
2. Maeda H et al (2001) Analysis of particle velocity and temperature distribution of struck surface in fine particle peening. Trans JSME(ver.C) 67(660):306–312 (in Japanese)
3. Ogawa K (1994) Measurement and analysis of shot velocity in pneumatic shot peening. Trans JSME(ver.C) 60:1120–1125 (in Japanese)
4. Ohta T et al (2019) Numerical analysis of shot peening with incident angle. J Jpn Soc Technol Plast 60(704):256–261 (in Japanese)
5. Ohta T et al (2020) Measurement of shot velocity using particle image velocimetry and numerical analysis of residual stress at two shot peening conditions. Mech Eng J, Belletin JSME 7(4)

Modelling of Dislocation Evolution in Nickel-Based Single Crystal Superalloy Under Shot Peening

Xinyu Yang, Yilun Xu, Siu Sin Quek, and Mark Hyunpong Jhon

Abstract Shot peening is a surface treatment technique that can improve the fatigue performance of metallic materials by introducing compressive residual stresses and modifying the microstructure via grain refinement, phase transformation, texture and/or dislocation structure changes. In this study, a multi-scale modelling framework is developed to understand the evolution of dislocation structures during shot peening of a nickel-based single crystal superalloy. Firstly, explicit finite-element (FE) simulations of the shot peening process are carried out using a Johnson–Cook model calibrated on reported mechanical properties of PWA1484 (Walston et al. In Superalloys 2004 (Tenth International Symposium). TMS, pp 15–24 [1], Cetel and Duhl In Superalloys 1988 (Sixth International Symposium). TMS, pp 235–244 [2]). Instead of using a systematically pre-defined distribution of shots, a random distribution of shot is adopted for these simulations. An Avrami-type equation is used to relate the surface coverage to the necessary peening time and corresponding number of shots (Kirk and Abyaneh in Shot Peen 9:28–30 [3], Pham et al. in Surf Eng 33:687–695 [4]). Various shot peening coverage is modeled by identifying the peening time required for 100%, then using a linear relationship between peening time and coverage. In order to study the effect of shot peening intensity, the initial velocities of shot are chosen according to experimentally measured average shot velocities applied at varying levels of intensity. Figure 1 shows the von Mises stress distribution on shot-peened surface with intensity of 0.15 mmA. During the shot peening process, specimens experience large plastic strain and strain gradient, which induced localized dislocation activity and

X. Yang (✉) · Y. Xu · S. S. Quek · M. H. Jhon
Institute of High Performance Computing (IHPC), Agency for Science, Technology and Research (A*STAR), 1 Fusionopolis Way, #16-16 Connexis, Singapore 138632, Republic of Singapore
e-mail: Yang_Xinyu@ihpc.a-star.edu.sg

Y. Xu
e-mail: Xu_Yilun@ihpc.a-star.edu.sg

S. S. Quek
e-mail: quekss@ihpc.a-star.edu.sg

M. H. Jhon
e-mail: jhonmh@ihpc.a-star.edu.sg

N. Maharjan and W. He (eds.), *Proceedings of the 3rd International Conference on Advanced Surface Enhancement (INCASE) 2023*, Lecture Notes in Mechanical Engineering, https://doi.org/10.1007/978-981-99-8643-9_35

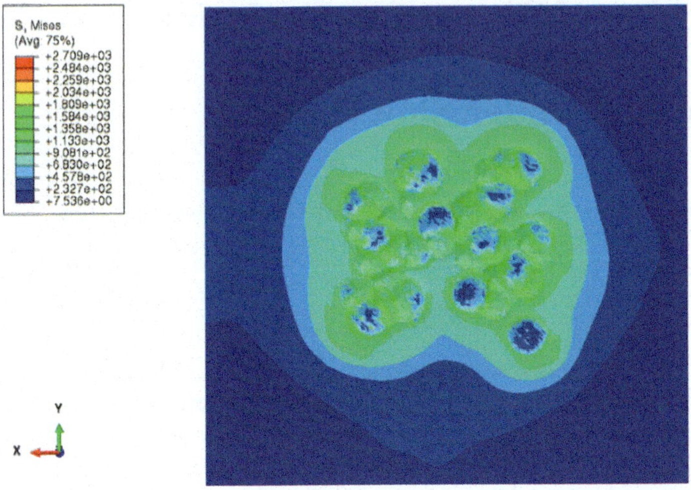

Fig. 1 Representative contour plot of von Mises stress distribution on shot-peened surface

complex dislocation pattern near the surface. A significant gradient in Geometrically Necessary Dislocations (GND) induced by shot peening from the surface to the bulk has been widely reported (Breumier et al. in Mater Sci Eng A 816 [5]), which is usually correlated to experimentally measured hardening gradient. Thus, in this study, the discrete dislocation plasticity (DDP) model (Giessen and Needleman in Model Simul Mater Sci Eng 3:689 [6]) is adopted in order to explicitly model the activities of individual dislocations in the region of substrate near the peened surface. The DDP model is coupled to the FE simulations of the shot peening process by applying boundary conditions corresponding to the displacement on the top surface, as shown in Fig. 2. The surface strengthening mechanisms of the nickel-based single crystal superalloy under different shot peening conditions are investigated by analyzing the resultant dislocation structures (i.e. Figure 3), dislocation density, slip field and lattice orientations.

Keywords Dislocation evolution · Single crystal · Shot peening

See Figs. 1, 2 and 3.

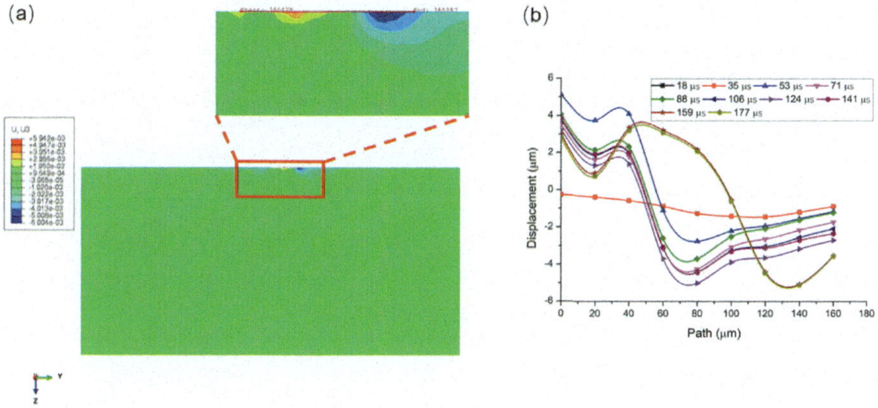

Fig. 2 Displacement history of the region on substrate near the peened surface obtained from FE simulations of shot peening process

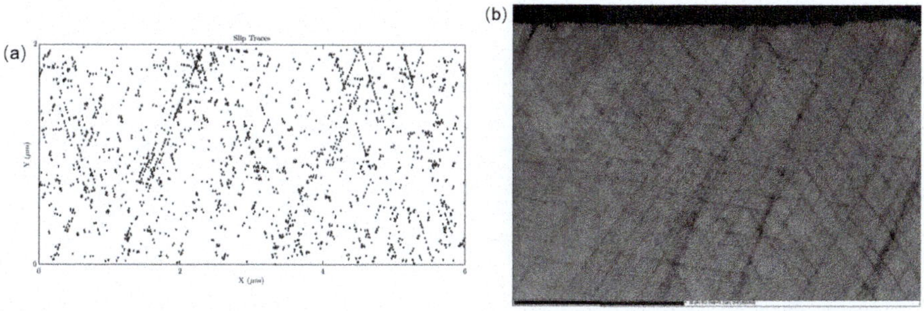

Fig. 3 Comparison between **a** DDP calculated (preliminary) and **b** experimentally observed slip traces

Acknowledgements The authors acknowledge support from A*STAR, Singapore via the Structural Metals and Alloys Programme (Grant No. A18B1b0061).

References

1. Walston S, Cetel A, MacKay R, O'Hara K, Duhl D, Dreshfield R (2004) Joint development of a fourth generation single crystal superalloy. In: Superalloys 2004 (Tenth International Symposium). TMS, pp 15–24
2. Cetel AD, Duhl DN (1988) Second-generation Nickel-base single crystal superalloy. In: Superalloys 1988 (Sixth International Symposium). TMS, pp 235–244
3. Kirk D, Abyaneh M (1995) Theoretical basis of shot peening coverage control. Shot Peen 9:28–30
4. Pham TQ, Khun NW, Butler DL (2017) New approach to estimate coverage parameter in 3D FEM shot peening simulation. Surf Eng 33:687–695

5. Breumier S, Adamski F, Badreddine J, Lévesque M, Kermouche G (2021) Microstructural and mechanical characterization of a shot peening induced rolled edge on direct aged Inconel 718 alloy. Mater Sci Eng A 816:141318
6. Van der Giessen E, Needleman A (1995) Discrete dislocation plasticity: a simple planar model. Model Simul Mater Sci Eng 3: 689–735

Inducing a Realistic Surface Roughness onto 3D Mesh Data Using Conditional Generative Adversarial Network (cGAN)

Bisma Mutiargo, Shan Lou, and Zheng Zheng Wong

Abstract In the age of machine learning, data-driven approaches with hybrid data (a mixture of real images and simulation images) are getting increasingly popular. One major issue with creating a realistic simulation for surface engineering is that the surface of the mesh model used in the simulation is smooth. Often, this mesh does not contain information on surface texture; thus, simulating an object based on these meshes may not represent an actual surface texture of a real component. This article presents a novel technique for introducing surface roughness onto a smooth mesh object to facilitate engineering simulation by using a conditional Generative-Adversarial Network (cGAN) that is trained on real height maps to generate random 2D height maps that represents a realistic texture of a typical upskin and downskin surface of an additively manufactured (AM) part. This approach extracted the past scans of AM components from the Focus Variation microscopy. The 3D surface deviation is extracted as height maps and used as the training data for the generative network. This paper will also discuss the structural similarities between the synthetic and real data using standard descriptors for surface texture characterisation, such as S_a, S_q and S_{dq}.

Keywords Surface roughness · Deep learning · GAN · Additive manufacturing · Synthetic data generation

1 Introduction and Related Work

One of the primary purposes of engineering simulations is to create a replicate and analyse physical responses as close to the real world as possible. Incorporating a realistic surface roughness into the digital model can mimic the irregularities and

B. Mutiargo (✉) · Z. Z. Wong
Advanced Remanufacturing and Technology Centre—A*STAR, Singapore, Singapore
e-mail: mutiargob@artc.a-star.edu.sg

B. Mutiargo · S. Lou
EPSRC Future Metrology Hub, University of Huddersfield, Huddersfield, UK

N. Maharjan and W. He (eds.), *Proceedings of the 3rd International Conference on Advanced Surface Enhancement (INCASE) 2023*, Lecture Notes in Mechanical Engineering, https://doi.org/10.1007/978-981-99-8643-9_36

imperfections of a real-world surface, enabling a more accurate prediction of how the simulated objects behave in practical applications. Such simulations include fluid turbulence analysis, advanced inspection procedure simulations [1], and AM post-processing design simulations in processes such as abrasive flow machining [2].

Incorporating surface roughness into a digital model improves simulation accuracy with fewer assumed parameters. Prior work by Eastwood et al. [3] shows that a Progressively Growing Generative Adversarial Network (PG-GAN) can create random AM surface roughness surface based on experimental data, which shows that machine learning can learn the features of an AM surface and make randomised predictions accurately.

This work presents a novel approach using a conditional-Generative Adversarial Network (cGAN) [4] to randomise surfaces based on the learned dataset and predict between the trained features. The main differing feature highlighted in this work is that c-GAN allows for GAN training with the class as its input, thus allowing for a controlled generation. It is known [5, 6] that many potential factors can affect the surface condition of an AM surface; in this paper, we are exploring only the build angle based on optimum printing parameters, in which a fully upskin surface is defined as 0° inclination and a fully down-skin surface is defined as 180°.

Goodfellow et al. [7] introduced the first concept of GAN, proposing a pair of "adversarial" networks that learn to deceive one another. The original GAN paper proved that this concept could generate an image from a random noise input to one with a considerable degree of realism using the two-player minimax concept. In a GAN network, a generator network G tries to create an image and fool the discriminator D network that the image it is creating is a "real" image. On the other hand, the discriminator network learns to adapt to the improvement of the generator and tries to predict better which image is real and which is not.

cGAN was first introduced by Mirza et al. as an extension of the original GAN [7]. cGAN architecture can be constructed by feeding data and corresponding labels to the generator and discriminator. The main difference between cGAN and the original GAN paper is that cGAN models are class encoded, which means that multiple classes can be trained. Latent space interpolation [8] is possible, which makes it possible to control the generation of the surface by keying in the desired angle as the input. The initial publication shows that this concept can generate MNIST [9] digits conditioned on class labels.

The primary reason for having the ability to interpolate between a limited number of trained angles is that gathering data points can be expensive and time-consuming. Using the class property of cGAN, combined with latent space interpolation, can be an optimal way to reduce the cost of developing a synthetic AM surface without compromising on the realism of the synthetic data.

2 CGAN Network Architecture

Adding class input in a cGAN allows the GAN to learn the critical features in each class. In this case, we define the classes as an angle checkpoint corresponding to the inclination deviation from the upskin surface, where 0° represents an upskin surface, and 180° represents a downskin surface.

The cGAn architecture explored in this paper is described in Fig. 1. The generator expects an input of 128 × 128 × *channel* × *class*. The *channel* in our work is one because the images are converted into a 16-bit grayscale. The *class* is the number of training points; in our case, we built six training points (0°, 30°, 60°, 90°, 120°, and 135°) for the model to learn based on the scanned data.

The first layer of the generator expects two inputs, a latent noise input of size 100 and the number of classes. The first layer's job is concatenating between these two inputs to create 106 dimensions. It is then passed onto a dense layer, which outputs a 1-dimensional tensor of 32,768, then reshaped onto an array of size (8,8) with 512 channels. From here on, standard convolutional upsampling is performed with batch normalisation performed for every Upsampling layer, with a rectified linear

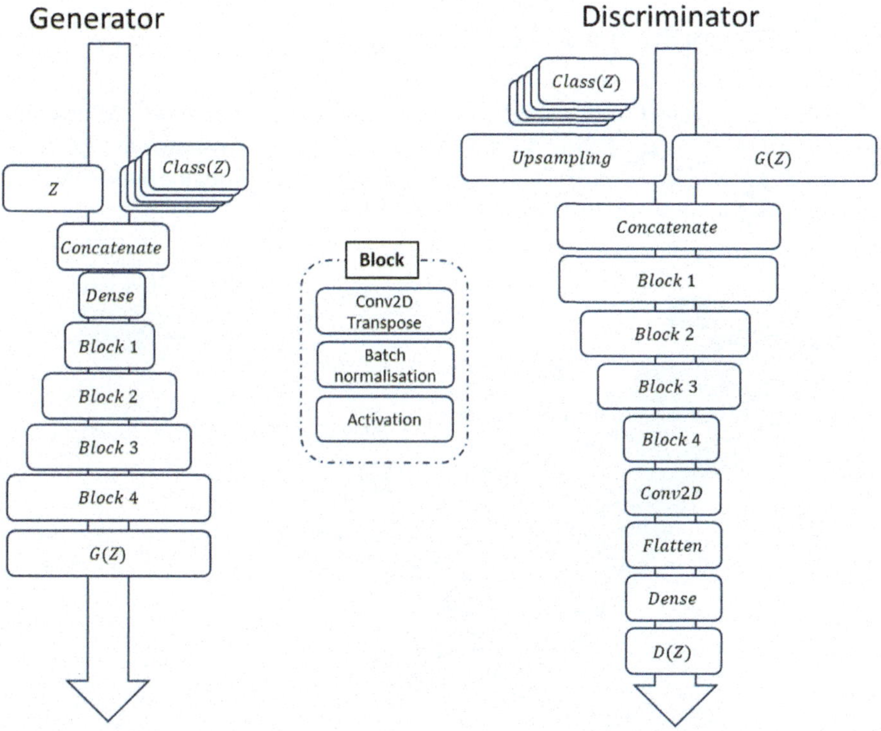

Fig. 1 Network architecture for cGAN (Left: Generator Right: Discriminator)

unit (ReLU) as the activation function to promote sparse activation and reduce the vanishing gradient.

The network's last activation layer is tanh, commonly used in a GAN architecture. Tanh was chosen over sigmoid for the generator because it can process negative values, thus reducing the chances of diminishing gradients. The loss function used in this experiment is binary cross-entropy, the original loss function associated with the initial cGAN implementation. Further analysis and recommendations on the appropriateness of this loss function will be discussed in the result section.

The discriminator network is displayed in Fig. 1 (right side). It expects two inputs, a class descriptor (6 in our case), reshaped into an array of shapes (1, 1, 6), and then upsampled into the same array size of the generator's output. An additional input for the "real" image of size 128 × 128 × 1 is concatenated into the first layer, creating an 8-channel image, which is then put into a standard down-sampling neural network with batch normalisation. The activation function in the generator is a leaky rectified linear unit (leakyReLU) instead of the standard ReLU to reduce the chances of a vanishing gradient in the discriminator network. Sigmoid is chosen as the last layer's activation function as proposed in the original cGAN paper.

3 Data Acquisition and Pre-Processing

The dataset is obtained from a single print with a Renishaw AM400 SLM machine, using 316L stainless steel alloy powder. The sample design is shown in Fig. 2, and the sample contains a surface of a different inclination angle from the build direction: from 0–135° (support structure used beyond 132°).

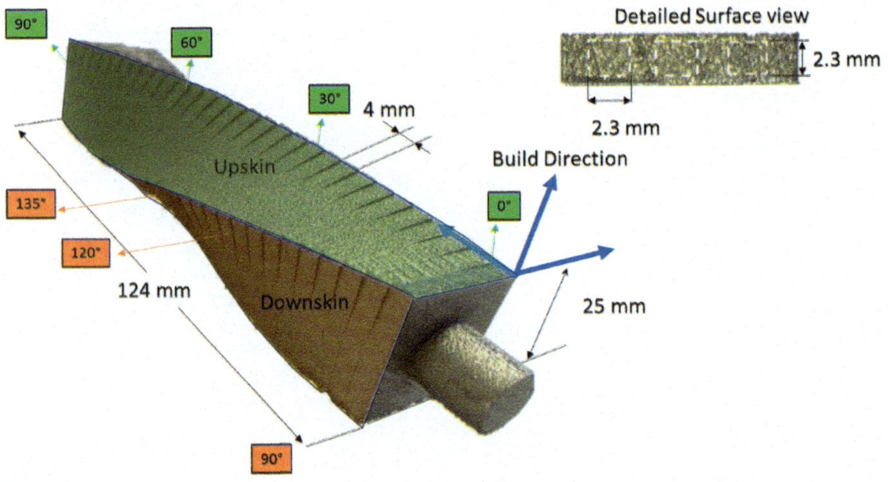

Fig. 2 Printed sample

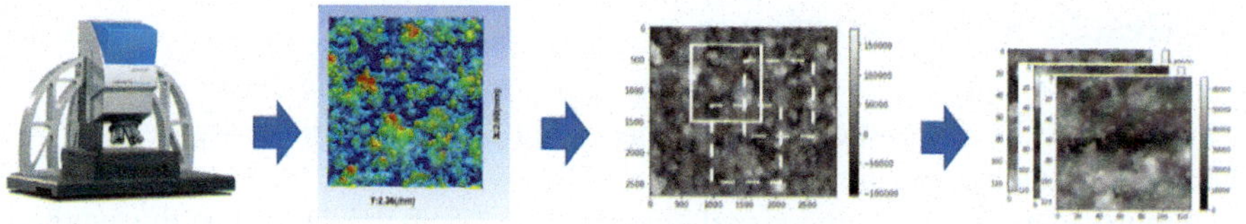

Fig. 3 Data pre-processing (Infinite focus image taken from [12])

The first step to obtaining the height map training data was to scan each angle inclination step in three regions using a focus variation [10] optical instrument over approximately 2.3 × 2.3 mm sampling size (x and y resolution of 0.878 μm). The data was then imported to Surstand software, and a surface filter was applied to level the surface and extract the medium and high-frequency surface. The data was then exported as an Excel file.

The x and y value in the Excel file was then converted into a pixel grid, and the z-value was converted as a normalised grey value heightmap in 16-bit with the OpenCV library [11]. These images were then cropped with a grid size of 1000 × 1000, with a random location, to generate 200 unique images from a single scanned heightmap. Moreover, since three scans were performed for each inclination angle, 600 images were derived as training data from the three scans (Fig. 3).

Six angles of a regular interval were chosen as the "classes" for the cGAN model, i.e., 0°, 30°, 60°, 90°, 120° and 135° which corresponds to classes 0, 1, 2, 3, 4 and 5, respectively. Each class is saved in a folder with its inclination angle as the folder name.

Data is loaded based on the name of each class; as memory management for the GPU and physical RAM, data is loaded in batches, and each epoch is split into steps with different images loaded into the model. Thus, the number of steps equals the total number of images divided by the batch number.

The data is trained for 950 epochs with binary cross-entropy as its primary loss function. The training took approximately 6 h with the system configuration described in Table 1.

Table 1 System configuration

Hardware	Configuration
CPU	Intel Xeon CPU @2.20 GHz
RAM	13 GB
GPU	NVIDIA Tesla T4, 16 GB GDDR6
Operating system	Linux 5.15.107 +
Deep learning python library	Keras 2.12.0

4 Results and Discussions

This section discusses the result of the training and the similarity between the generated data, against the actual data.

Figure 4 further shows the visual comparison of actual data against the synthetic data for 0° (downskin) and 135° (upskin). The visual comparison shows a remarkable consistency between the real data and the synthetic data for the trained classes, and the increasing dominance of the particle features can be observed.

Figure 5 shows an STL file containing inclination angles of 0°, 30°, 60°, 90°, 120° and 135°. A Python script was developed to create a synthetic surface height map for every 2 × 2 mm grid based on the deviation of its average vector against the z-axis, which mimics the build direction in this case. One of the advantages of cGAN

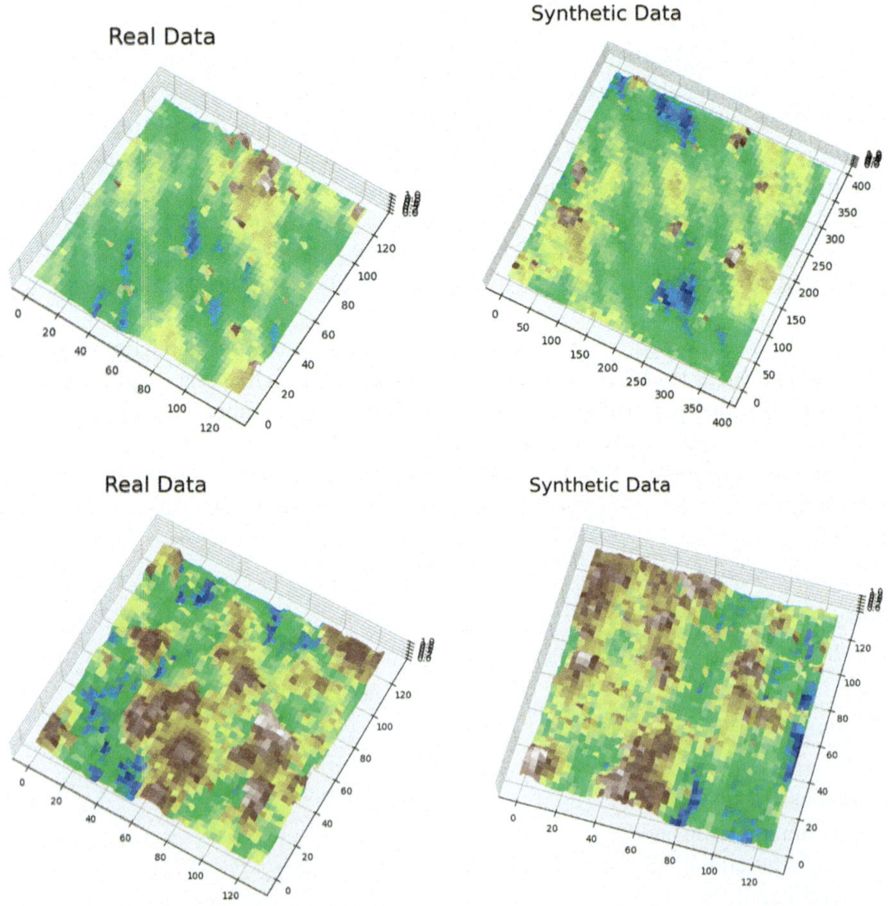

Fig. 4 Visual comparison of real heightmap obtained from infinite focus scan and GAN-generated heightmap

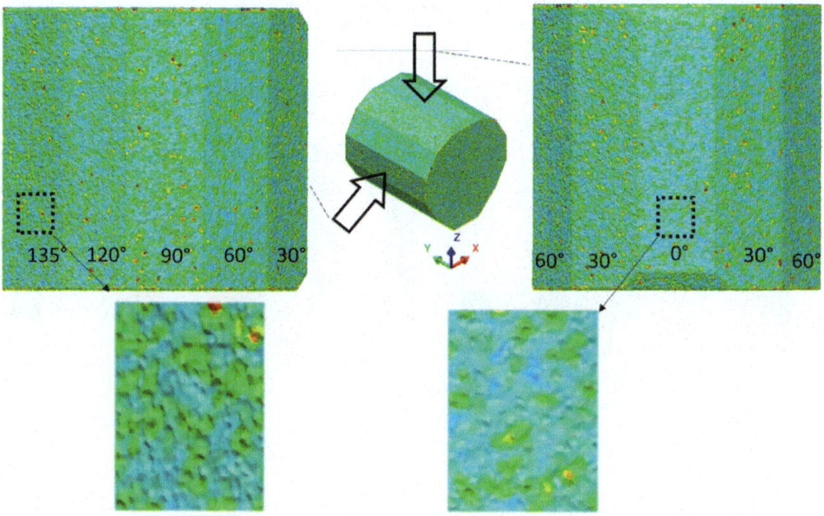

Fig. 5 Incorporating synthetic surface roughness onto an STL file

is that because it can interpolate between its trained angles, an infinite number of interpolated heightmaps can be generated for non-uniform surfaces such as a circle or spherical design.

Some immediate observation points out that the "staircase" feature is not visible on the surface; the reduced lateral triangle resolution in the mesh causes this. In a scenario such as an X-ray CT simulation, an overly high-resolution triangle may cause choking to the memory and cause the computing cost to increase significantly [13]. Ideally, one should decide the trade-offs between having an extremely high-resolution mesh and a reasonable simulation speed, and it is based on application. In the case of a typical X-ray Monte Carlo simulation, a mesh size not exceeding 1,000,000 triangles is usually desirable.

Characterising a surface texture is a field of study with already established guidelines. The most accepted approach is the work published in ISO 25178–2:2012 [14] on surface numeric descriptors. We use the same ISO surface descriptors used in Subharavaanan et al. [15] to perform the quantitative analysis in our study. These surfaces are the surface arithmetic mean height (S_a), Surface root mean square average height (S_q) and root mean square gradient (S_{dq}) to analyse the surface.

Figure 6 shows the S_a value of 100 randomly selected data and 100 generated images. It can be observed that the surface S_a value of the generated surface matches within less than one standard deviation to that of the real surface, both in the distribution characteristics and the absolute value, with 90° as an exception. Overall, the surface arithmetic average of the generated surface closely matches that of the real surface except for the 60° and 90° class, which sees slightly larger surface arithmetic mean for 60°, followed by a lower arithmetic mean for 90°. Despite that, the deviation of the surface arithmetic average between the generated data and the real is

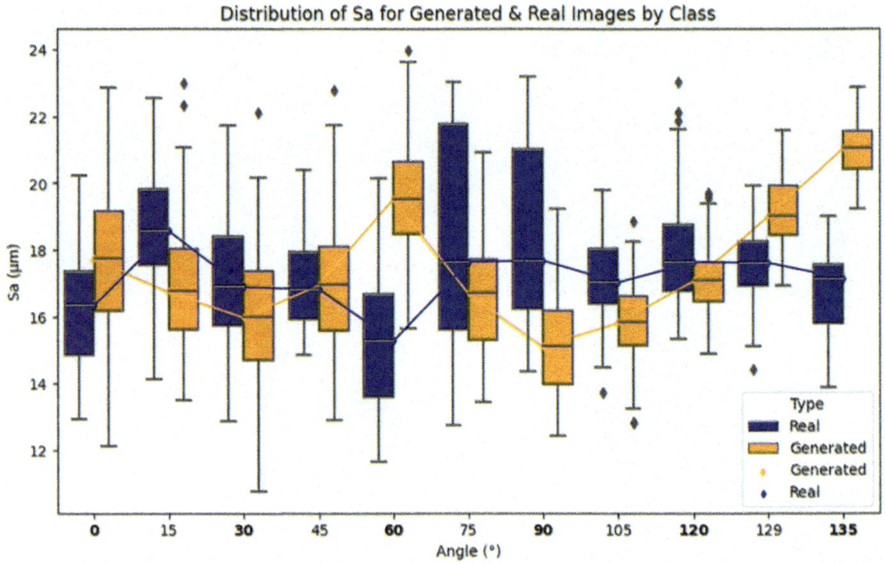

Fig. 6 Surface arithmetic average (S_a) distribution of real and generated surface, with interpolated angles (*Note* Trained angles are in highlighted **bold**, and ◆ in the graph represents outlier)

observed to be remarkably similar (within < 5 μm) and all the distribution overlaps with one another.

This observation suggests that the network is learning to recreate the average surface arithmetic average, creating an accurate representation of a real surface. It is important to note that due to the normalising function in the data pre-processing, the height map value is stretched to a 16-bit integer for memory management, which may cause a minor data loss; converting the data into a 32-bit integer without stretching the data could potentially improve the surface arithmetic average even further.

Further observations can also be made from the surface arithmetic figure that a class that yields a higher deviation from the real value will cause a mis-interpolation between classes, and this is evident in the interpolated angle between 60° and 90° where the surface arithmetic mean of the interpolated class (75°) is almost a linear approximation of the nearest trained classes. Moreover, because of the slightly higher distribution in the 60° class and the lower distribution in the 90° class, the error tends to average out in the interpolated class. Similarly, it can be observed that due to the generation error in the class 135°, the interpolation is also erroneous in the surface arithmetic mean value of 129°.

Surface rms (S_q) represent the root-mean-square of the surface heights. The S_q is a more straightforward descriptor from S_a that observes the standard deviation of the height distribution from its normal arithmetic average; it can be observed that generated surface follows the standard deviation of the real surface. Compared to the arithmetic average (S_a), the S_q in the generated surface tends to be much more prominent in distribution and absolute value. However, it can also be observed that the generated surface's root mean square height measurement still falls within the standard deviation error range of the real images.

A potential cause of this is the "checkerboard" artefact [16], a common occurrence in the GAN network that causes it to create a checker feature. Checker artefacts can be reduced by adjusting the stride value of the network; the current stride value is set to 1 as it is a default value for a cGAN network. Checker artefacts can also be further reduced by increasing the number of epochs for the training. In this instance, the model training was performed on 950 epochs, which may not be sufficient to remove the checker artefact from the image, as evidenced by observing the S_q distribution (Fig. 7).

S_{dq} is a surface parameter representing the standard deviation of the gradient given a fixed sampling area. It can be observed that the generated surface has a lower absolute S_{dq} value, but it maintains the same pattern throughout the angle. For example, in the real data, the S_{dq} drops slightly before peaking at 90°, where the value linearly decreases. This pattern is also observed in the generated images, meaning the generated heightmap learns an equivalent pattern between the classes.

The S_{dq} analysis also gives an insight into the non-linear nature of the relationship between the inclination angle and the surface texture and that the cGAN network can capture this non-linear relationship remarkably close to an actual surface. Some direct observations can be made that further improvement to the cGAN model can be made by additional training so that the S_{dq} absolute value of the generated model matches that of the real data. Moreover, since a clear movement pattern between the increasing inclination angle and the S_{dq} value can be observed, S_{dq} can be incorporated together

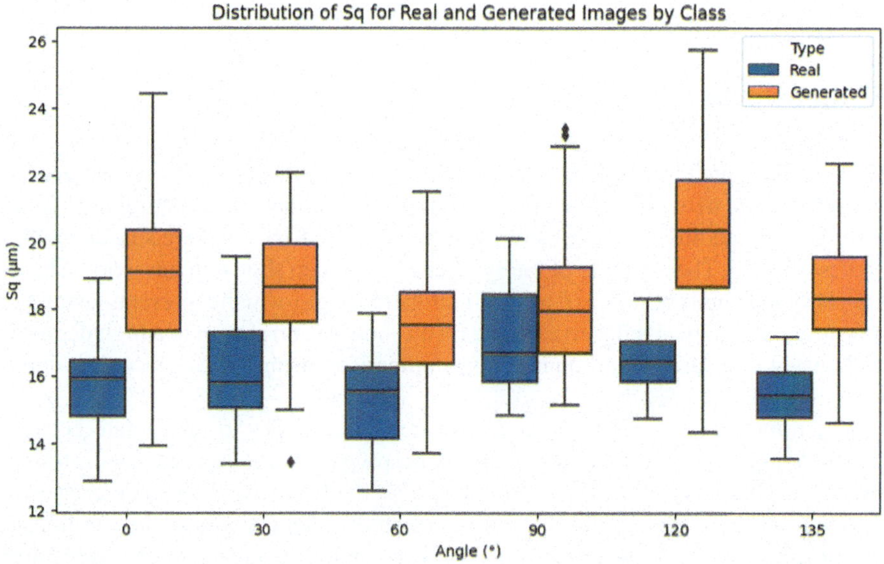

Fig. 7 Comparison of real and generated surface root mean square height (S_q) distribution over trained surfaces

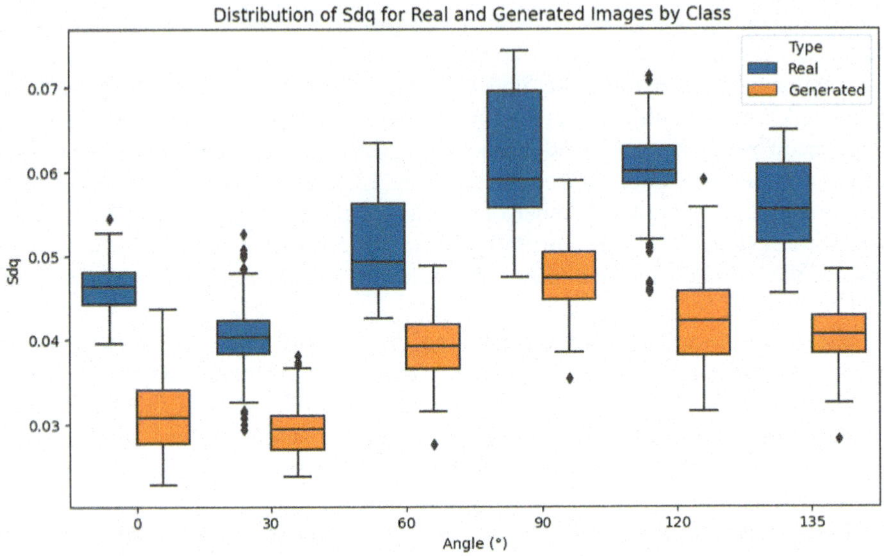

Fig. 8 Comparison of total surface area (S_{dq}) in real and generated surfaces in trained angles

with binary cross entropy as a hybrid loss function, with S_{dq} carrying a slightly higher weight than that of a binary cross-entropy function. The weighted loss function will reward the model for learning the parameters better, and this method is commonly known as physics-based learning (Fig. 8).

5 Conclusion

We have presented the implementation of cGAN, a variance of the original generative adversarial network, to generate a synthetic heightmap representing the surface texture of an additively manufactured surface printed by a sintered laser melting process (SLM). The example of the synthetic heightmap shown in our work is based on six trained points between 0° to 135° inclination angle, which represents an upskin, transition, and a downskin surface. The result shows a correlation to a high degree in both visual comparison and quantitative comparison using ISO surface descriptors such as S_a, S_q, and S_{dq}.

The trained cGAN model also demonstrates the ability to interpolate between trained points using extrapolation of the latent spaces. It was observed that the trained angles' accuracy is critical to accurately predict the latent areas between classes, as the relationship in each surface descriptor of all the latent spaces is close to linear.

Some drawbacks to the model's current state include evidence of checkered artefacts seen in the higher S_q absolute figures compared to the S_a value. These checker artefacts should be reduced as much as possible in future work by adjusting the strides, adjusting the learning rate, and increasing the training epoch of the model. The current model is evaluated at 950 epochs.

Further improvement to the model can also be made through physics-based learning. The current state of the cGAN demonstrated a very promising capability to learn the features that affect the S_{dq}, where the pattern increment is almost identical, even though the absolute value in the generated data tends to show a systemic error slightly below the real value. Training the model with a weighted hybrid loss function and changing the primary loss function to categorical cross-entropy instead of binary may significantly improve the model's performance.

The ability to generate a synthetic heightmap and apply it digitally is essential to create a better engineering simulation, especially in areas of engineering where surface texture impacts the overall result, such as additive manufacturing and fluid dynamics. This paper has shown that cGAN is an excellent neural network for generating a realistic surface roughness map that can be controlled.

This paper demonstrates a sound architecture that can interpolate between classes with minimal training data points. One of the biggest problems with performing a deep learning-based generation is the massive dataset required for a good classification or generation. Due to the non-linear nature of the surface texture in an AM surface, it is possible to extract a large amount of data from a single scan by cropping in random locations and performing image augmentation, such as 180° rotation to mirror the images. The result is a GAN network that can generate arbitrary surfaces with a remarkably high realism compared to real data.

References

1. Liu M, Cheung CF, Senin N, Wang S, Su R, Leach R (2020) Machine surface defect detection using light scattering and deep learning. J Opt Soc Am A 3(9):B53–B59
2. Kum CW, Wu CH, Wan S, Kang CW (2020) Prediction and compensation of material removal for abrasive flow machining of additively manufactured metal components. J Mater Proc Technol 282
3. Eastwood J, Newton L, Leach R, Piano S (2022) Generation and categorisation of surface texture data using a modified progressively growing adversarial network. Precis Eng 74:1–11
4. Mehdi M, Osindero S (2014) Conditional generative adversarial nets. arXiv preprint
5. Surface roughness considerations in design for additive manufacturing (2021) Proceedings of the design society, vol 1, pp 2481–2850
6. Udroiu R, Braga IC, Nedelcu A (2016) Evaluating the quality surface performance of additive manufacturing systems: methodology and a material jetting case study. Materials (Basel) 12(6):662
7. Goodfellow I, Pouget-Abadie J, Mirza M, Xu B, Warde-Farley D, Ozair S, Courville A, Bengio Y (2020) Generative adversarial networks. Commun ACM 63(11):139–144
8. Pham Van T, Nguyen TM, Tran NN, Nguyen HV, Doan LB, Dao HQ, Minh TT (2020) Interpreting the latent space of generative adversarial networks using supervised learning. In: International conference on advanced computing and applications. Vietnam
9. Deng L (2012) The mnist database of handwritten digit images for machine learning research. IEEE Sig Proc Mag 29(6):141–142
10. Danzl R, Helmli F, Scherer S (2011) Focus variation—a robust technology for high resolution optical 3D surface metrology. Strojniski Vestnik 245–256

11. Bradski G (2000) The OpenCV library. Dr. Dobb's J Softw Tools
12. Bruker A. G-series infinite focus application note—application: safety washer measurement, Taken from https://www.alicona.com/fileadmin/user_upload/Alicona/Downloads/Application_Note_Safety_Washers.pdf
13. Zhenga K, Qiu B, Wang Z, Li X, Li J, Gao K (2020) Image-based numerical study of three-dimensional meso-structure effects on damage and failure of heterogeneous coal-rock under dynamic impact loads. Particuology 51:132–141
14. International Standard Organisation (2012) ISO 25178-2 geometrical product specifications (GPS)—surface texture: areal—part 2: terms, definitions and surface texture parameters, Geneva, Switzerland
15. Shubhavardhan RN, Liu W, Zeng W, See TL, Scott P, Jiang XQ (2021) Surface texture characterization of metal selective laser melted part with varying surface inclinations. J Tribol 143(5):1–37
16. Odena A, Dumoulin V, Olah C (2016) Deconvolution and checkerboard artifacts. Distill

Fluid/Material Coupled Numerical Simulation of a Bubble Collapse Near a Wall for Laser Cavitation Peening

Yuka Iga, Chieko Kuji, Hirotoshi Sasaki, and Hitoshi Soyama

Abstract An impact of a bubble induced by a submerged pulsed laser is utilized for improvement of fatigue strength of metallic materials. As the bubble induced by the pulsed laser behaves like a cavitation bubble, the laser induced bubble is called as "laser cavitation". The mechanical surface treatment using the laser cavitation impact is named as "laser cavitation peening". At laser cavitation peening, the impact induced by laser cavitation collapse strongly depends on the bubble geometry. There are two typical mode at the bubble collapse. One mode is "microjet mode", at which bubble develops near the target and is collapsed with generating a microjet in the bubble. The other mode is "hemispherical mode", at which a hemispherical bubble develops on the target surface and is collapsed on the surface. As the bubble collapse of microjet mode is interesting phenomenon, a lot of researchers investigate "microjet mode". However, impact induced by "hemispherical mode" is significantly larger than that of "microjet mode". In the present paper, to optimize laser cavitation peening condition, a fluid/material coupled numerical simulation of a bubble collapse near a wall was carried out changing with standoff distance from wall. It was revealed that the equivalent stress induced by hemispherical mode was larger than that of microjet mode.

Keywords Cavitation · Bubble · Numerical simulation · Laser · Peening

---

Y. Iga (✉)
Institute of Fluid Science, Tohoku University, 2-2-1 Katahira, Sendai 980-8577, Japan
e-mail: yuka.iga.b4@tohoku.ac.jp

C. Kuji · H. Soyama
Department of Finemechanics, Tohoku University, 6-6-01 Aoba-Ku, Sendai 980-8579, Japan
e-mail: c-kuji@tohoku.ac.jp

H. Soyama
e-mail: soyama@mm.mech.tohoku.ac.jp

H. Sasaki
Tokai Head Office, The Japan Atomic Power Company, 1-1 Shirakata, Tokai 319-1198, Japan

N. Maharjan and W. He (eds.), *Proceedings of the 3rd International Conference on Advanced Surface Enhancement (INCASE) 2023*, Lecture Notes in Mechanical Engineering, https://doi.org/10.1007/978-981-99-8643-9_37

1 Introduction

In the case of laser peening, which is also called as laser shock peening, there are two types. One is laser peening with water film [1]. The other is laser peening under water, i.e., submerged laser peening [2]. For both laser peening, it has been thought that peening mechanism to introduce local plastic deformation is caused by laser ablation [3]. Namely, it is thought that inertial force caused by shock wave at laser ablation is contained by water, and it produces plastic deformation into metallic materials [1, 3]. However, in the case of submerged laser peening, a bubble which behaves like a cavitation bubble is generated after the laser ablation, and the bubble impact can be utilized for peening [4] (see Fig. 1a). In the present paper, the bubble is called as "laser cavitation", and a peening method using laser cavitation impact is named as "laser cavitation peening" [5–7].

When pressure wave in water was measured by a submerged shock wave sensor, the amplitude of laser ablation was larger than that of laser cavitation collapse as shown in Fig. 1b [4]. However, an impact induced of laser cavitation collapse measured by a PVDF sensor in target was larger than that by laser ablation (see Fig. 1c) [4]. Namely, the laser cavitation impact is useful for the laser cavitation peening.

As the improvements of fatigue properties of metallic materials such as additive manufactured titanium alloy Ti6Al4V [8] and magnesium alloy AZ31 [9] by laser cavitation peening were reported, it is worthwhile to optimized and enhance peening intensity of laser cavitation peening.

In the case of research area on cavitation, i.e., bubble dynamics, a bubble collapse with a microjet, i.e., "microjet mode", was an attractive phenomenon in both experimental and numerical analysis since early 1970s [10, 11]. When peening effects such as introduction of compressive residual stress was compared experimentally, hemispherical bubble collapse, i.e., "hemispherical mode" was more aggressive

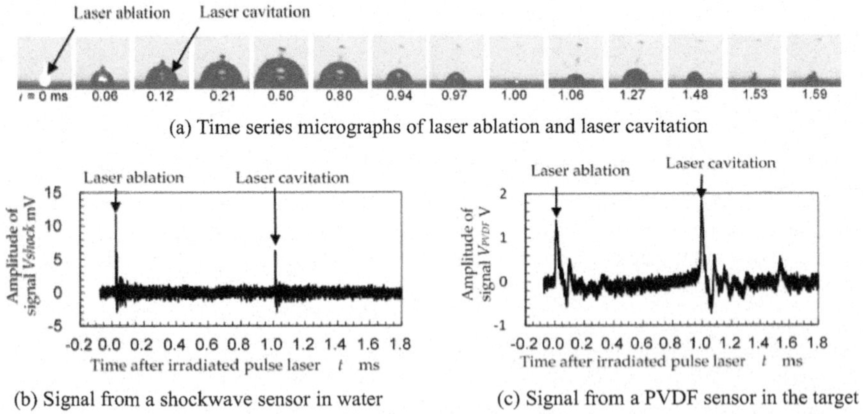

(a) Time series micrographs of laser ablation and laser cavitation

(b) Signal from a shockwave sensor in water

(c) Signal from a PVDF sensor in the target

Fig. 1 Laser ablation and laser cavitation produced by a pulsed laser [4] comparing with "microjet mode" [7]. Iga and Sasaki demonstrated numerically by a fluid/material coupled numerical simulation that effects of bubble collapse into material surface strongly depended on the bubble shape.

In the present paper, in order to optimize laser cavitation peening, the difference between "microjet mode" and "hemispherical mode" was investigated numerically by the fluid/material coupled numerical simulation.

2 Numerical Method

For fluid domain, the locally homogeneous compressible gas–liquid two-phase medium model [12] was used for capturing the bubble interface behavior. In the present model, compressibility is considered in the liquid phase, and then pressure, which propagates not only in the gas phase but also in the liquid phase, can be reproduced numerically. The surface tension on the bubble surface is calculated by using the CSF (Continuum surface force) model, and phase change of evaporation of liquid phase and condensation of vapor phase are considered.

For material domain, because the purpose of the present simulation of material is to analyze the propagation of stress waves and history and stress distribution, rather than material deformation, the material is assumed as an elastic body. The governing equations are equations of motion and the time-differential constitutive equations of homogeneous isotropic elastic medium.

We can use the same numerical scheme of fluid calculation for the calculation of material. The cell-centered finite volume formulation was used to discretize the governing equations in both the fluid and material domains. The fluid and material domains are coupled by two-way with considering the reflection and transmission of pressure and stress with an acoustic impedance in each domain. The detail numerical and coupling schemes were reported our previous study [13].

Figure 2 shows the calculation area and initial condition of calculation. Duralumin was simulated as target material. In order to investigate the collapsing mode of the bubble, $\gamma = L/R_{max}$ is defined by the distance between the center of the bubble and the wall surface, L, and the maximum bubble radius R_{max}. Here, $\gamma = 0$ and $\gamma = 1.2$ correspond to "hemispherical mode" and "microjet mode", respectively [13].

3 Results

Figure 3 reveals the color contour at bubble collapse Fig. 3a hemispherical mode and Fig. 3b microjet mode. As bubble shape, pressure in water and stress in material were changing with time, the time of maximum pressure in water at each mode was chosen for Fig. 3, respectively. In Fig. 3, black lines show the bubble shape changing with time from $R = R_{max}$ to the collapse. The color contour of upper left of each figure reveals void fraction. The void fraction is almost zero at the moment in Fig. 3 because

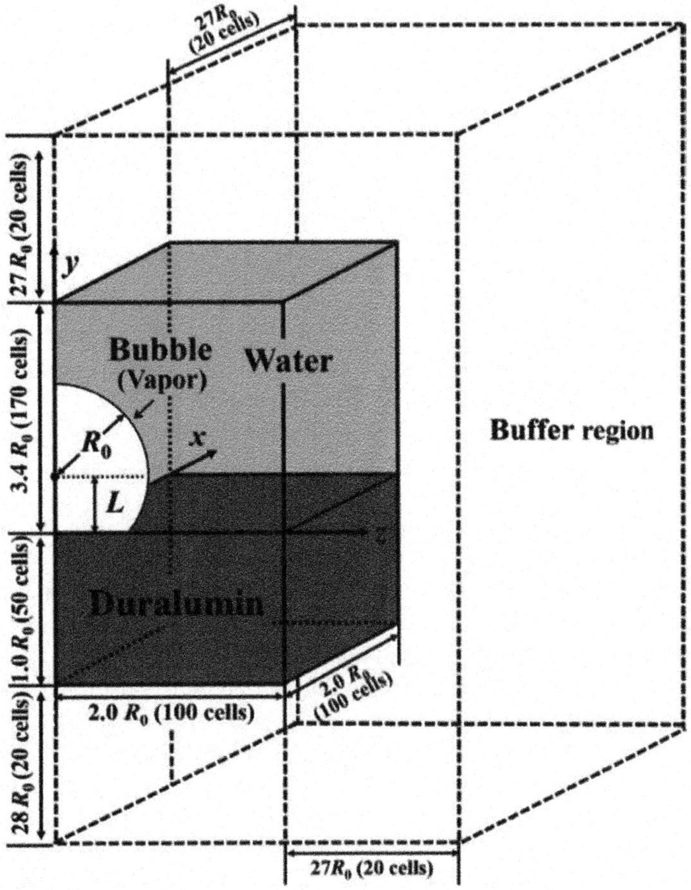

Fig. 2 Calculation area and initial condition of calculation

the moment of maximum pressure is final stage of the bubble collapse, although the bubble will rebound after this moment. The upper right illustrates pressure and velocity vectors in the fluid domain. The equivalent stress in the material domain was shown in lower part.

Figure 4 shows the maximum equivalent stress in material domain $(\sigma_{eq})_{max}$ during the collapse. In the case of microjet mode, the pressure in water reached about 450 MPa due to water hammer pressure of the microjet, however, $(\sigma_{eq})_{max}$ was less than 100 MPa. On the other hand, $(\sigma_{eq})_{max}$ of hemispherical mode reached over 2.5 GPa. And also, the volume of stress influential area and the depth of stress influential area of hemispherical mode were larger than those of microjet mode [13]. These results well correspond to the experimental result, in which the introduced compressive residual stress by hemispherical mode was larger and deeper than that of microjet mode [7].

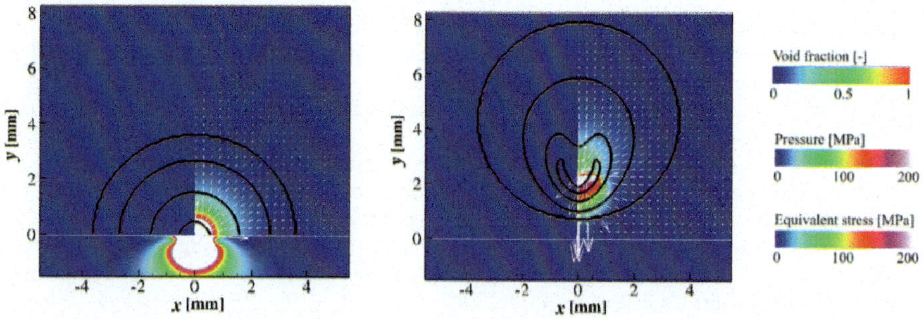

(a) Hemispherical bubble on the wall (hemispherical mode)

(b) Spherical bubble near the wall (microjet mode);

Fig. 3 Fluid/material coupled numerical simulation of collapse of a bubble near a wall (upper left of each sub-figure: void fraction, upper right: pressure and velocity vectors, lower: equivalent stress in the material)

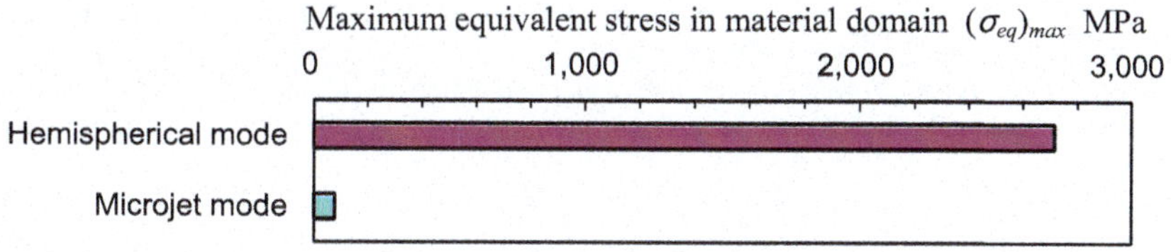

Fig. 4 Difference between hemispherical mode and microjet mode on maximum equivalent stress in material domain

4 Conclusions

In order to demonstrate the difference on peening intensity of laser cavitation peening between microjet mode and hemispherical mode, fluid/material coupled numerical simulation was carried out. It was concluded that maximum equivalent stress induced by hemispherical mode was larger than that of microjet mode.

Acknowledgements This research was partly supported by JSPS KAKENHI grant numbers 22KK0050 and 23H01292.

References

1. Peyre P, Fabbro R, Merrien P et al (1996) Laser shock processing of aluminium alloys. Application to high cycle fatigue behaviour. Mater Sci Eng A 210:102–113
2. Sano Y, Obata M, Kubo T et al (2006) Retardation of crack initiation and growth in austenitic stainless steels by laser peening without protective coating. Mater Sci Eng A 417:334–340
3. Sano Y, Akita K, Sano T (2020) A mechanism for inducing compressive residual stresses on a surface by laser peening without coating. Metals 10:816

4. Soyama H (2019) Comparison between the improvements made to the fatigue strength of stainless steel by cavitation peening, water jet peening, shot peening and laser peening. J Mater Process Technol 269:65–78
5. Soyama H (2021) Laser cavitation peening and its application for improving the fatigue strength of welded parts. Metals 11:531
6. Soyama H, Korsunsky AM (2022) A critical comparative review of cavitation peening and other surface peening methods. J Mater Process Technol 305:117586
7. Soyama H, Iga Y (2023) Laser cavitation peening: a review. Preprint (wwwpreprintsorg)
8. Soyama H, Kuji C (2022) Improving effects of cavitation peening, using a pulsed laser or a cavitating jet, and shot peening on the fatigue properties of additively manufactured titanium alloy Ti6Al4V. Surf Coat Technol 451:129047
9. Soyama H, Kuji C, Liao YL (2023) Comparison of the effects of submerged laser peening, cavitation peening and shot peening on the improvement of the fatigue strength of magnesium alloy AZ31. J Magnes Alloy 11, in press
10. Plesset MS, Chapman RB (1971) Collapse of an initially spherical vapour cavity in neighbourhood of a solid boundary. J Fluid Mech 47:283–290
11. Lauterborn W, Bolle H (1975) Experimental investigations of cavitation-bubble collapse in neighborhood of a solid boundary. J Fluid Mech 72:391–399
12. Ochiai N, Iga Y, Nohmi M et al (2011) Numerical analysis of nonspherical bubble collapse behavior and induced impulsive pressure during first and second collapses near the wall boundary. J Fluid Sci Technol 6:860–874
13. Iga Y, Sasaki H (2023) Relationship between a non-spherical collapse of a bubble and a stress state inside a wall. Phys Fluids 35:023312

Advanced Coating Materials Design Synthesis and Industry Applications

Electrochemical Performance of Synthesized CoCuFeMnNi High Entropy Alloy Nanoparticles in OER Applications

Hamzah Kamaruddin, Lu Yu, Mingzhen Xiu, Huang Yizhong, and Wei Yuefan

Abstract High entropy alloy nanoparticles (HEA NPs) have recently found tremendous popularity as effective catalysts in water splitting applications, such as oxygen evolution reactions (OERs). HEA NPs that contain expensive and less sustainable noble metals, such as platinum and iridium, as constituent elements are extensively investigated. With consideration of environmental friendliness, we propose a novel and facile approach for the fabrication of CoCuFeMnNi HEA NPs upon conventional carbon cloth (CC) substrates using laser irradiation. Noble metal-free CoCuFeMnNi HEA NPs were synthesized on CC substrates. Characterization using scanning electron microscopy (SEM), transmission electron microscopy (TEM) and X-ray diffraction (XRD) confirmed the successful synthesis of the HEA NPs on CC. Preliminary electrochemical results show that CoCuFeMnNi HEA NPs on CC display promising catalytic performance of lower overpotentials during OER tests. These results are competitive with traditionally used RuO_2, IrO_2, and other novel nanocatalysts in research today. This work can provide evidence that CoCuFeMnNi

H. Kamaruddin (✉) · L. Yu · M. Xiu · H. Yizhong
School of Materials Science and Engineering, Nanyang Technological University, 50 Nanyang Avenue, Singapore 639798, Singapore
e-mail: HAMZ0002@e.ntu.edu.sg

L. Yu
e-mail: luyu0010@e.ntu.edu.sg

M. Xiu
e-mail: xium0002@e.ntu.edu.sg

H. Yizhong
e-mail: yzhuang@ntu.edu.sg

W. Yuefan
Advanced Remanufacturing and Technology Centre (ARTC), Agency for Science, Technology and Research (A*STAR), 3 Cleantech Loop, #01/01 CleanTech Two, Singapore 637143, Republic of Singapore
e-mail: wei_yuefan@artc.a-star.edu.sg

N. Maharjan and W. He (eds.), *Proceedings of the 3rd International Conference on Advanced Surface Enhancement (INCASE) 2023*, Lecture Notes in Mechanical Engineering, https://doi.org/10.1007/978-981-99-8643-9_38

HEA NPs synthesized by laser irradiation can still perform comparably to conventional noble metal catalysts. It also provides insights into using more sustainable and cheaper nanomaterials as electrocatalysts.

Keywords High entropy alloys · Catalysis · Nanomaterials

1 Introduction

The transition towards greener forms of energy in our future necessitates the development of cost-effective and efficient forms of energy storage and conversion technologies. In recent years, fuel cells and novel nano-sized catalysts have gained significant traction, attributed to their high energy conversion capabilities which posits these technologies as potential alternative, renewable energy sources. The oxygen evolution reaction (OER) is one such important application for fuel cells and water-splitting to generate oxygen. High Entropy Alloys nanoparticles (HEA NPs) have surfaced as a class of materials with significant potential for use in OER because of their unique structural and catalytic abilities [1, 2]. HEAs are multi-metallic materials with five or more constituent transition metal elements which may vary stoichiometrically, consequently resulting in the formation of a disordered solid solution with higher system entropy [3]. Additionally, there is a great degree of tunability to enhance the performance of HEA NPs by adjusting the respective stoichiometries of different constituents.

However, in the design of HEA NPs for OER currently, existing research has been largely focused on the use of noble metals as constituents in HEA NPs since noble metals, including ruthenium and iridium for OER [4, 5], are known to display excellent catalytic performance in these applications. The production of noble metals, however, is known to be a greatly unsustainable process while the commercialization of these noble metal-containing HEA NPs becomes exceedingly limited by the high cost associated with the production of these materials [6]. Novel research on metal–organic frameworks (MOF) as a template to synthesize non-noble metal HEA NPs have been reported to be an effective strategy to produce binary and ternary structures like NiFe [7–9] and CoFeNi respectively [10–12], with carbon supports having displayed as complementing the electrochemical performance of these systems.

Noble metal-free HEAs were first reported to possess excellent thermocatalytic performances by Yao et al. [13, 14]. Recently, their potential applications in electrochemical catalysis have attracted vast attention. For example, Qiu et al. synthesized AlNiCoFeX (X: Mo, Nb, Cr) HEA catalysts by using a top-down method and confirmed that the synthesized HEA catalyst has good OER performance [15]. A high entropy catalyst of AlFeCoNiCr fabricated by Fang et al. also shows a low overpotential towards OER when it is employed as an anodic electrode [16]. However, most conventional methods have mainly been focused on the production of HEAs with bulk sizes instead of nano-sized structures. Acquisition of nano-HEAs without impure phases involves either a complicated process or is subjected to harsh conditions. Therefore, much emphasis has recently been placed on using more streamlined synthesis methods of nanocatalysts with less strenuous conditions, and the use of constituent non-noble metals in the HEA.

In this work, we report on the facile and quick synthesis of CoCuFeMnNi HEA NPs synthesized upon industrial carbon cloth (CC) using laser irradiation in inert conditions, which can be a potential electrocatalyst for OER applications. The pulsed irradiation process requires a comparatively short time to complete which is crucial for upscaling and commercialization compared to other existing synthesis methods. CoCuFeMnNi upon CC substrates were benchmarked in terms of OER electrochemical performance against traditional RuO_2 on CC and IrO_2 on CC catalysts and displayed promising results of lower overpotentials to achieve a set current density and lower charge transfer resistance, coupled with the ease of synthesis using laser irradiation. Furthermore, CoCuFeMnNi contains no noble metal constituents and is composed of only transition metal alloy constituents yet performs as competitively compared to other well-known precious metal catalysts, such as IrO_2 and RuO_2.

2 Methods

2.1 Preparation of Acid-Treated CC

Industrial-grade carbon cloth with plastic supports interwoven between fibers was immersed into concentrated HNO_3 (69% purity) for 9 h. Subsequently, the acid-treated carbon cloth was triply washed with ethanol and further sonicated in DI water for 30 min prior to drying in an oven at 60 °C for 1 h. This methodology was similarly used by Thamilselvan et al. [17] as a means of chemically activating the surfaces of the carbon fibers. Furthermore, as reported by Mei et al. [18], oxygen-containing groups such as carboxylic groups will be introduced onto the surfaces of the carbon fibers while also making the fibers more hydrophilic. This is meant to facilitate further reduction in nanoparticle sizes and increased size uniformity upon nanoparticle synthesis [13].

2.2 Synthesis of CoCuFeMnNi on CC Substrates

The process of synthesizing CoCuFeMnNi HEA NPs onto CC firstly involves the creation of a stock solution composed of the constituent metal chloride precursor solutions and 37% fuming hydrochloric acid was created, following the stoichiometric proportions as stated in Table 1. Each constituent metal chloride powder was dissolved in absolute ethanol to produce a precursor solution of 0.1 mol L^{-1}, prior to combining with the other respective precursors to create the stock solution. Secondly,

Table 1 List of stock solution constituent components

Stock Solution Constituents	Molarity (mmol)
$CoCl_2 \cdot 6H_2O$	0.055
$CuCl_2 \cdot 2H_2O$	0.008
$FeCl_3$	0.040
$MnCl_2$	0.041
$NiCl_2 \cdot 6H_2O$	0.055
37% Fuming HCl	0.136

Table 2 List of laser irradiation parameters

Laser Parameters	
Output power	3 W
Scan speed	40 mm s^{-1}
Wavelength	1064 nm
Frequency	900 kHz
Pulse width	2 ns

equally sized sectioned pieces of industrial grade carbon cloth were drop-casted with varying volumes of 200, 50, 40, 30 and 20 μL of CoCuFeMnNi stock solution homogeneously throughout the entire surface of the carbon cloth. These samples were denoted as HEA@CC-200, HEA@CC-50, HEA@CC-40, HEA@CC-30 and HEA@CC-20 respectively. The drop-casted samples were then dried in an oven for 20 min at 60 °C. Thirdly, the sample was placed into a vacuum box which was initially evacuated of air using a vacuum pump and subsequently subjected to an inert atmosphere of nitrogen gas. Laser irradiation was subsequently performed on samples using a commercially available JPT-M7-100W pulsed laser with maximum power rating of 100W. The laser spot size was 0.03 mm and was set to scan along a zig-zag hatch pattern covering the entire sample with 0.03 mm spacing between each track. The laser irradiation parameters are also listed in Table 2. Finally, the samples were washed with DI water and fully dried in an oven for 20 min at 60 °C.

2.3 *Synthesis of RuO_2@CC and IrO_2@CC Benchmarks*

As in the synthesis of HEA@CC, the stock solutions of RuO_2 and IrO_2 were synthesized from respective powders that were dissolved in absolute ethanol to produce a salt solution of 0.1 mol L^{-1} respectively. 200 μL of RuO_2 and 50 μL of IrO_2 salt solution, identified as the optimum loading volume for each, was then drop-casted onto equally sized acid-treated carbon cloth substrates homogeneously, dried, then subjected to laser irradiation in air using the same parameters listed in Table 2. Then, the samples were washed with DI water and dried as in the synthesis of HEA@CC.

3 Results and Discussion

3.1 SEM and XRD Analyses

The morphology of pure CC and HEA@CC was imaged using a JEOL JSM-7600F field-emission scanning electron microscope (FESEM). FESEM micrographs clearly show the surfaces of pure as-received carbon cloth, with fibers of 7 μm in width, appear to have a smooth and clear appearance which is free of any particles, prior to addition of HEA NPs (Fig. 1a). After concentrated nitric acid treatment, drop-casting with stock solution, and drying, the carbon fibers appear to not only be homogeneously coated with unmelted salts which formed a thin layer above each CC fiber before any laser irradiation, but also has a slightly rougher surface (Fig. 1b). After the laser irradiation process however, the surfaces of carbon fibers appear to be homogeneously and densely covered with small, uniformly sized, and spherical particles on all fibers. Upon further magnification, the particles appear to have sizes of around 15 nm (Fig. 1c, inset). The particles also have a generally uniform size distribution with marginal agglomeration of particles (Fig. 1c, inset). This is further supported by the particle size distribution analysis in which the average particle size is found to be around 15.634 ± 0.540 nm and the distribution of particle sizes follows a Gaussian trend (Fig. 1d). This suggests the successful synthesis of the CoCuFeMnNi HEA NPs on CC.

XRD analysis was performed using an XRD Shimadzu XRD-6000 with Cu-Kα radiation. Pure carbon fibers produced a spectrum of three major peaks positioned at 25.8°, 43.6° and 53.5°, which is attributed to the reflections of the (0 0 2), (1 0 1) and (0 0 4) facets of the carbon cloth respectively (Fig. 2a) [19]. As reported in existing research [20–22], CoCuFeMnNi HEA has a single-phased face-centered cubic (FCC) lattice microstructure which is apparent in the XRD spectra in Fig. 3b, whereby the peaks are likely attributed to the reflections of the (1 1 1), (2 0 0) and (2 2 0) facets of CoCuFeMnNi HEA NPs at 43.9°, 53.0° and 79.6° (Fig. 2b). When compared to bare CC and laser irradiated CC, it is apparent that HEA@CC-30 produces more intense peaks which is likely due to the formation of the FCC phases of the HEA NPs. This comparison further demonstrates that the peaks originate from the HEA NPs themselves and not from the effects of laser irradiation on CC as there are insignificant changes between the XRD spectra of bare CC and laser-irradiated CC. This further supports the SEM findings and suggests strongly that the HEA NPs have been successfully synthesized.

The formation of extremely small particles can be attributed to the laser irradiation synthesis strategy employed. The mixed metal salts in the stock solution which was drop-casted onto the carbon cloth experienced ultra-fast heating and cooling, leading to the decomposition of the constituent precursors of CoCuFeMnNi HEA NPs by laser interaction with CC [23]. Additionally, as described above, in carbothermal shock synthesis experiments, nitric acid treatments on carbon cloth are essential as the introduction of carboxylic groups on the surfaces of carbon fibers encourages the fission and fusion processes of the catalyst nanoparticles, which ultimately results in

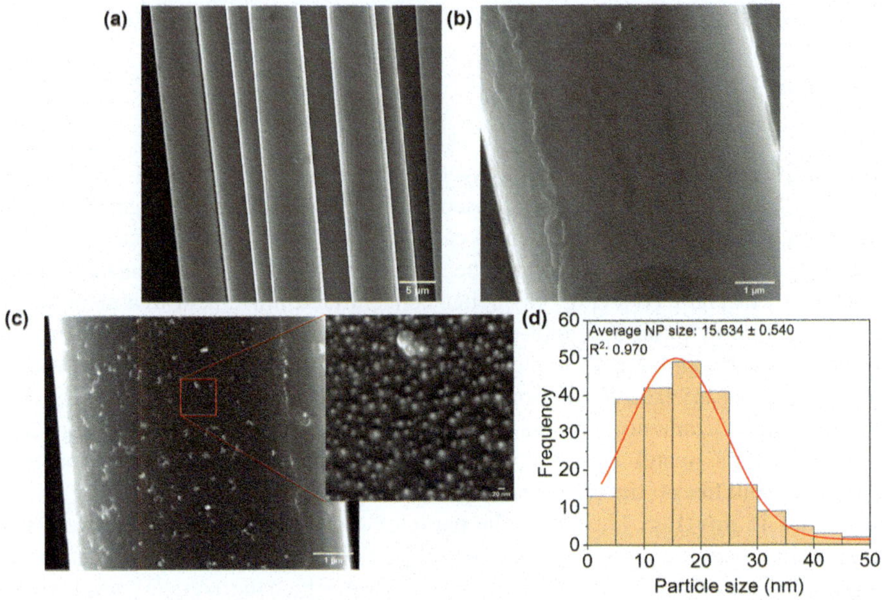

Fig. 1 **a** Bare as-received CC fibers, **b** acid-treated CC fiber after drop-casting with HEA stock solution, **c** drop-casted acid-treated CC fiber after laser irradiation (inset: magnified view of the surface of CC), and **d** HEA NP size distribution

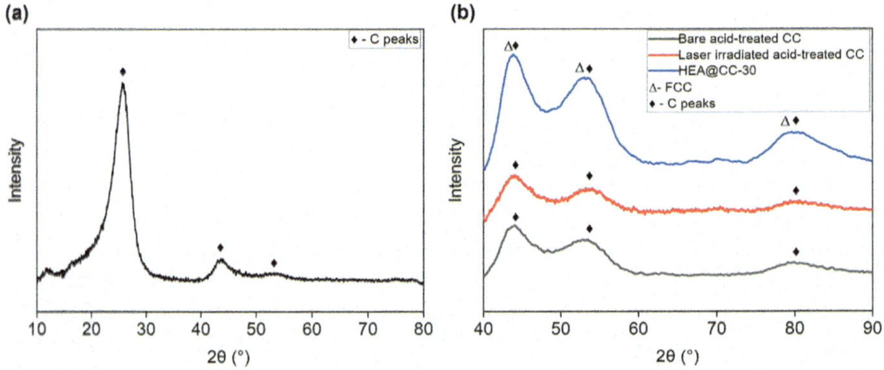

Fig. 2 XRD spectra of **a** bare acid-treated CC with carbon peaks demarcated, and **b** HEA@CC-30 when compared to bare CC and laser irradiated CC with both FCC peaks and carbon peaks demarcated

a further decrease in particle size [18, 24]. In the case of HEA@CC, the nitric acid treatment was essential in the fission of the respective larger-sized metal chloride particles, and the further fusion of these particles with other respective metallic particles into the small HEA particles observed in Fig. 1c.

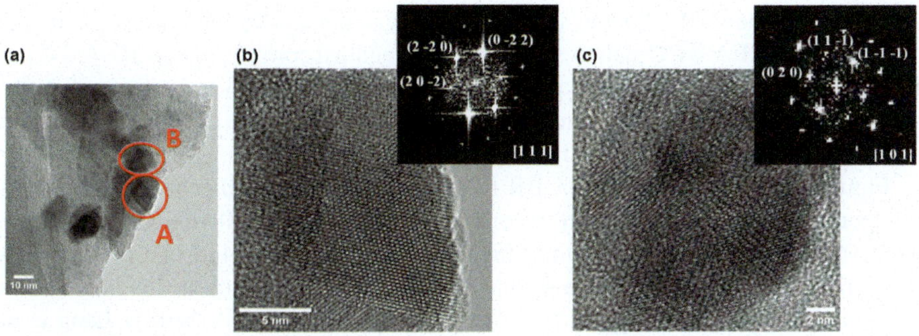

Fig. 3 HRTEM analysis of HEA@CC-30 **a** bright-field HRTEM image at lower magnification, **b** magnified image of the particle A showing the atomic lattices (inset: FFT image), and **c** magnified image of the particle B showing the atomic lattices (inset: FFT image)

3.2 HRTEM Analysis

The JEOL-2100F field emission high-resolution transmission electron microscope (HRTEM) was employed to image and analyze the HEA@CC-30 particles. Figure 3a shows a bright-field HRTEM image of HEA@CC-30 whereby circular particles are observed to be embedded onto carbon cloth remnants. The particles appear to be spherical with sizes of around 15 nm, which greatly corroborates with the information retrieved from SEM analysis. Two specific areas of well-oriented particles are highlighted in Fig. 3a as areas A and B, referring to Fig. 3b, c respectively. Figure 3b reveals that the particle A conforms to an FCC structure as evidenced by its Fast Fourier Transformation (FFT) image (inset in Fig. 3b) showing that the zone axis is [1 1 1], with the diffraction planes being {2 2 0}. This corroborates well with the results of the XRD analyses which shows a (2 2 0) peak at $2\theta = 79.6^\circ$. In Fig. 3c, the particle B similarly conforms to an FCC structure with its FFT image (inset in Fig. 3c) showing that the zone axis to be [1 0 1]. Three planes can be discerned from the FFT image, namely (1 1 − 1) or (− 1 − 1 1), (− 1 1 1) or (1 − 1 − 1), and (0 2 0) or (0 − 2 0). Essentially, these planes belong to the {1 1 1} and {2 0 0} family of planes respectively which supports strongly the XRD results which shows a (1 1 1) peak at $2\theta = 43.9^\circ$ and a (2 0 0) peak at $2\theta = 53.0^\circ$. In all, the HRTEM analysis adds weight to the evidence already provided by SEM and XRD analyses that CoCuFeMnNi HEA NPs were indeed successfully synthesized with the correct crystallographic orientations.

3.3 Electrocatalytic Performance Analysis

Electrochemical measurements were performed using an Autolab PGSTAT 302N electrochemical workstation with a standard three electrode configuration. Linear sweep voltammetry (LSV) was utilized to analyze the electrocatalytic performance of the varying HEA@CC samples benchmarked against conventional RuO_2@CC and IrO_2@CC. The HER and OER performance of all catalysts was performed in O_2-saturated 1.0 M KOH electrolyte using an AgCl/Ag reference electrode, a graphite counter electrode, and the working electrode being the sample.

Figure 4a, b prove that, for OER reactions, HEA@CC-30 required the lowest overpotential of 293 mV to produce a current density of 10 mA cm^{-2} whereas RuO_2@CC and IrO_2@CC required higher overpotentials of 410 mV and 299 mV respectively to produce the same current density. The variation of the loading volume can influence the electrochemical performance of NPs as described by Jiang et al. [25]. An appropriate amount of loading is essential to produce optimum performance in the OER reaction. A higher amount of loading likely would result in the increase in electron transfer resistances between the catalyst and substrate, thus resulting in lower overpotentials [25, 26]. This finding is supported by the results of the OER overpotentials of HEA@CC-200 to HEA@CC-30 where the decrease in loading volume produced lower overpotentials (Fig. 4b). However, for HEA@CC-20, the loading volume was likely too low to produce a better overpotential value as there is an absence of active sites created with such a low volume of drop-casting [25]. This alludes to the loading volume of 30 μL as the optimum loading volume for the system. Comparing the Tafel slope between the three systems in Fig. 4c, HEA@CC-30 displayed the strongest electrocatalytic reaction kinetics compared to the two other noble metal catalysts with a smaller Tafel slope of 55 mV dec^{-1} against 113 mV dec^{-1} for RuO_2@CC and 63 mV dec^{-1} for IrO_2@CC. This indicates that the HEA@CC can, with the increase in potential, stimulate the rapid increase in OER current density.

Electrochemical impedance spectroscopy (EIS) experiments provided further insights into the catalytic capabilities of the three systems. In Fig. 4d, the Nyquist plot is composed of two adjoining semicircles which describe two separate electrochemical processes occurring with the electrode during the OER process, in which the equivalent electrical circuit shown in Fig. 4e is employed to study these two processes. In Fig. 4e, R_s refers to the electrolyte solution resistance, R_{ct} refers to the charge transfer resistance while C_{dl} refers to the double layer capacitance. It is found that HEA@CC-30 displayed a low R_{ct} value of only 1.37 Ω while RuO_2@CC and IrO_2@CC displayed higher values of 11.37 Ω and 1.43 Ω respectively. This suggests that the charge transfer capabilities of HEA@CC far outweigh the capabilities of noble metal catalysts.

The OER experiments with HEA@CC demonstrate that with the contribution of nitric acid treatment to the CC substrate, the tuning of the amount of stock solution loading to the substrate, and the optimization of the stoichiometric ratios of the various elements, CoCuFeMnNi on CC is demonstrated to have performed as well as, or even better, than commercial noble metal catalysts. Furthermore, for the case of OER, this reaction involves the complicated and energy-intensive transfer of four electrons [27]. It is therefore necessary to introduce higher energy into the system in the form of higher overpotential values to induce a specific current density for the OER reaction which is the reason even noble metals such as RuO_2 and IrO_2 systems, as in this work, are shown to be less efficient than HEA NPs. To further

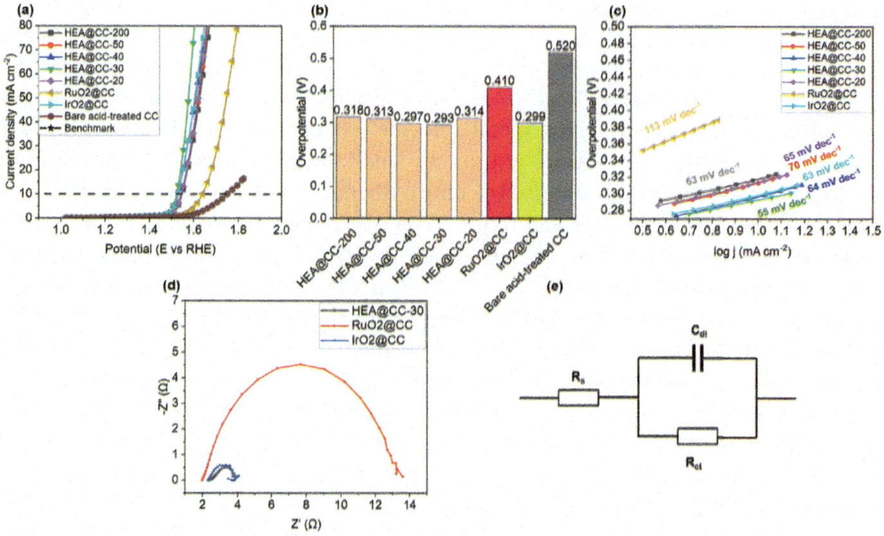

Fig. 4 Electrochemical performances of HEA@CC compared to RuO_2 and IrO_2. **a** OER LSV curves, **b** OER overpotentials to achieve current density of 10 mA cm^{-2}, **c** OER Tafel slopes, **d** EIS spectra, and **e** equivalent electrical circuit assumed during EIS testing

provide evidence that CoCuFeMnNi HEA NPs are as competitive as RuO_2 and IrO_2 catalysts reported in other established works, and even other novel nanocatalysts in recent work, the OER performance of HEA@CC was benchmarked against these studies in Table 3.

Table 3 List of nanocatalyst benchmarks

Catalyst	Solvent	OER overpotential, η_{10} (mV)	References
This work	1.0 M KOH	293	–
RuO_2	1.0 M NaOH	380 ± 20	[28]
IrO_2	1.0 M NaOH	380 ± 10	[28]
NiCoP microflowers	1.0 M KOH	314	[29]
FeNi/N-doped CNT	1.0 M KOH	300	[30]
FeCoNi	0.5 M KOH	400	[31]
Co_3O_4 nanosheets	1.0 M KOH	367	[32]
$MnCo_2O_4$ microspheres	0.1 M KOH	400	[33]

4 Conclusion

In this work, we report the synthesis of CoCuFeMnNi HEA NPs by way of laser irradiation of drop-casted mixed metal chloride precursors onto CC substrates. The HEA NPs are observed to have extremely small sizes of around 15 nm and are homogeneously coated throughout the entire fibers of the CC. These particles have a single-phase FCC structure and required an overpotential of 293 mV for OER in 1.0 M of KOH to achieve a current density of 10 mA cm^{-2}. The charge transfer resistance was also small at only 1.37 Ω. The Tafel slope produced is also only 55 mV dec^{-1} for the OER reaction. Compared to the overpotentials produced by RuO_2@CC, IrO_2@CC, and other similar benchmarks, the CoCuFeMnNi HEA NPs synthesized on CC performed comparably or even better than many existing catalysts in research today. In summary, CoCuFeMnNi HEA NPs have enormous potential as an alternative to existing, less sustainable noble metal catalysts. With the assistance of laser irradiation, many similar non-noble metal HEA catalysts can be synthesized in an extremely fast and cost-effective way.

References

1. Fu M, Ma X, Zhao K, Li X, Su D (2021) High-entropy materials for energy-related applications. iScience24:102177
2. Sun Y, Dai S (2021) High-entropy materials for catalysis: a new frontier. Sci Adv **7**:eabg1600
3. Miracle DB, Senkov ON (2017) A critical review of high entropy alloys and related concepts. Acta Mater 122:448–511
4. Huang X et al (2022) Noble-metal-based high-entropy-alloy nanoparticles for electrocatalysis. J Energy Chem 68:721–751
5. Ostovari Moghaddam A, Trofimov EA (2021) Toward expanding the realm of high entropy materials to platinum group metals: a review. J Alloy Compd 851:156838
6. Rho Y-J, Yoo YJ, Ryu W-H (2023) Research trends on minimizing the size of noble metal catalysts for Li-CO2 batteries: from nanoparticle to single atom. Korean J Chem Eng 40:461–472
7. lv Q et al (2022) Controlled direct electrodeposition of crystalline NiFe/amorphous NiFe-(oxy)hydroxide on NiMo alloy as a highly efficient bifunctional electrocatalyst for overall water splitting. Chem Eng J 446:137420
8. Qin X et al (2023) Magnetic field enhancing OER electrocatalysis of NiFe layered double hydroxide. Catal Lett 153:673–681
9. Chen Y-F et al (2023) Constructing robust NiFe LDHs–NiFe alloy gradient hybrid bifunctional catalyst for overall water splitting: one-step electrodeposition and surface reconstruction. Rare Met. https://doi.org/10.1007/s12598-022-02249-x
10. Chen D et al (2021) Developing nitrogen and Co/Fe/Ni multi-doped carbon nanotubes as high-performance bifunctional catalyst for rechargeable zinc-air battery. J Colloid Interface Sci 593:204–213
11. Chen D et al (2022) CoFeNi/N-codoped carbon nanotubes with small diameters derived from spherical Prussian blue analog as bifunctional oxygen electrocatalysts for rechargeable Zn-air batteries. J Alloy Compd 910:164964
12. Han C, Li W, Li W, Yang L, Huang Z (2023) CoFeNi based trifunctional electrocatalysts featuring in-situ formed heterostructure. Inorg Chem Commun 149:110402

13. Yao Y et al (2018) Carbothermal shock synthesis of high-entropy-alloy nanoparticles. Science 359:1489–1494
14. Xie P et al (2019) Highly efficient decomposition of ammonia using high-entropy alloy catalysts. Nat Commun 10:4011
15. Qiu H-J et al (2019) Noble metal-free nanoporous high-entropy alloys as highly efficient electrocatalysts for oxygen evolution reaction. ACS Mater Lett. 1:526–533
16. Fang G et al (2020) Multi-component nanoporous alloy/(oxy)hydroxide for bifunctional oxygen electrocatalysis and rechargeable Zn-air batteries. Appl Catal B 268:118431
17. Thamilselvan A, Nesaraj AS, Noel M, James EJ (2015) Effect of chemically treated/untreated carbon cloth: potential use as electrode materials in the capacitive deionization process of desalination of aqueous salt solution. J Electrochem Sci Technol 6:139–145
18. Mei T, Gao M, Wang Y, Huang Y, Chu PK (2022) Effects of acid treatment and plasma micromachining on the surface properties of carbon fibers. Appl Surf Sci 592:153261
19. Ungár T, Gubicza J, Ribárik G, Pantea C, Zerda TW (2002) Microstructure of carbon blacks determined by X-ray diffraction profile analysis. Carbon 40:929–937
20. Fiocchi J, Mostaed A, Coduri M, Tuissi A, Casati R (2022) Enhanced cryogenic and ambient temperature mechanical properties of CoCuFeMnNi high entropy alloy through controlled heat treatment. J Alloy Compd 910:164810
21. MacDonald BE et al (2019) Influence of phase decomposition on mechanical behavior of an equiatomic CoCuFeMnNi high entropy alloy. Acta Mater 181:25–35
22. Huang K et al (2020) Exploring the impact of atomic lattice deformation on oxygen evolution reactions based on a sub-5 nm pure face-centred cubic high-entropy alloy electrocatalyst. J Mater Chem A 8:11938–11947
23. Hoffman J et al (2014) The effect of laser wavelength on the ablation rate of carbon. Appl Phys A 117:395–400
24. Lu Y et al (2022) Atomically dispersed intrinsic hollow sites of M—M 1—M (M 1 = Pt, Ir; M = Fe Co, Ni, Cu, Pt, Ir) on FeCoNiCuPtIr nanocrystals enabling rapid water redox. Adv Funct Materials 32:2110645
25. Jiang W, Lehnert W, Shviro M (2023) The Influence of loadings and substrates on the performance of nickel-based catalysts for the oxygen evolution reaction. ChemElectroChem 10
26. Stevens MB, Trang CDM, Enman LJ, Deng J, Boettcher SW (2017) Reactive Fe-Sites in Ni/Fe (Oxy)hydroxide are responsible for exceptional oxygen electrocatalysis activity. J Am Chem Soc 139:11361–11364
27. Li H, Lai J, Li Z, Wang L (2021) Multi-Sites Electrocatalysis in High-Entropy Alloys. Adv Funct Materials 31:2106715
28. Jung S, McCrory CCL, Ferrer IM, Peters JC, Jaramillo TF (2016) Benchmarking nanoparticulate metal oxide electrocatalysts for the alkaline water oxidation reaction. J Mater Chem A 4:3068–3076
29. Chen T et al (2021) Nanosheet self-assembled NiCoP microflowers as efficient bifunctional catalysts (HER and OER) in alkaline medium. Int J Hydrogen Energy 46:29889–29895
30. Tao Z, Wang T, Wang X, Zheng J, Li X (2016) MOF-derived noble metal free catalysts for electrochemical water splitting. ACS Appl Mater Interfaces 8:35390–35397
31. Saha S, Ganguli AK (2017) FeCoNi alloy as noble metal-free electrocatalyst for oxygen evolution reaction (OER). ChemistrySelect 2:1630–1636
32. Zhang S et al (2021) Oxygen vacancy-based ultrathin Co3O4 nanosheets as a high-efficiency electrocatalyst for oxygen evolution reaction. Int J Hydrogen Energy 46:5286–5295
33. Zeng K et al (2021) Multilayer hollow MnCo2O4 microsphere with oxygen vacancies as efficient electrocatalyst for oxygen evolution reaction. Chem Eng J 421:127831

Facile Tuning of Black Nickel Electroplating Bath for Optimal Performance

Kee Seng Joseph Wong, Yajuan Sun, Yujie Zhou, and Yong Teck Tan

Abstract Black nickel plating is a decorative coating typically electrodeposited from an electroplating bath containing nickel and zinc cations. To ensure coating quality, chemical composition and pH value of plating bath should be closely monitored and maintained within the working ranges, especially when an inert anode is used. With plating by using an inert anode, both pH and bath composition will change and eventually drift out of the working range, leading to coating defects. In this paper, we propose a new methodology to tune the black nickel plating bath back to its usable range, by identifying the critical parameters and their limits, and give a recommended sequence for facile tuning to obtain optimal performance in terms of uniform black appearance. The recommended sequence of tuning is to first adjust the pH value, then the zinc concentration, and finally the nickel concentration when they reach their critical limits. The deposited black nickel coatings are characterized using sphere spectrophotometer and scanning electron microscopy equipped with energy dispersive X-ray spectroscopy where the ratio of deposited zinc to nickel is found to be critical for obtaining a uniform and consistent black appearance.

Keywords Coating · Electroplating · Black nickel plating · Bath analysis · Bath tuning

---

K. S. J. Wong · Y. Sun · Y. Zhou · Y. T. Tan (✉)
Singapore Institute of Manufacturing Technology (SIMTech), Agency for Science, Technology and Research (A*STAR), 2 Fusionopolis Way, Innovis #08-04, Singapore 138634, Republic of Singapore
e-mail: tanyongt@simtech.a-star.edu.sg

K. S. J. Wong · Y. Zhou · Y. T. Tan
Advanced Remanufacturing & Technology Centre (ARTC), Agency for Science, Technology and Research (A*STAR), 3 CleanTech Loop, #01-01 CleanTech Two, Singapore 637143, Republic of Singapore

N. Maharjan and W. He (eds.), *Proceedings of the 3rd International Conference on Advanced Surface Enhancement (INCASE) 2023*, Lecture Notes in Mechanical Engineering, https://doi.org/10.1007/978-981-99-8643-9_39

1 Introduction

There are various types of electrochemical baths employed for black nickel plating, such as thiocyanate-type bath containing ammonium nickel sulfate hexahydrate, zinc sulfate heptahydrate and sodium thiocyanate that produces NiS-ZnS deposits for a black coating appearance [1–8]. Tuning of a thiocyanate-type black nickel plating bath requires attentive replenishment especially when an inert anode is used, as zinc and nickel are plated out of the plating bath at different rates [9]. When electroplating using an inert anode, both pH value and bath composition of the black nickel plating bath will change and eventually drift out of the working range, leading to coating defects and/or undesired coating appearance. Bath analysis is required to check the bath composition on a regular basis and then do chemical additions to tune the electroplating bath back to within its working limits to ensure good coating quality, which can be a time-consuming process.

There are recommended procedures in industry for periodic testing of processing solutions, such as SAE ARP 4992 for aerospace industry, which recommends keeping to between 25 and 75% of the permissible concentration range, and a frequency of analysis that can be as frequent as just before each production load or before each working shift [10]. The bath maintenance process typically includes bath sample collection, chemical analysis, concentration limits verification followed by chemical addition/tuning. Due to the time required to execute all these steps, the test results when obtained may not reflect the actual conditions of the bath as the bath would have further changed between bath sample collection and the availability of the analysis results due to continuous bath usage.

In this work, we propose a new methodology to tune a boric acid-free black nickel electroplating bath back to its usable range, that has the advantage of being facile to implement, that reduces the number of tunings that need to be done and hence less time-consuming, by identifying the key parameters, their critical values and a logical sequence of tuning. This is especially important for this boric acid-free formulation, since boric acid acts as a pH buffer to keep the pH value stable and without it the electroplating bath drifts out of its working limits more readily.

In a previous work, we identified a critical pH value (i.e. pH 4) below which good quality black nickel coating cannot be achieved even when zinc and nickel concentrations are relatively high when using the formulation given in MIL-P-18317 and also observed that zinc is deposited at a much faster rate compared to nickel [9]. We further observed that the sulfur content in the deposited coating is relatively constant at around 14 wt.% (considering only Zn, Ni, S and O for the composition of the deposited coating) independent of bath pH variations from pH 6 to pH 3 [9].

Therefore, we hypothesize that a bath maintenance routine according to the following sequence may be helpful in reducing the number of tunings required by focusing on the key parameters in the following sequence, starting from the most critical: bath pH, zinc concentration, and nickel concentration. The thiocyanate concentration is proposed to be tuned during the pH adjustment step as needed. This work is to test out the above hypothesis and to identify the critical values for the key parameters of zinc concentration and nickel concentration.

2 Methodology

2.1 *Preparation of Copper Panels for Electroplating*

For the black nickel plating experiments, copper panels with dimensions of 100 × 65 × 3 mm were used as cathode and round platinized titanium mesh as anode. The exposed cathode area was controlled to be between 0.90 dm^2 and 0.95 dm^2.

Before electroplating, each panel was first degreased in an alkaline cleaning solution at 80 °C for 10 min. Then the copper panel was activated in 10% sulfuric acid solution to remove any oxide film to avoid non-uniform electroplating. Water break test was done prior to immersion into the electroplating bath to check that the copper panel surface was prepared properly.

2.2 *Black Nickel Electroplating*

Table 1 shows the bath composition of the black nickel electroplating formulation used in this work, which follows that given in Military Specification MIL-P-18317 [8]. The electroplating process parameters used are given in Table 2. A pH value of 6.0 was used for all the electroplating experiments, by adjusting with 28% ammonia solution as needed. The current density used was 0.15 A/dm^2, and the applied current was based on the sample's exposed area.

Table 1 Black nickel plating bath formulation used in this work [8]

Bath constituents	Quantity (g/L)
Ammonium nickel (II) sulphate hexahydrate ($(NH_4)_2SO_4.NiSO_4.6H_2O$)	60
Zinc sulphate heptahydrate ($ZnSO_4.7H_2O$)	7.5
Sodium thiocyanate (NaSCN)	15
Deionized water	Balance

Table 2 Black nickel plating process parameters

Process Parameters	Operating condition
Current density	0.15 A/dm^2
Electroplating duration	20 min
Temperature	Ambient temperature (~ 22 °C)
Bath pH	6.0

2.3 Characterization of the Black Nickel Coating

After electroplating, the panels were dried and visually inspected to check for any coating defects such as localized discoloration and silvery (instead of black) appearance. Next, the surface appearance of the black nickel coating was characterized in a quantifiable manner by using a handheld spectrophotometer (X-rite Ci60) to identify variations in colour and gloss with three coordinates of L*A*B*, F2/10 and SPIN. Lightness (L) value of 40 and below corresponds to a visually black appearance.

The black nickel coating samples were imaged with JEOL IT300LV scanning electron microscope (SEM) at high magnifications with an accelerating voltage of 20 kV and the chemical composition of the coating was characterized by energy dispersive X-ray spectroscopy (EDX) using an Oxford Instruments X-MaxN EDX detector attached to the SEM machine. The ratio of deposited zinc to nickel atoms was calculated using the atomic percentage of the coating composition from the EDX results.

3 Results and Discussion

In an earlier work, a critical pH value of 4 and above was identified as necessary to achieve a good black nickel coating using the formulation investigated in this work [9]. Furthermore, the sulfur content in the deposited coating was observed to be relatively constant at around 14 wt.% independent of bath pH variations from pH 6 to pH 3 [9]. The amount of thiocyanate in the bath should not be a major limiting factor since the amount of sulfur from thiocyanate in the bath is around 3.5 times the amount of zinc in the bath by mass while the amount of sulfur in a good black nickel coating is only around 0.3 times the deposited zinc by mass. This means that if the bath pH is tuned back to pH 6 before reaching pH 4, and during this step we also tune the thiocyanate concentration back to its original level, a good black nickel coating should be deposited.

Based on our hypothesis, the next step was to identify the critical zinc concentration and we identified this critical value by using a plating bath at pH 6 made up with 90% of the recommended nickel concentration and 10% of the recommended zinc concentration from the formulation of MIL-P-18317. After each cycle of electroplating, the zinc concentration was increased by 5% until the zinc concentration was

90% of the recommended zinc concentration. Figure 1 shows that as the zinc concentration increased, the lightness of coated panels decreased gradually until reaching a transition zone between 50 and 65% zinc concentration in which there is a sharp drop in the lightness value i.e. the coating appearance changed from silvery to black. Optical images of coatings with different lightness values are shown in Fig. 2 for comparison.

Using EDX to study the ratio of deposited zinc to nickel atoms on the coated samples, as the L value decreased to below 40, the ratio of deposited zinc to nickel increased to more than 2. The critical value of zinc concentration was identified to be 65% of the recommended bath formulation to achieve L value ≤ 40 and deposited zinc to nickel ratio value ≥ 2.00.

To identify the critical nickel concentration, a plating bath made up of 90% of the recommended zinc concentration and 10% of the recommended nickel concentration from the formulation of MIL-P-18317 was used. After each cycle of electroplating, the nickel concentration was increased by 10% to the bath in each subsequent plating

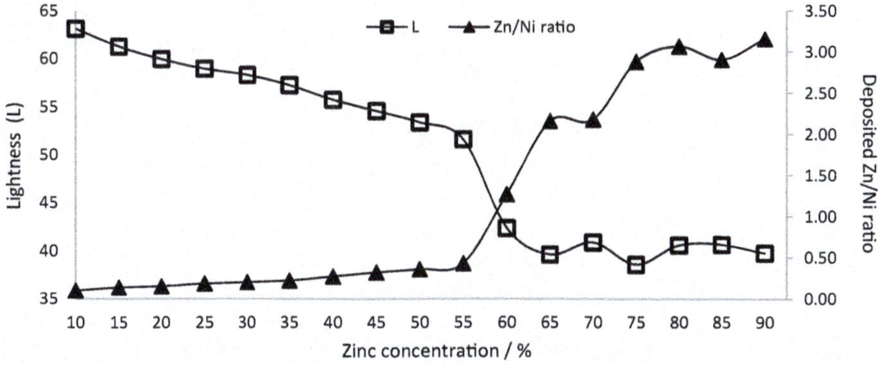

Fig. 1 Lightness value of coating (left axis) and deposited zinc to nickel ratio (right axis) against zinc concentration (as a percentage of the recommended amount in MIL-P-18317)

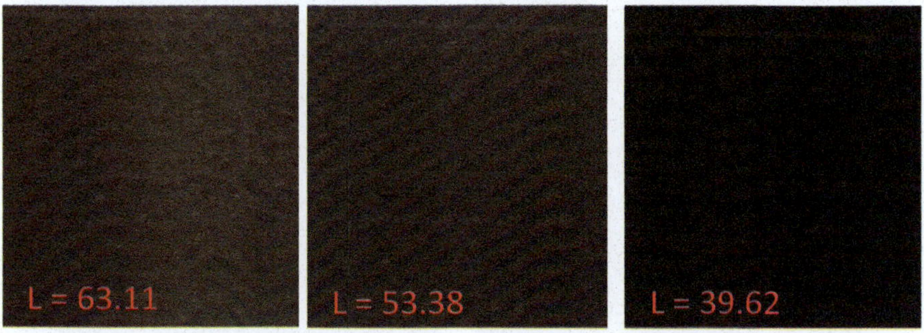

Fig. 2 Optical images of coatings with different lightness value. A lightness value lower than 40 corresponds to a visually black surface

until it is 90% of the recommended nickel concentration. In Fig. 3, the first four electroplating experiments (curves in red) resulted in uneven and powdery coatings as shown in Fig. 4. To achieve uniform non-defective black nickel coating, the critical value of nickel concentration was identified to be 50% of the recommended nickel concentration given in MIL-P-18317.

After identifying the critical zinc and nickel concentration values, plus having the critical value of pH from the earlier work, we validated that a black nickel plating bath at the critical limit for both zinc and nickel concentrations can give good black nickel coating. A plating bath at pH 6 with concentration of 65% of the recommended zinc concentration and 50% of the recommended nickel concentration was used to electroplate a copper panel and the resultant uniform black nickel coating achieved an L value of 35 and deposited zinc to nickel ratio more than 2. Figure 5 shows the

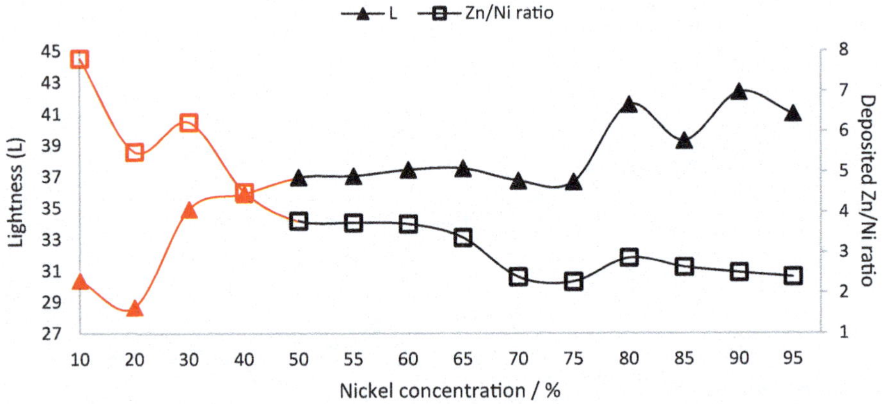

Fig. 3 Lightness value of coating (left axis) and deposited zinc to nickel ratio (right axis) against nickel concentration (as a percentage of the recommended amount in MIL-P-18317)

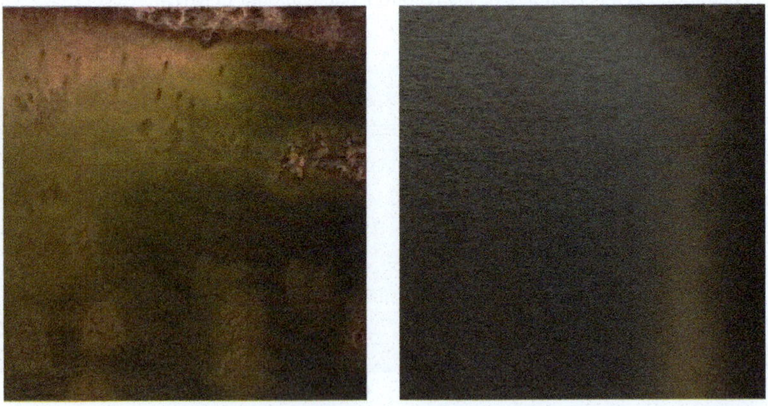

Fig. 4 Visual images of uneven and powdery black nickel coatings (i.e. defective coatings)

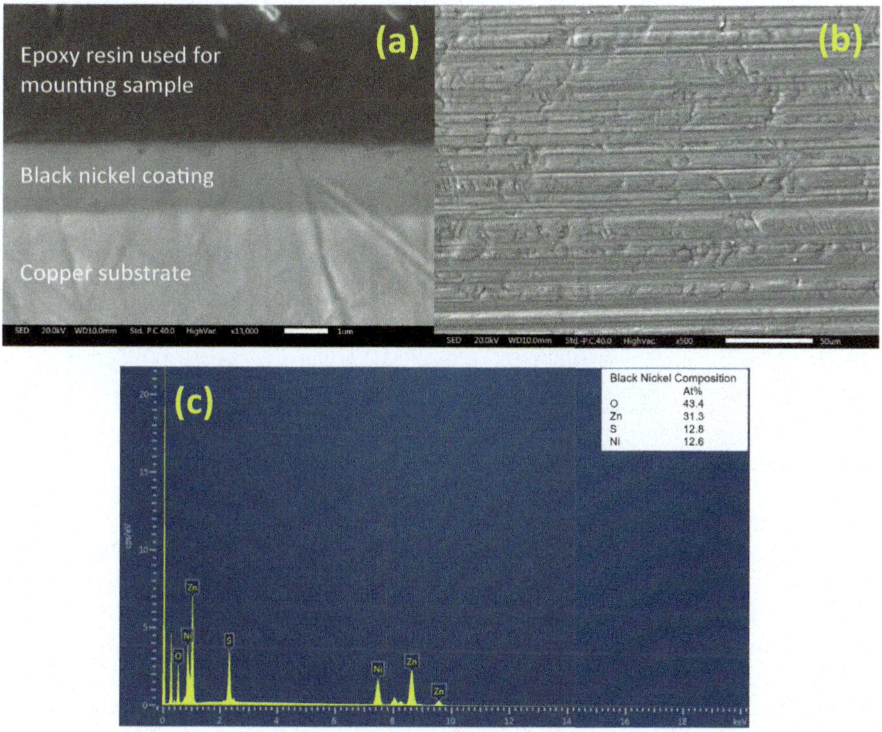

Fig. 5 SEM images of **a** cross-section and **b** surface topography of black nickel coating electroplated on copper substrate, and **c** EDX result of the elemental composition of a good black nickel coating

SEM images of the cross-sectional view, the surface topography and the EDX results for the elemental composition of a good black nickel coating.

Therefore, in this work we proposed and validated a facile tuning methodology for the black nickel plating bath based on the formulation of MIL-P-18317 as shown in the process flowchart in Fig. 6. Firstly, the bath pH should be tuned back to pH 6 before it reaches pH 4. This is the first step as the pH drop results in silvery coating instead of black coating, even when the zinc and nickel concentrations are still quite high. In addition, pH monitoring can be done relatively fast and conveniently when using a pH electrode sensor. The thiocyanate concentration is tuned during this pH adjustment step as required. After the pH value and thiocyanate concentration are adjusted and maintained at the appropriate level, the zinc concentration is the second key parameter to be tuned before it reaches its critical value (65%). This is because the zinc concentration depletes faster than the nickel concentration; for every 1 nickel atom deposited on the substrate, more than 2 zinc atoms are deposited. Furthermore, the starting zinc concentration is lower than the nickel concentration. Finally, the nickel concentration is tuned before its concentration depletes to its critical value (50%) due to continuous usage.

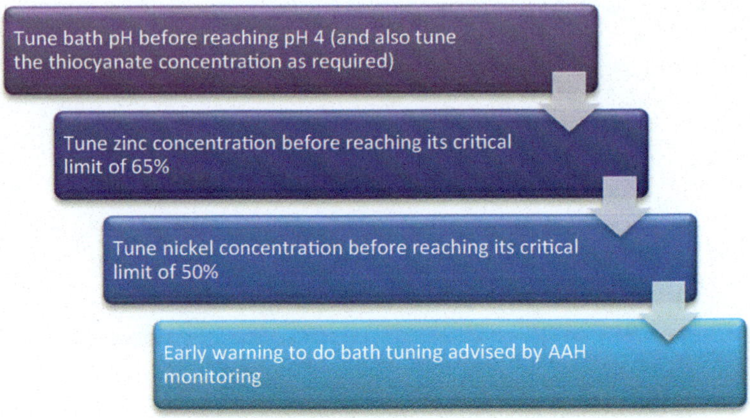

Fig. 6 Process flowchart for facile tuning of black nickel electroplating bath

The zinc and nickel concentration remaining in the tank can be estimated using Faraday's Law, by monitoring the Accumulated Ampere Hour (AAH), having an empirical value for the cathode efficiency and by assuming a ratio of 2:1 zinc atoms deposited to nickel atoms deposited. The bath can then be analyzed for its zinc and nickel concentrations, only required at certain timings (with a buffer time to account for assumptions made on the cathode efficiency and ratio of zinc and nickel deposited) advised by the AAH estimations and tuning actions can be taken before any key parameter falls below their critical value. In this way the key parameters are monitored for early warning to do bath tuning, and productivity can remain high as the plating bath is maintained before going out of its working range.

4 Conclusions

Chemical analysis and maintenance of electroplating baths can be time-consuming and lagging behind plating operations, which can lead to lower productivity and more rework. In this work, a facile tuning methodology for a black nickel electroplating bath was proposed and validated which gives a recommended sequence of tuning and the critical values for key parameters. The key parameters have been identified in sequence as pH (to be more than pH 4), zinc concentration (to be more than 65% of the recommended amount), nickel concentration (to be more than 50% of the recommended amount). The thiocyanate concentration can be tuned during the pH adjustment step as required. Monitoring of Accumulated Ampere Hour can help provide estimations of timings when the critical values will be reached as an early warning sign so that bath tuning can be done before it drifts out of its working range. This will ensure good black nickel coating with lightness value of less than 40 and a deposited zinc to nickel ratio more than 2.

Acknowledgements This work is supported by A*STAR Career Development Fund project—"Continuous Process Optimization and Prescriptive Quality Control for Electrochemical Process Using Machine Learning Approach" (Project No. C210112027).

References

1. Somasundaram S, Pillai AM, Rajendra A, Sharma AK (2015) High emittance black nickel coating on copper substrate for space applications. J Alloys Compd 643:263–269
2. Shalini VM, Arockiasamy P, Rani RU, Sharma AK (2012) Flat absorber phosphorus black nickel coatings for space applications. J Mater Sci Technol 28(2):118–124
3. Kafle BP, Basnet B, Timalsina B, Deo A, Malla TN, Acharya N, Adhikari A (2022) Optical, structural, and thermal performances of black nickel selective coatings for solar thermal collectors. Sol Energy 234:262–274
4. Mennucci M, Montes R, Bastos A, Monteiro A, Oliveira P, Tedim J, Ferreira M (2021) Nanostructured black nickel coating as replacement for black Cr(VI) finish. Appl Sci 11
5. Estrella-Gutierrez MA, Lizama-Tzec FI, Arés-Muzio O, Oskam G (2016) Influence of a metallic nickel interlayer on the performance of solar absorber coatings based on black nickel electrodeposited onto copper. Electrochim Acta 213:460–468
6. Takadoum J (2010) Black coatings: a review. Eur Phys J 52(3)
7. Lizama-Tzec FI, Herrera-Zamora DM, Arés-Muzio O, Gómez-Espinoza VH, Santos-González I, Cetina-Dorantes M, Vega-Poot AG, García-Valladares O, Oskam G (2019) Electrodeposition of selective coatings based on black nickel for flat-plate solar water heaters. Sol Energy 194:302–310
8. Military Specification MIL-P-18317: Plating, Black Nickel (Electrodeposited) on Brass, Bronze, or Steel (1954)
9. Teo ZHA, Sun Y, Tan YT (2022) Electroplating of black nickel: bath age and its effect on coating quality. Int Conf Asian Soc Precision Eng Nanotechnol 312–315
10. SAE ARP4992, Revision B: Periodic test for processing solutions. SAE International

High Phosphorus Nickel Alloy Deposition by Brush Plating Process

Yujie Zhou and Zhaohong Huang

Abstract Nickel (Ni) alloy coating with high phosphorus (P), P > 10wt.%, is usually deposited by electroless and electrolytic plating processes and has been widely used for corrosion and wear resistance applications. However, the deposition rate is low and plating tanks with temperature control are required, and not viable for on-site coating and coating repair. The tank-free brush plating is usually controlled by constant voltage and featured with high deposition rate due to high current density (>100 A/dm^2), while the current density of conventional tank plating is usually lower than 10 A/dm^2. However, the high current density of brush plating leads to a challenge to deposit a high phosphorus (Hi-P) Ni alloy, as the P content in the nickel phosphorus (NiP) alloy coating is inversely proportional to the operating current density based on the tank plating practices. In this paper, we report a brush plating process to deposit high phosphorous nickel thin film on carbon steel coupons. Design of experiments (DOE) was used to optimize the brush plating process, including plating bath formation, and plating parameters. It is found that low pH value and high phosphorous acid (H_3PO_3) concentration in the plating bath are beneficial for high P content in the NiP deposit. High pH value and low 1-hydroxyethylidene-1,1-diphosphonic acid (HEDP) concentration are beneficial for high current efficiency. By using the optimized bath chemistry and plating parameters, NiP coating with P content of 13.7 wt% was deposited on carbon steel coupons with a current efficiency as high as 19.4%.

Keywords Brush plating · Nickel alloy plating · High phosphorus nickel plating · Nickel bath formulation · High current density plating

---

Y. Zhou (✉) · Z. Huang
Singapore Institute of Manufacturing Technology (SIMTech), Agency for Science, Technology and Research (A*STAR), 2 Fusionopolis Way, Innovis #08-04, Singapore 138634, Singapore
e-mail: yjzhou@simtech.a-star.edu.sg

Advanced Remanufacturing and Technology Centre (ARTC), Agency for Science, Technology and Research (A*STAR), 3 Cleantech Loop, #01-01, Cleantech Two, Singapore 637143, Singapore

Z. Huang
e-mail: zhhuang@simtech.a-star.edu.sg

N. Maharjan and W. He (eds.), *Proceedings of the 3rd International Conference on Advanced Surface Enhancement (INCASE) 2023*, Lecture Notes in Mechanical Engineering, https://doi.org/10.1007/978-981-99-8643-9_40

1 Introduction

NiP alloy coating has found its applications in various industry sectors such as aerospace, oil & gas (O&G) and marine & offshore (M&O) due to the excellent performance in wear and corrosion resistance. The physical and chemical properties of NiP alloy coatings are significantly influenced by the phosphorus content in the coating and the post heat treatment. Based on the P content, NiP alloy coatings are generally grouped into three categories, namely low phosphorus (LP), medium phosphorus (MP), and high phosphorus (HP) with P content of 2–5, 6–9 and 10–13 wt%, respectively [1]. As a general guideline, LP coating shows high hardness and better wear resistance. With increasing P content, the hardness of NiP coating reduces, but the corrosion resistance drastically increases.

To deposit NiP alloy coatings, both electroless and electrolytic plating processes are widely used. Since developed by Brenner and Riddell [2], the electroless plating process is well known for the excellent corrosion and wear resistance of the resulted coating, in particular, it is comparable with that of hard chromium in hardness after heat treatment [3]. However, electroless NiP alloy coating processes suffer many drawbacks, such as an expensive reducing agent required, a high operating temperature, and more bath control than chromium plating. Furthermore, the deposition rate of electroless plating is very low. To overcome these difficulties, Brenner et al. [4] proposed electrolytic deposition of NiP coating.

Compared with electroless nickel (EN) plating, electrolytic approach for NiP alloy coating has attracted more attentions recently because of its easy maintainability, high deposition rate, low operation cost, as well as a possibility of changing the properties and adapting them to many applications.

However, like EN coating process, the tank-based electrolytic deposition process is also not viable for plating on large work pieces such as those in M&O and O&G industries. To address this challenge, brush plating has been proposed, which also known as selective plating, or swab plating, is a very useful and portable method of contact plating [5–8].

In its simplest form, brush plating process resembles painting, as shown in Fig. 1. Brush plating equipment includes power packs, solutions, plating tools, anode covers, and auxiliary equipment. The power pack has two leads. One is connected to the plating tool and the other is connected to the workpiece to be plated. The anode is covered with an absorbent material which holds the solution. The operator dips the plating tool in the solution and then brushes it against the surface of the workpiece that is to be finished. When the anode touches the work surface a circuit is formed and an electrodeposit is produced. Plating occurs only where the anode contacts the workpiece. During the brush plating process, the plating tool is always kept in motion whenever it is in contact with the work surface. The movement will apply even plating on the entire area being plated and ensure a quality finish.

The tank-free brush plating is usually controlled by constant voltage and featured with high deposition rate due to high current density (>100 A/dm^2), while the current density of conventional tank plating is usually lower than 10 A/dm^2. The high current

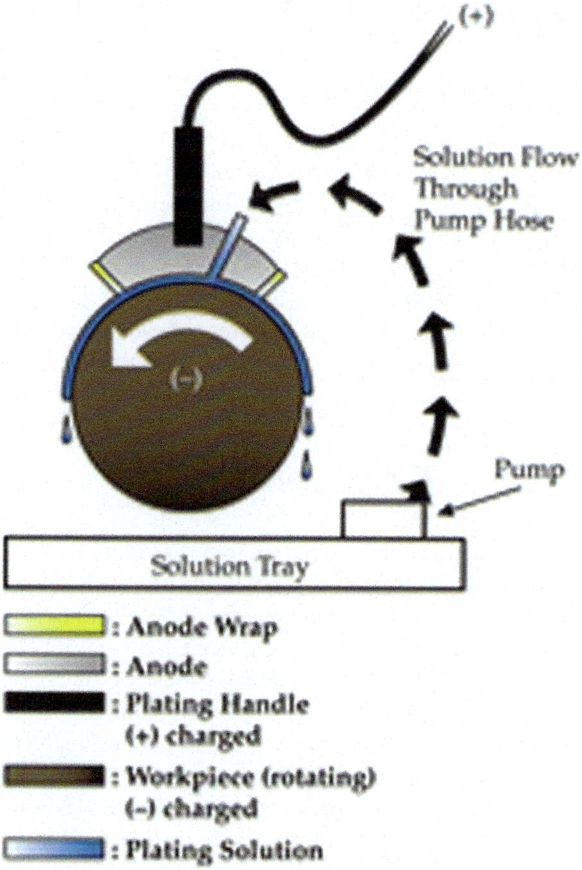

Fig. 1 Illustration of brush plating on outer surface of round work piece [8]

density of brush plating leads to a high deposition rate, but also results in a challenge to deposit a high phosphorus (>10 wt.%) nickel alloy, as the phosphorus content in the NiP alloy coating is inversely proportional to the operating current density based on the tank plating practices. In this paper, we report a brush plating process, including bath chemistry, and plating parameters, such that uniform NiP coating with a higher P content more than 10 wt.% can be deposited with a current efficiency of about 20%.

2 Experimental

2.1 Methodology

The formulation of plating bath and the operating voltage are identified as the critical factors which have significant influence on the P content in NiP coating, in particular, with high current efficiency. At the first, the compounds for making the plating solution were chosen basing on the demands, experiences and references. Secondly, orthogonal experiment was used to determine the optimized content of each chemical in the plating solution and the effect of each ingredient was investigated. Thirdly, the operating voltage was varied to obtain higher P content and current efficiency.

2.2 Bath Formulation

The chemical bath formulation for brush plating of high phosphorus nickel alloy was carried out based on nickel sulfate and phosphorous acid with complexing agents. For example, nickel sulfate hexahydrate ($NiSO_4.6H_2O$) was chosen as the source of Ni in brush plating. Nickel chloride ($NiCl_2$) was not being considered, as Cl^- is a well-known element causing the tensile stress of the coating. Phosphorous acid (H_3PO_3) was chosen as the source of P in brush plating. Phosphoric acid (H_3PO_4) was chosen as a pH buffer rather than a source of P because it was too difficult to be reduced into P under the plating condition applied in this study. A further and important role of the H_3PO_4 is to suppress the oxidation of H_3PO_3 on the anode and weaken the reduction ability of Ni ions on cathode to benefit increasing P content in the deposit. It also can act partly as complexing agent. Hypophosphorous acid (H_3PO_2) was not being considered, because it is relatively unstable in brush plating. An organic phosphorus compound, HEDP was chosen as buffering and complexing agent as well, because it has 4 hydrolysable H^+ and strong complexing due to the chelation. Boric acid (H_3BO_3) was not being considered, because it is prone to precipitation at room temperature. So, the chosen ingredients were $NiSO_4.6H_2O$, H_3PO_4, H_3PO_3 and HEDP. The pH value was adjusted by potassium hydroxide (KOH) and sulfuric acid (H_2SO_4).

2.3 Design of Experiments (DOE)

In order to make the design of orthogonal experiments more accurate, some preliminary experiments were conducted in the basis of experience and references to test the boundary of the bath ingredient concentration and pH value (refer to Table 1).

After that, orthogonal DOE (refer to Table 2) was designed with five factors and four levels based on the results of the above preliminary experiments. These

Table 1 The formulation of plating baths and plating test results at plating voltage of 12 V

Exp. No.	$NiSO_4.6H_2O$ (g/L)	H_3PO_4 (g/L)	H_3PO_3 (g/L)	HEDP (g/L)	pH	P content (wt%)	Current efficiency (%)
1	200	10	20	30	0.91	7.7	21.8
2	200	10	50	0	0.96	8.5	10
3	200	10	50	20	1.35	6.6	19.8
4	200	10	60	0	1.50	5.6	12.5
5	200	50	60	0	1.35	10.3	15.1
6	200	50	40	0	1.50	7.3	17.1
7	200	50	60	0	1.50	7.4	17.9
8	200	50	80	0	1.50	9.9	17.4
9	200	50	100	0	1.50	11	7.6

Table 2 Design of orthogonal experiments with five factors and four levels

Level / Factors	1	2	3	4
$NiSO_4.6H_2O$ (g/L)	100	150	200	250
H_3PO_4 (g/L)	10	30	50	70
H_3PO_3 (g/L)	30	50	70	90
HEDP (g/L)	0	20	40	60
pH	1	1.3	1.6	2

experiments were expected to find out the optimized content of each ingredient in the plating solution and the pH value for higher P content in the coating with higher current efficiency. The optimized bath formulation was further tested at different plating voltages.

2.4 *Materials and Experimental Methods*

Carbon steel was used as the substrate material of testing coupons in this work. The electrolytes for surface cleaning and activating were obtained from a chemical supplier (Sterling Impreglon Asia Pte Ltd). The samples were electro-cleaned by 15 V for 0.043 Ah, and then activated using activator-2 under -10 V for 0.01 Ah and activator-3 under -15 V for 0.02 Ah. Brush plating was conducted under 12 V for 0.5 Ah. The P content was detected by EDS. The current efficiency was calculated based on the measurement of sample mass change.

3 Results and Discussion

Preliminary experiments with the formulations of different compositions and pH value, and resulted P contents in NiP coating, as well as current efficiency were summarized in Table 1.

The experiments 1–5 were conducted based on literature, and the results do not meet the expectation of the project, i.e., higher P content than 10 wt% and current efficiency of about 20%. H_3PO_3 was the source of P in brush plating. The experiments 6–9 were used to test the effect of H_3PO_3 concentration on P content and current efficiency. The results are shown in Fig. 2.

As shown in Fig. 2, with the increase of H_3PO_3 concentration, the P content in NiP coating increases to 11 wt%, and the current efficiency decreases down to 7.6%. As described above, H_3PO_3 was the only source of phosphorus for NiP deposit in this study, therefore, the increased H_3PO_3 concentration would promote the deposition of phosphorus. On the other hand, H_3PO_3 is a weak complexing agent of Ni^{2+} due to the existence of P-O- species. The increase of H_3PO_3 in the plating bath would suppress the deposition of Ni, and therefore, lower down the current efficiency. The results show that an optimization of the bath formulation is required. In this study, DOE (orthogonal experiments) was used for this purpose, as shown in Table 2.

Furthermore, the H_3PO_4 was used to suppress the oxidation of H_3PO_3. HEDP was used as pH buffering agent and complexing agent of Ni^{2+}. It is noted that low pH value and high H_3PO_3 are beneficial for high P content. On the contrast, high pH value and low HEDP were beneficial for high current efficiency. The design of orthogonal experiments, DOE results and scores are shown in Tables 2, 3, and 4, respectively.

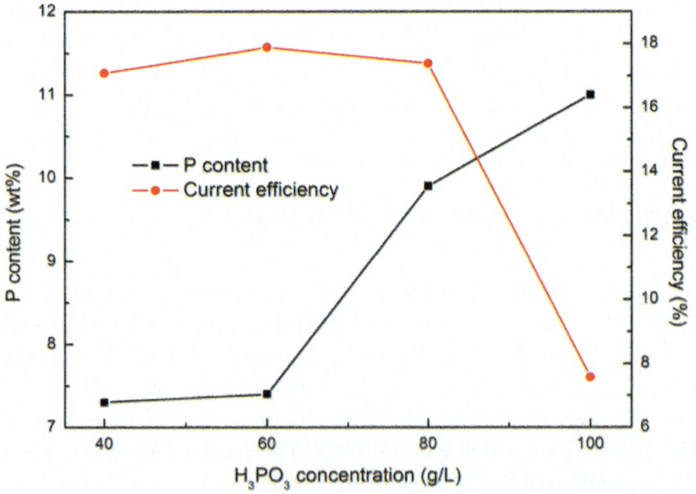

Fig. 2 The effect of H_3PO_3 concentration on P content and current efficiency

Table 3 Testing results of orthogonal experiments

Level / Exp. No.	$NiSO_4 \cdot 6H_2O$	H_3PO_4	H_3PO_3	HEDP	pH	P content (wt%)	Current efficiency (%)	Score
1	1	1	1	1	1	7.0	18.8	0 + 2 (2)
2	1	2	2	2	2	7.8	13.7	0 + 1 (1)
3	1	3	3	3	3	13.0	4.5	3 + 0 (3)
4	1	4	4	4	4	15.9	-2.4	3 + 0 (3)
5	2	1	2	3	4	11.2	19.9	2 + 2 (4)
6	2	2	1	4	3	10.2	8.7	1 + 0 (1)
7	2	3	4	1	2	15.5	2.3	3 + 0 (3)
8	2	4	3	2	1	16.4	-7.7	3 + 0 (3)
9	3	1	3	4	2	12.4	9.6	3 + 0 (3)
10	3	2	4	3	1	15.3	-16.5	3 + 0 (3)
11	3	3	1	2	4	5.8	18.0	0 + 2 (2)
12	3	4	2	1	3	9.3	21.6	0 + 3 (3)
13	4	1	4	2	3	10.7	23.5	1 + 3 (4)
14	4	2	3	1	4	10.5	31.8	1 + 3 (4)
15	4	3	2	4	1	9.4	-2.2	0 + 0 (0)
16	4	4	1	3	2	7.4	22.0	0 + 3 (3)

Table 4 The score of orthogonal experiments

P content (wt.%)	< 10	10–11	11–12	> 12
Score	0	1	2	3
Current efficiency (%)	<10	10–15	15–20	> 20
Score	0	1	2	3

According to the results given in Table 3, the pH value and H_3PO_3 concentration have greater impact on P content and current efficiency. Low pH value and high H_3PO_3 concentration were beneficial for high P content. High pH value and low HEDP concentration were beneficial for high current efficiency. Based on the score of orthogonal experiments given in Table 4, three optimal bath formulations were selected and shown in Table 5, which meet the requirements for phosphorous content (>10 wt%) and current efficiency (>19%).

Table 5 shows that HEDP may not be a necessary ingredient in the plating bath. For example, high phosphorus (10.5 wt%) and high efficiency (31.8%) was obtained simultaneously from bath formula 14, which contains zero HEDP. It should be noted that this chemical bath contains more H_3PO_4 than other two, which implies that HEDP and H_3PO_4 can be exchangeable to some extent. Further study is needed

Table 5 The optimized formula of orthogonal experiments

Exp. No.	$NiSO_4.6H_2O$ (g/L)	H_3PO_4 (g/L)	H_3PO_3 (g/L)	HEDP	pH	P content (wt%)	Current efficiency (%)
5	150	10	50	40	2	11.2	19.9
13	250	10	90	20	1.6	10.7	23.5
14	250	30	70	0	2	10.5	31.8

to understand the effects of the two ingredients on the current efficiency and the phosphorus content in the resulted coating.

Repeatability test for the optimal bath formulations were carried out, and the testing results are given in Table 6. Compared with Table 5, it is found that the repeatability is very high. All the three bath formulations can be used to deposit NiP coating with high phosphorus and high current efficiency. However, the max. phosphorus obtained with a current efficiency of about 20% is still less than 12 wt%. To further increase the phosphorus content in NiP coating, the effect of plating voltage was investigated. With decreasing the operating voltage from 12 to 10 V both P content and current efficiency were changed, as shown in Table 7. It is found that by using bath formula 14 and plating with 10 V, NiP coating containing phosphorus of 13.7 wt.% was deposited with a current efficiency of 19.4%. It is obvious that the decreased plating voltage significantly increases the P content of NiP coating, with a loss of current efficiency.

Table 6 The repeatability evaluation of some tests in orthogonal experiments

Exp. No	$NiSO_4.6H_2O$ (g/L)	H_3PO_4 (g/L)	H_3PO_3 (g/L)	HEDP (g/L)	pH	P content (wt%)	Current efficiency (%)
5	150	10	50	40	2	11.6	19.7
13	250	10	90	20	1.6	10.5	19.7
14	250	30	70	0	2	11.0	25.2

Table 7 The result for good formula at 10V

Exp. No.	$NiSO_4$ $6H_2O$ (g/L)	H_3PO_4 (g/L)	H_3PO_3 (g/L)	HEDP (g/L)	pH	P content (wt%)	Current efficiency (%)
5	150	10	50	40	2	8.6	12.8
13	250	10	90	20	1.6	15.1	13.9
14	250	30	70	0	2	13.7	19.4

4 Conclusion

DOE was used to develop brush plating process, including plating bath formulation and plating parameters. More specifically, by varying pH value from 1 to 2, the bath chemistry was optimized by using DOE with five factors and four levels, and the effect of each ingredient was discussed. The effect of operating voltage (10 V and 12 V) on P content was also studied. It is found that low pH value and high H_3PO_3 concentration in the plating bath are beneficial for high P content in the NiP deposit. High pH value and low HEDP concentration are beneficial for high current efficiency. By using the optimized bath chemistry and plating parameters, NiP coating with P content of 13.7 wt% was deposited on carbon steel coupons with a current efficiency as high as 19.4%.

Acknowledgements We acknowledge intern student Haiyang Nan in brush plating experimental work. This work is supported by A*STAR (Project No. C14-S-027).

References

1. Parkinson R Properties, and applications of electroless nickel. Nickel development institute. Retrieved from https://www.nickelinstitute.org/media/1769/propertiesandapplicationsofelectrolessnickel_10081_.pdf
2. Brenner A, Riddell GE (1946) Nickel plating on steel by chemical reduction. J Res Natl Bur Stand 37:31–34
3. Ruff AW, Lashmore DS (1982). In: Baye RG (ed) Proceeding 62nd ASTM symposium, December 1981. American society for testing and materials, Philadelphia, PA, pp 134–156
4. Brenner A, Couch DE, Williams EK (1950) Electrodeposition. of phosphorous with nickel or cobalt. J Res Natl Bur Stand 44:109
5. Gabel MK, Donovan DM (1980) Wear resistant coatings and treatments. In: Peterson MB, Winer WO (eds) Wear control handbook, ASME, pp 343
6. Norris JC (1995) Brush plating. Metal Finish 93(1):349–361
7. Vanek DL (2002) An updating on brush plating. Met Finish 100(7):18
8. http://www.brushplate.com

Hardness Investigation of the Electroplated Ni–Co Alloy

Wei He

Abstract Electroplated hard metal coatings are sought after for their anti-wear properties and cost-effective deposition process. However, traditional choices like hard chrome and tungsten present challenges in terms of fabrication and environmental concerns. To address this, we explored the electroplating process under a binary sulfamate system to produce a hard Ni-Co alloy coating on metal surfaces. Our findings revealed that the electroplating current density predominantly influenced the hardness of the Ni–Co alloy, with higher current densities resulting in lower Co concentrations. Surprisingly, the composition of the alloy remained consistent regardless of the current density and showed little sensitivity to the Co^{2+} concentration. The highest hardness value of 608.1 HV0.05, with a Co concentration of 34%, was achieved at a current density of 3 ASD. These results highlight the potential of electroplated Ni–Co alloy coatings as alternatives to traditional options, offering controlled hardness properties and stable compositions. Further research is needed to assess their mechanical properties and applicability in various industries.

Keywords Electroplated Ni–Co alloy · Binary sulfamate system · Current density · Hardness

---

W. He (✉)
Advanced Remanufacturing and Technology Centre (ARTC), Agency for Science, Technology and Research (A*STAR), 3 CleanTech Loop, #01-01 CleanTech Two, Singapore 637143, Republic of Singapore
e-mail: hewei@simtech.a-star.edu.sg

Singapore Institute of Manufacturing Technology (SIMTech), Agency for Science, Technology and Research (A*STAR), 2 Fusionopolis Way, Innovis #08-04, Singapore 138634, Republic of Singapore

N. Maharjan and W. He (eds.), *Proceedings of the 3rd International Conference on Advanced Surface Enhancement (INCASE) 2023*, Lecture Notes in Mechanical Engineering, https://doi.org/10.1007/978-981-99-8643-9_41

1 Background

Metal coatings with high hardness are of great importance in various industrial applications where surfaces are subjected to high stress and wear. High hardness metal coatings are essential for producing enhanced durability, corrosion resistance, increased service life, and improved performance. High hardness coatings offer superior resistance to wear and abrasion compared to softer coatings [1]. This makes them ideal for use in applications where surfaces are exposed to high levels of mechanical stress, such as in the aerospace [2], automotive [3], and other manufacturing industries [4].

Some of the most common methods to produce hard metal coatings include thermal spray [5] and physical vapor deposition (PVD) [6]. Thermal spray involves melting metal or ceramic powders and then spraying them onto a surface, where they solidify and form a hard coating. This process is commonly used to produce coatings with high wear resistance, corrosion resistance, and thermal insulation properties. PVD is another popular method for producing harder metal coatings. This process involves evaporating metal atoms in a vacuum chamber and then condensing them onto a surface, where they form a thin, hard coating with excellent adhesion and wear resistance properties. One limitation of thermal spray is its high porosity which makes its coating susceptible to corrosion. Many efforts are devoted to improving its anti-corrosion property [7]. PVD also has some limitations, particularly in terms of the size and complexity of the parts that can be coated. The process is generally better suited for producing thin coatings on flat or simple geometries, as more complex shapes can be difficult to coat uniformly [8]. Additionally, PVD metal coatings can be brittle and prone to cracking under high stress, limiting their usefulness in applications where the coating is subjected to mechanical stress [9]. In addition to these limitations, both thermal spray and PVD can be relatively expensive and time-consuming processes, requiring specialized equipment and expertise.

Electroplating is particularly useful in producing metal coatings because it produces a uniform and dense metal layer on top of a complex-shaped substrate. The internal stress can be well managed (especially in the sulfamate system [10]) to prevent the cracking phenomenon. Its equipment investment is relatively smaller than the other processes. This makes it possible to produce coatings with specific properties, such as increased hardness and wear resistance, as required by different applications.

Examples of the use of electroplating to produce harder metal coatings include the production of hard chrome coatings on engine components in the automotive industry [11]. Hard chrome [12] coatings provide increased durability and wear resistance, making them ideal for use in high-stress environments such as engine components. However, despite its promising properties, hexavalent chrome is heavily used to produce hard chrome plating. Hexavalent chrome is a known carcinogen that affects multiple key organs, such as the heart, kidneys, and lungs. Other hard metals such as tungsten [13] tungsten alloy [14] are difficult to electroplate. Jovic [15] and Zamani [16] have explored the synthesis of Ni-Co alloy material through an electrochemical process previously. However, the relationship between electroplating current and composition under the sulfamate system is not well studied. The reason to study the sulfamate system is that it has shown to give less internal stress for the electroplated Ni system [10].

In this paper, we reported the detailed investigation of the electroplating process in the sulfamate system to produce Ni-Co alloy coating and studied the relation between the electroplating condition and the hardness and composition profile of the Ni-Co alloy coating. The effect of the plating current density is the focus of this study.

2 Experimental Conditions

The electroplating condition is listed in Table 1. The electroplating setup is shown in Fig. 1 (left). Sacrificial sulfur-activated Ni balls were used as the anode, and a binary sulfamate system (Ni Sulfamate and Co sulfamate) was used as the electrolyte. The Cu coupon was masked with tape to create a plating window of 20 × 20 mm (Fig. 1, middle). By controlling the plating window, the current density can be precisely calculated. We produced multiple samples with current densities ranging from 2 ASD (ampere per square dm) to 10 ASD. The hardness of the Ni–Co coating was measured using a Micro-Vickers hardness testing machine (HM-210). The thickness of the Ni–Co coating on the Cu coupon is approximately 120 μm, which makes it easy for us to use the microhardness machine. The composition of the Ni–Co coating is obtained from the cross-section (Fig. 1, right) using a Field Emission Scanning Electron Microscope (FESEM) with Energy Dispersive X-ray Spectroscopy (EDS). The purpose of the EDS is to investigate the composition changes during the electroplating process.

Table 1 Electroplating conditions

Bath formulation	
Anode	Sulphur activated Ni ball inside Ti cage
Ni sulfamate ($Ni(SO_3NH_2)_2$)	55 g/L
Co sulfamate ($Co(SO_3NH_2)_2$)	20 g/L
Boric acid (H_3BO_3)	35 g/L
Saccharin ($C_7H_5NO_3S$)	1.5 g/L
sodium dodecyl sulfate ($NaC_{12}H_{25}SO_4$, SDS)	0.15 g/L
pH	4
Temperature	50 ^{0}C
Beaker size	1 L
Stirrer speed	500 rpm
Current density	2–10 ASD
Targeted thickness	120 μm

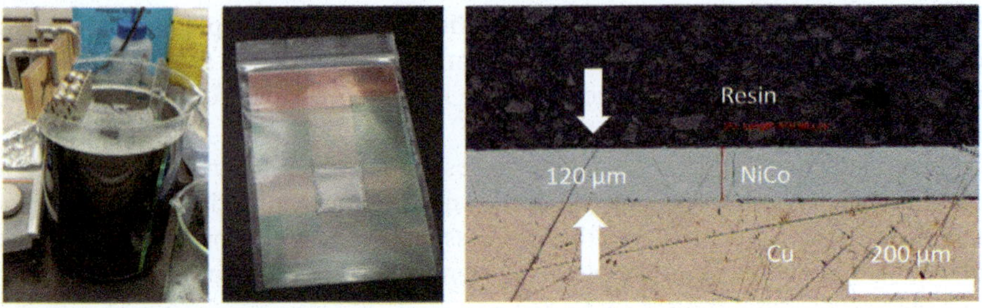

Fig. 1 Setup for Ni–Co alloy electroplating process (left) and masked Cu coupon with electroplated Ni–Co coating (middle) and its cross-section (right)

3 Results and Discussions

The relationship between hardness and composition with respect to the electroplating current density was recorded in Fig. 2, plotted against the current density. The hardness values were obtained from five locations throughout the surface of the Ni–Co coating for each sample, and the deviation of the hardness value is shown as the error bar. The composition value chosen represents the number obtained at the surface of the Ni–Co alloy close to the substrate.

From the orange line, it can be observed that higher current density results in lower Co composition in the Ni–Co alloy, and vice versa. Regarding the hardness value (blue line), the maximum hardness (608.1 HV0.05) is achieved at a composition of 34% Co, which is produced with a current density of 3.0 ASD. In comparison, electroplated pure Ni only has a hardness of 356.1 HV0.05.

The advantage of using a higher current density is that it shortens the production time. For example, to electroplate a cantilever with a thickness of 75 µm, with a

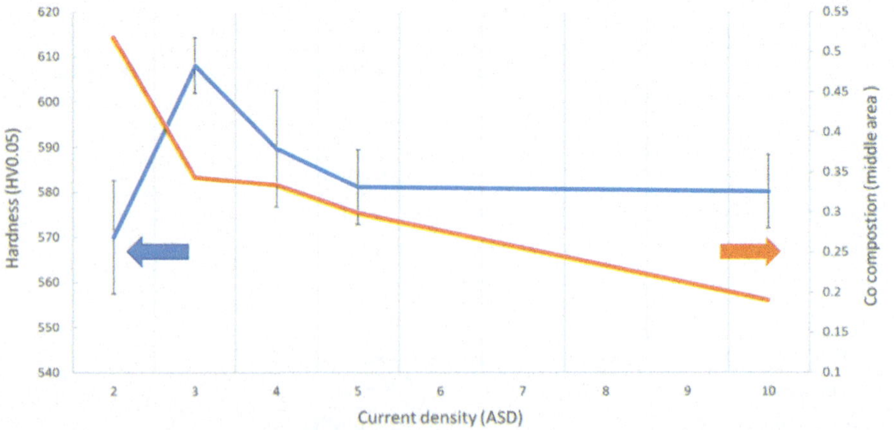

Fig. 2 Relation between current density and hardness/composition of the NiCO alloy

current density of 2 ASD, it can be produced within 2.6 h. However, with a current density of 10 ASD, it only requires 0.52 h. If hardness is the determining factor, a higher current density can be used to reduce the production time.

The processing methodologies and material form have a significant impact on the properties of Ni–Co alloys. When compared to bulk Ni–Co alloys produced through low-temperature chemical reduction, spark plasma sintering, and arc melting [17], higher Co composition (>60%) is associated with higher hardness. In the case of other electroplated Ni–Co films [16], it has been observed that a Co composition of 45% results in the highest values for hardness, yield strength, and ultimate tensile strength. This trend could potentially be attributed to a decrease in the grain size of cobalt within the alloy film [18]. Additionally, the presence of a higher percentage of Ni fcc (111) phase within the alloy film may also contribute to these properties [18] However, the specific reason why the electroplated Ni–Co alloy film with 34% Co exhibits the highest hardness is still being investigated.

Another important research issue is the significance of Co^{2+} concentration in relation to the composition of electroplated Ni–Co alloy. In other words, since the Co^{2+} ions in the electrolyte are not replenished like the Ni^{2+} ions, it is crucial to understand the sensitivity of the composition to changes in Co^{2+} concentration. We obtained the composition distribution along the direction shown in Fig. 1 (right) using the same cross-section sample. The starting point is at the interface between Cu and Ni–Co. We have listed the Co composition distribution for samples produced at 10 ASD (Fig. 3a), 5 ASD (Fig. 3b), and 2 ASD (Fig. 3c) current densities. For each electroplating process, approximately 1.3 g of Ni–Co alloy was coated on the surface, with a maximum Co composition of 50% (2 ASD). In that process, 0.65 g of Co was deposited onto the copper surface, corresponding to a consumption of 14% Co sulfamate. Based on the results, we believe that the composition of the Ni–Co alloy is not significantly affected by the depletion of Co^{2+} in the electrolyte but is primarily sensitive to the plating current density. This finding contradicts the behavior observed in other Ni–Co electroplating systems (using nickel sulfate and cobalt sulfate) [16] where the Co^{2+} concentration typically determines the Co composition in the electroplated Ni–Co alloy film. This finding is significant as it allows for better quality control of the electroplated alloy without the need to constantly monitor the Co^{2+} concentration, as the Co is continuously consumed within the process.

4 Conclusion

In our study, we utilized the electroplating process (sulfamate system) to produce a hard Ni–Co alloy coating on a metal surface to investigate its hardness and composition. We found that the hardness of the Ni–Co alloy is primarily determined by the Co composition within the alloy, which is predominantly influenced by the electroplating current density. Higher current density results in a lower concentration of Co. The highest hardness value (608.1 HV0.05) was achieved at a current density of 3 ASD, corresponding to a Co concentration of 34% in the Ni–Co alloy.

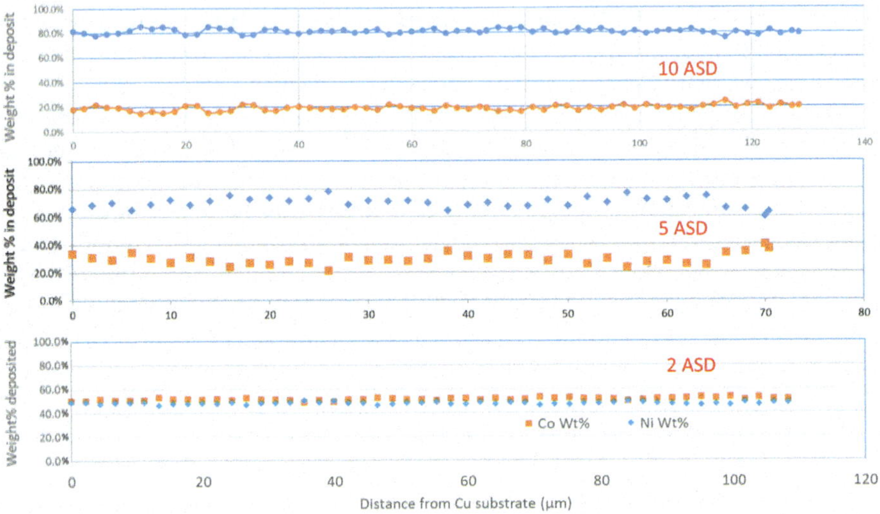

Fig. 3 Composition distribution on the cross section of Ni–Co alloy with different current density (**a** 10 ASD; **b** 5 ASD; **c** 2 ASD)

Interestingly, the composition of the Ni–Co alloy remained consistent throughout the electroplating process, regardless of the electroplating current density, and was not found to be sensitive to the Co^{2+} concentration (like in the sulfate system). Further investigations are needed to explore the impact of temperature on the hardness of the alloy. Additionally, a detailed study of the mechanical performance of the Ni–Co coating with a 34% Co composition is planned as the next step. Stress analysis of this alloy coating is also important for its potential industrial applications.

Overall, our research highlights the significance of electroplating parameters, such as current density, in controlling the composition and hardness of Ni–Co alloy coatings, and underscores the need for further characterization and analysis to fully understand their mechanical properties and performance in various industries.

Acknowledgements This research was supported by A*STAR (Agency for Science, Technology and Research) under grant number C19-S2-034/SC26/19-111034. We would like to express our gratitude for the valuable contributions made by Mr. Lee Xiaohua in the areas of electroplating and sample analysis. His expertise and dedication were instrumental in the success of this research project.

References

1. Bai H, Zhong L, Kang L, Liu L, Zhuang W, Lv V (2021) A review on wear-resistant coating with high hardness and high toughness on the surface of titanium alloy. J Alloy Compd 882:160645

2. Schopphoven T, Gasser A, Wissenbach K, Poprawe R (2016) Investigations on ultra-high-speed laser material deposition as alternative for hard chrome plating and thermal spraying. J Laser Appl 28(2):022501
3. ChandraYadaw R, Singh S, Chattopadhyaya S, Kumar S, Singh R (2018) Comparative study of surface coating for automotive application—a review. Int J Appl Eng Res 13(6):17
4. Gupta P, Tenhundfeld G, Daigle E, Schilling P (2005) Synthesis and characterization of hard metal coatings by electro-plasma technology. Surf Coat Technol 200(5–6):1587
5. Noble N, Radhik N, Sathishkumar M, Saleh B (2023) Characterization and property evaluation of high entropy alloy coating on 316 l steel via thermal spray synthesis. Tribol Int 108525
6. Kazlauskas D, Jankauskas V, Jankauskas X (2020) Research on tribological characteristics of hard metal WC-Co tools with TiAlN and CrN PVD coatings for processing solid oak wood. Coatings 10(7):632
7. Sadeghi E, Markocsan N, Joshi X (2019) Advances in corrosion-resistant thermal spray coatings for renewable energy power plants. Part I: effect of composition and microstructure. J Therm Spray Technol 28:1749
8. Eberhardt M, Hodge A, Branicio P (2022) Atomistic modeling of physical vapor deposition on complex topology substrates. Comput Mater Sci 203:111111
9. Valiev R, Selivanov K, Smyslova M, Dyblenko Y, Savina Y, Valiev R (2022) Enhanced Erosion Resistance of an Ultrafine-Grained Ti Alloy with a PVD Coating. Metals 12(5):818
10. Kelly J, Goods S, Talin A, Hachman J (2006) Electrodeposition of Ni from low-temperature sulfamate electrolytes: I. Electrochemistry and film stress. J Electrochem Soc 153(5):C318
11. Dhinakaran V, Jagadeesha T (2021) Tribological characterization of electrolytic hard chrome and WC-Co coatings. Mater Today: Proceed 46:9176
12. Merlo A, Léonard G (2021) Magnetron sputtering vs. electrodeposition for hard chrome coatings: a comparison of environmental and economic performances. Materials 14(14):3823
13. Obradović M, Stevanović J, Stevanović R, Despić A (2000) A contribution to the knowledge of electrochemical deposition of W induced by Ni. J Electroanal Chem 491(1–2):188
14. Hoar T, Bucklow I (1954) On the electrodeposition of tungsten-cobalt alloys from aqueous solutions. Transact IMF 32(1):186
15. Jović V, Jović B, Pavlović M (2006) Electrodeposition of Ni, Co and Ni–Co alloy powders. Electrochim Acta 51(25):5468
16. Zamani M, Amadeh A, Baghal S (2016) Effect of Co content on electrodeposition mechanism and mechanical properties of electrodeposited Ni–Co alloy. Transact Nonferr Metals Soc China 26(2):484
17. Tan LP, Padhy SP, Tsakadze Z, Chaudhary V, Ramanujan RV (2022) Accelerated property evaluation of NieCo materials libraries produced by multiple processing techniques. J Mark Res 20:4186
18. Karpuz A, Kockar H, Alper M, Karaagac O, Haciismailoglu M (2012) Electrodeposited Ni–Co films from electrolytes with different Co contents. Appl Surf Sci 258:4005

The Challenges of Thermosetting Polymers Metallisation via Low-Pressure Cold Spray

Davide Verdi, David Merino-Millan, Alicia Salazar, and Pedro Poza

Abstract Metallisation is the process of applying a metallic coating on the surface of a non-metallic component for structural, functional, or decorative purposes. It can be used to replace metallic parts with metallic-coated polymers with the objective of weight and cost reduction, or to extend the field of application of polymeric components granting them thermal or electrical conductivity. For example, it is used in aerospace applications to provide conductivity as protection against lighting strikes. Cold Spray was demonstrated to be suitable for the deposition of metallic coatings on polymer-based substrates without any prior surface treatment. It can be considered as an environmentally friendly alternative for the traditional electroplating metallisation practices and thermal spray coatings technologies as it is a non-combustion process and any unadhered powder residues can be easily disposed of while the metallisation via Cold Spray of thermoplastic polymers was reported by various authors in the last few years, the coating of thermosetting presents additional challenges that have yet to be fully resolved. In this work, a low-pressure Cold Spray system with a maximum pressure of 6 bar was employed to investigate the challenges associated with the metallisation of an epoxy resin.

Keywords Cold spray · Metallisation · Thermosetting · Polymers

D. Verdi (✉) · D. Merino-Millan · A. Salazar · P. Poza
DIMME—Durability and Mechanical Integrity of Structural Materials, Escuela Superior de Ciencias Experimentales y Tecnología, Universidad Rey Juan Carlos, C/ Tulipán S/N, 28933 Móstoles, Madrid, Spain
e-mail: davide.verdi@urjc.es

A. Salazar
Instituto de Tecnologías para la Sostenibilidad, Universidad Rey Juan Carlos, C/ Tulipán S/N, 28933 Móstoles, Madrid, Spain

N. Maharjan and W. He (eds.), *Proceedings of the 3rd International Conference on Advanced Surface Enhancement (INCASE) 2023*, Lecture Notes in Mechanical Engineering, https://doi.org/10.1007/978-981-99-8643-9_42

1 Introduction

Metallisation is a practice commonly used in different industrial fields. It is employed to add a metallic layer on the surface of a metallic or non-metallic component with the objective to provide structural or functional improvements or for aesthetic purposes. Multiple metallisation methods can be listed, such as electroplating process and the different coating technologies. Metallic-coated polymers can be used to replace metallic parts with the objective of reducing weight and costs, or to extend the field of application of polymeric components, for example by providing them with thermal or electrical conductivity. The application of polymer metallisation was used for example in the medical field to improve the biocompatibility of polymeric implants [1]. Another field of application is in the aerospace industry, where metallisation is used to provide lighting strike protection to aircrafts [2]. Metallisation is also used in the automotive sector, for functional purposes such as in the headlamp reflectors, commonly coated with aluminium via physical vapour deposition (PVD) process, or in the case of decorative purposes, where Cr-electroplating is used [3].

Due to their different nature, the adhesion between polymers and metals can be challenging. For this reason, pre-treatments are commonly required to improve the adhesion without compromising the bulk characteristics of the polymer. This requires an additional step within the metallisation process, and on some occasion may need to involve hazardous substances, wastes, or emissions. For example, a common practice to metallise the acrylonitrile butadiene styrene (ABS), a very commonly used polymer, involves the use of chromic acid which contains hexavalent chromium (Cr-VI), a well-known toxic substance [4]. These issues can be avoided by using thermal spray process. However, even with these technologies, the requirement of activating the substrate to improve the adhesion remains as a challenge [5]. The most common process in this case is grit blasting, in which a hard medium (e.g., crushed alumina) is projected through a stream of compressed air against the target surface generating a superficial erosion. An alternative could be the inclusion of ceramic or metallic powder particles on the surface of the polymers during their curing phase. However, this solution is not compatible with all the polymers manufacturing processes.

The Cold Spray (CS) process is a unique solid-state thermal spray technology in which the feedstock powder is accelerated up to supersonic velocities (400–1200 m/s) through a pressurised and preheated carrier gas which expands in a converging–diverging (de Laval) nozzle. Compared to other thermal spray processes such as flame spray, plasma spray, high velocity oxygen fuel, etc., the carrier gas temperature in the CS process is always below the melting point of the sprayed material. CS is based on the kinetic energy of the powder particles impacting the substrate. Thanks to its characteristics, CS is suitable for the deposition of coatings onto materials that are sensitive to high temperatures (e.g., polymers). In fact, in the last decade, the direct deposition on polymers via CS without a previous surface treatment was demonstrated by different authors [6, 7]. CS can also be considered as an environmentally friendly alternative to the most conventional metallisation processes as it is a non-combustive method and its residues (i.e., non-adhered powder) can be easily collected and disposed of [8].

In recent times, CS has gained increased acceptance across various industrial sectors especially since the introduction in the market of compact and portable Low-Pressure Cold Spray (LPCS) systems [9]. The LPCS can be considered a branch of the CS technique. The main differences are related to the type of carrier gas, air, nitrogen, or a combination of them, and the injection of the feedstock powder that is fed downstream, perpendicularly to the diverging part of the nozzle ("radial injection"). In addition, the gas heating temperature and pressure are substantially low, with values between room temperature and 650 °C and below 6 bar respectively, compared to the ~1100 °C and 70 bar of a conventional CS system.

Even with the development of LPCS, the deposition on polymers remains challenging. While on metals, the coating deposition depends on two main mechanisms, namely the formation of shear instabilities generated by the plastic deformation of the feedstock powder particles impacting the substrate or previously deposited material, and the mechanical interlocking on polymers only the second one is possible [10]. Additionally, most of the kinetic energy of the powder particles is consumed in the deformation of the polymeric substrate. As a result, the metallic particles will be embedded in the polymer. Issues may then arise with the growth of the coating on top of the first layer of embedded particles as the energy of the incoming powder may generate erosion or the extrusion of the previously deposited particles [11]. To obtain a proper coating, the polymer deformed by the impact from the first metallic particles must create lips that will act as binder and facilitate the adhesion of subsequent incoming powders. This mechanism is however intrinsic to only thermoplastic polymers as they undergo a reversible thermal softening when the glass transition temperature (T_g) is reached. On the other hand, thermosetting polymers possess higher T_g with respect to thermoplastics (i.e., > 200 °C, against 30–80 °C) and their degradation temperatures are close to their T_g, making the thermal softening process irreversible. The high glass transition temperature of thermosetting polymers in comparison with thermoplastics guarantees its dimensional stability and brittleness for a wider range of high temperatures. This brittleness is responsible of substrate destruction independently of CS parameters [11]. This is the main additional issue for thermosetting polymers metallisation via CS or LPCS.

To avoid the thermosetting polymer substrate erosion, several authors suggested the application of a buffer layer. Ganesan et al. [12], sprayed a Sn buffer layer which is a low melting point metal to generate a metallic interlayer subsequently coated with Cu. Fallah et al. [13], electroplated carbon fibre-reinforced polymers (CFRP) with Cu before spraying Cu powder. Rubino et al. [14], added Cu and Ti powder on the surface of glass fibre-reinforced polymers (GFRP) during the impregnation step to facilitate the adhesion of the CS Sn-Al powder mixture. Affi et al. [15], deposited Al onto CFRP using two spray technologies. The authors first generated a buffer layer using plasma spray and then coated it using CS. Another solution was suggested by Viscusi et al. [16], who applied a 30 μm thick thermoplastic film (i.e., polylactic acid, PLA) on a basalt-epoxy surface and heated the whole sample under a press plate to help the PLA-composite coupling. In all the cases, the nature and properties of the buffer layer employed differ from those of the base or the coating materials.

In this document, the preliminary results achieved from the direct LPCS deposition of a metallic coating on thermosetting polymers without buffer layer are reported. Some key challenges were observed as expected. A feasible solution is provided by applying a thin layer of thermosetting epoxy resin before the LPCS process. The effect of the epoxy curing conditions on the LPCS process are reported.

2 Methods

A Titomic D523 system (Titomic, Australia) was used for the experiments discussed in this document. This LPCS equipment works with pressures between 4.0 and 6.0 bar and with temperatures below 600 °C. To reduce costs and environmental impact, compressed air was used as carrier gas, although nitrogen may also be employed. The deposition gun was mounted on a six-axis robotic arm, KUKA KR6 R90 (KUKA, Germany), responsible of providing the demanded standoff distance (SOD) and deposition speed.

The feedstock powder used is a 50:50 mix of pure Al and pure Zn particles. This product was supplied by Titomic under the commercial name A-20-10, and it is characterised by particles with sizes that range between 5 μm and 30 μm.

Epoxy resin samples of dimensions 25 mm (length) × 15 mm (width) × 6 mm (thick) were used as substrate. They were additively manufactured via masked stereolithography (MSLA), using a Creality LD-006 Resin 3D Printer (Creality, China). A Monocure 3D Rapid Model Resin was used (Monocure 3D, USA). The 3D printing parameters were selected based on the recommendations that can be found on the Monocure 3D webpage [17]. The printed samples were washed with isopropyl alcohol and subsequently cured for 60 s with ultraviolet (UV) light using an Anycubic Wash& Cure Plus Machine (Anycubic, USA).

After some preliminary trials, LPCS tests were performed by varying the carrier gas temperature and the nozzle-to-substrate distance (i.e., SOD) while keeping other process parameters fixed, such as carrier gas pressure (4.5 bar), powder mass flowrate (26 g/min), and robot scanning speed (50 mm/s). Three values of temperature (250, 400, and 550 °C) and three SODs (40, 60, 80 mm) were tested for a total of nine trials. The process toolpath consisted in two tracks deposited in the same direction on the same location. This was to analyse the effect of the second pass that may generate the build-up of the coating, the detachment of previously deposited particles, or the erosion of the substrate.

Additional tests were carried out applying a thin (~50 μm) layer of Akasel Aka-Resin Liquid Epoxy (Akasel, Denmark), mixed with the corresponding amount of Akasel Aka-Cure Slow epoxy curing agent, using a small brush. In this case, the temperature of the carrier gas was fixed at 550 °C and the SOD at 80 mm. The curing conditions (time and temperature) of the epoxy resin were varied.

In addition to a naked-eye observation of the tested specimens, some of them were sectioned, mounted in resin, and polished using diamond suspensions down to 0.25 μm, for scanning electron microscopy (SEM) analysis. A Hitachi S-3400 microscope (Hitachi, Japan) was employed to capture secondary (SE) and backscattered (BSE) electron images. Energy dispersive X-ray microanalysis (EDX) were also carried out to analyse the feedstock material and the obtained samples.

Finally, a multimeter was used to determine if a continuous electrical path was formed in the deposited coatings.

3 Results and Discussion

Figure 1 reports the results obtained from the SEM analysis of the Al-Zn feedstock powder. As can be seen from the BSE-SEM image in Fig. 1a, the powder particles are characterised by a spherical morphology. The presence of satellites and clusters can be appreciated. Two different phases can be clearly discerned. The darker particles, marked with "1" in Fig. 1a, were identified as aluminium while the white ones, marked with "2" in Fig. 1a, were recognised to be zinc, as can be seen in the EDX spectra reported in Fig. 1b, c, respectively. Additionally, it was observed that while most of the powder particles range between 5 and 30 μm, some exceed these limits. Moreover, it was observed that the Zn particles do not exceed the 15 μm and the bigger particles are all made of Al.

The combinations of carrier gas temperatures and SODs used in the first set of tests are listed in Table 1. A picture of the superficial appearance of the tested samples is showed in Fig. 2. The position of the LPCS tracks is highlighted with a couple of dashed lines in each picture. Erosion can be appreciated in almost all the tested conditions where the LPCS tracks appear translucent, like the colour of the epoxy resin substrate, with exception to tests number 04, 07, and 08, where not even signs of deposition can be observed. Remaining of the sprayed powder can be observed on the sides of the deposition tracks (grey coloured areas).

These initial results agree with the observations reported by other authors that highlighted the challenges related to the metallisation via CS of thermosetting polymers. Due to their intrinsic behaviour, these polymers are more prone to erode or fracture during the CS process. As in thermoplastic polymers metallisation a key role is played by their capability to become softer when their temperature exceeds the T_g, the proposed solution to metallise thermosetting polymers is to perform the coating process when the substrate is not completely cured. For this reason, a thin layer of epoxy resin was painted on top of the 3D printed coupons used for the second sets of experiments. The samples were then cured at ambient temperature (~20 °C), and 40 °C to accelerate the process. Different curing times were investigated, as reported in Table 2.

The surface aspect of the tested samples is illustrated in Fig. 3. In this case, no erosion was observed, but the main event was the dislodging of the painted epoxy resin when it was too fluid to adhere to the substrate. An exception can be observed

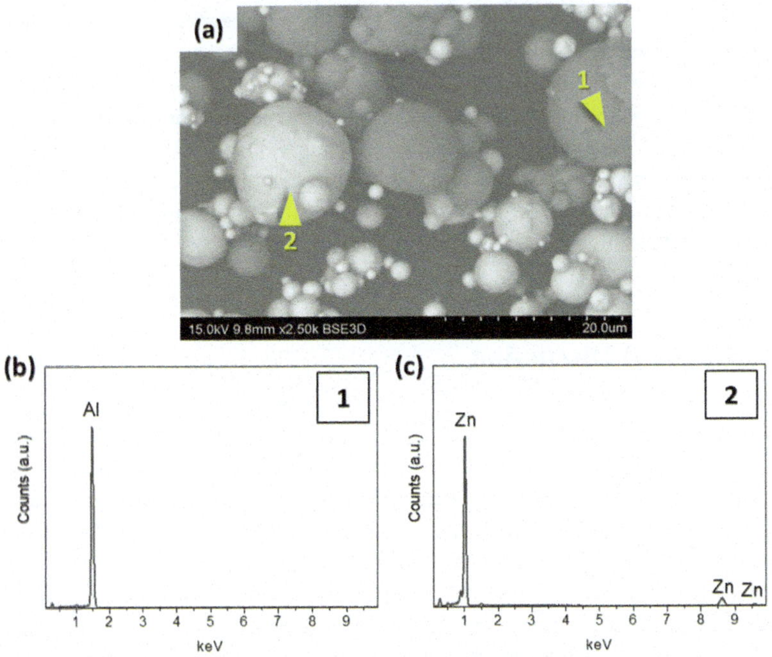

Fig. 1 SEM analysis of the Al-Zn feedstock powder. **a** BSE-SEM image; **b** EDX spectrum of the phase marked with 1 in figure (**a**); and **c** EDX spectrum of the phase marked with 2 in figure (**a**)

Table 1 Carrier gas temperature and SOD combinations for the first set of tests

Test (#)	Carrier gas temperature (°C)	SOD (mm)
01	250	40
02	400	40
03	550	40
04	250	60
05	400	60
06	550	60
07	250	80
08	400	80
09	550	80

for sample number 17, cured at 40 °C for 120 min. In this case, the incorporation of metallic powder can be observed. This sample was further analysed via SEM.

SEM images of the top surface of sample number 17 are reported in Fig. 4. It is possible to observe how the metallic particles adhered to the painted epoxy resin and in between them. Little to no deformation of the smaller particles can be seen, while the biggest Al ones undergone a certain degree of deformation. In addition to that, a

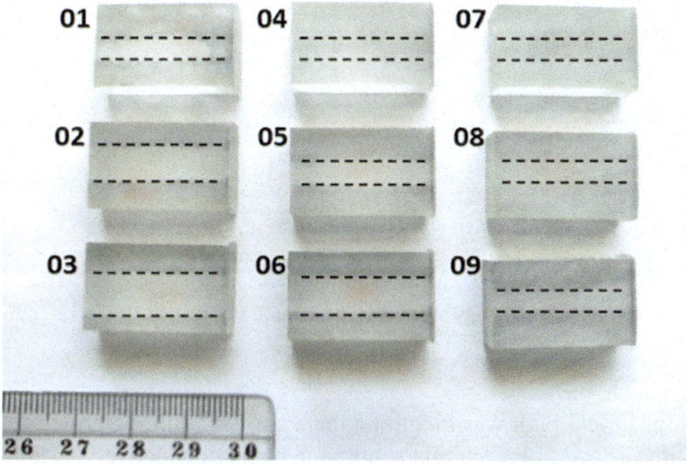

Fig. 2 First set of tests: samples' surface appearance

Table 2 Curing temperature and time combinations for the second set of tests

Test (#)	Curing temperature (°C)	Curing time (min)
10	T_{amb}	0
11	40	30
12	T_{amb}	30
13	40	60
14	T_{amb}	60
15	40	90
16	T_{amb}	90
17	40	120
18	T_{amb}	120

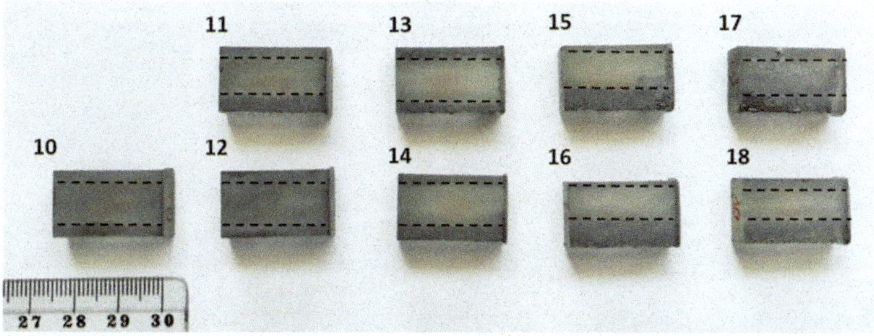

Fig. 3 Second set of tests: samples' surface appearance

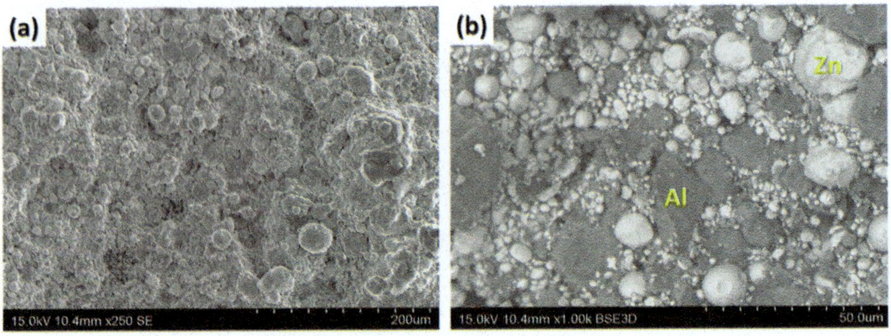

Fig. 4 Top surface view after LPCS process of the sample cured at 40 °C for 120 min. **a** SE-, and **b** BSE-SEM images

continuous electrical path was identified thanks to the multimeter. However, it was not the entire surface of the sample resulted in being conductive.

From a cross section analysis of the sample, it was possible to appreciate a continuous layer of Al and Zn particles embedded in resin all over the 3D printed substrate, with a thickness that varies between ~30 and ~60 μm (Fig. 5a). From a closer observation of the superficial layer, it can be seen that Al and Zn particles of different sizes were embedded in the painted resin that act as a binder (Fig. 5b). At the bottom of the deposited layer, the metallic particles embedded resulted to be mainly Zn. This can be connected to the smaller dimension of these particles with respect to the Al ones, as illustrated in Fig. 1a. The smaller the particle size, the higher is the particle velocity during the CS process.

It is also possible to speculate that the heat provided by the carrier gas may also help to complete the curing process of the deposited layer during the LPCS process.

Thanks to these initial encouraging results, more tests will be performed to optimise the curing conditions of the painted epoxy layer (i.e., to identify the mechanical behaviour and properties that allow the incorporation of metallic particles). Following

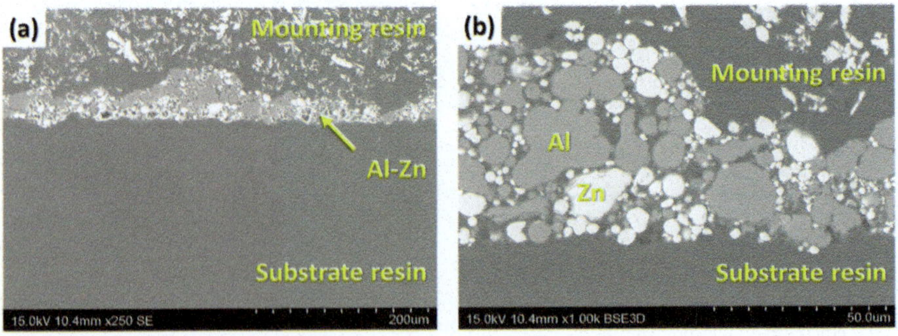

Fig. 5 Cross section of the sample cured at 40 °C for 120 min after LPCS process. **a** SE, and **b** BSE-SEM images

that, more experiments will be carried out to study the possibility to grow a full Al-Zn metallic coating on top of this first mixed layer. Another solution that will be tested is to pre-mix the metallic powder with the epoxy resin before painting it on the 3D printed epoxy substrate. A proper distribution of the metallic particles in the resin must be achieved.

4 Conclusions

Metallisation of polymers via LPCS is a potentially eco-friendly alternative to other conventional practices. Different challenges are still to be fully tackled to allow a wider industrial acceptance of this process. One particular challenge arises in the intrinsic nature of the polymers, where their low hardness hinders the mechanism of plastic deformation of the sprayed metallic particles. To add on, once cured, the fragile nature of thermosetting polymers provides an additional challenge. For this reason, the authors suggested that the deposition should be carried out before the thermosetting resin fully cures.

A layer of epoxy resin was applied to the surface of a 3D printed epoxy coupon before undergoing metallisation via LPCS. By curing the resin at 40 °C for 120 min, it was possible to achieve the incorporation of Al-Zn particles. This solution can be considered an alternative to the addition of Sn or thermoplastic polymers as buffer layer for the metallisation of thermosetting polymers. Additional experiments and analysis must be carried out to fully understand and optimise the viability of this solution.

Acknowledgements The authors would like to thank the Spanish Ministry of Universities for the financial support through the "María Zambrano" grant.

References

1. Guarnieri V, Biazi L, Marchiori R, Lago A (2014) Platinum metallization for MEMS application. Biomatter 4:e28822. https://doi.org/10.4161/biom.28822
2. PSM R, Cauchy X, Gagne ME, Klemberg-Sapieha J, Sirois F, Therriault D (2015) Wet chemical metallization of aerospace composites as a lightning protection strategy. In: Engineering solutions for sustainability. John Wiley & Sons, Inc., Hoboken, NJ, USA, pp 187–192. https://doi.org/10.1002/9781119179856.ch18
3. Melentiev R, Yudhanto A, Tao R, Vuchkov T, Lubineau G (2022) Metallization of polymers and composites: state-of-the-art approaches. Mater Des 221:110958. https://doi.org/10.1016/j.matdes.2022.110958
4. European Commission Directorate-General for Employment Social Affairs and Inclusion, Hartwig A, Klein C, Kromhout H et al (2017) SCOEL/REC/386 Chromium VI compounds: recommendation from the scientific committee on occupational exposure limits. https://doi.org/10.2767/165340

5. Gonzalez R, Ashrafizadeh H, Lopera A, Mertiny P, McDonald A (2016) A review of thermal spray metallization of polymer-based structures. J Therm Spray Technol 25:897–919. https://doi.org/10.1007/s11666-016-0415-7
6. Lupoi R, O'Neill W (2010) Deposition of metallic coatings on polymer surfaces using cold spray. Surf Coat Technol 205:2167–2173. https://doi.org/10.1016/j.surfcoat.2010.08.128
7. Zhou XL, Chen AF, Liu JC, Wu XK, Zhang JS (2011) Preparation of metallic coatings on polymer matrix composites by cold spray. Surf Coat Technol 206:132–136. https://doi.org/10.1016/j.surfcoat.2011.07.005
8. Vilardell AM, Cinca N, Concustell A, Dosta S, Cano IG, Guilemany JM (2015) Cold spray as an emerging technology for biocompatible and antibacterial coatings: state of art. J Mater Sci 50:4441–4462. https://doi.org/10.1007/s10853-015-9013-1
9. Cavaliere P (2018) Cold-spray coatings. Springer International Publishing, Cham. https://doi.org/10.1007/978-3-319-67183-3
10. Ganesan A, Affi J, Yamada M, Fukumoto M (2012) Bonding behavior studies of cold sprayed copper coating on the PVC polymer substrate. Surf Coat Technol 207:262–269. https://doi.org/10.1016/j.surfcoat.2012.06.086
11. Melentiev R, Yu N, Lubineau G (2021) Polymer metallization via cold spray additive manufacturing: a review of process control, coating qualities, and prospective applications. Addit Manuf 48:102459. https://doi.org/10.1016/j.addma.2021.102459
12. Ganesan A, Yamada M, Fukumoto M (2013) Cold spray coating deposition mechanism on the thermoplastic and thermosetting polymer substrates. J Therm Spray Technol 22:1275–1282. https://doi.org/10.1007/s11666-013-9984-x
13. Fallah P, Alidokht SA, Vo P, Chromik RR, McDonald A, Yue S (2022) Properties of multilayered metallic coatings deposited on carbon fiber-reinforced polymers (CFRPS) through electrochemical and cold spray processes. J Therm Spray Technol. https://doi.org/10.1007/s11666-022-01475-0
14. Rubino F, Tucci F, Esperto V, Perna AS, Astarita A, Carlone P, Squillace A (2020) Metallization of fiber reinforced composite by surface functionalization and cold spray deposition. Procedia Manuf 47:1084–1088. https://doi.org/10.1016/j.promfg.2020.04.353
15. Affi J, Okazaki H, Yamada M, Fukumoto M (2011) Fabrication of aluminum coating onto CFRP substrate by cold Spray. Mater Trans 52:1759–1763. https://doi.org/10.2320/matertrans.T-M2011807
16. Viscusi A, Antonucci V, Carrino L, Della Gatta R, Lopresto V, Papa I, Perna AS, Ricciardi MR, Astarita A (2020) Manufacturing of an innovative composite structure: design, manufacturing and impact behaviour. Compos Struct 250:112637. https://doi.org/10.1016/j.compstruct.2020.112637
17. Monocure 3D Rapid Model Resin. Retrieved from https://monocure3d.com.au/product/3d-rapid-model-resin/

Surface Engineering and Protective Coating Towards Advanced Manufacturing and Remanufacturing Applications

Hongfei Liu, Tzee Luai Meng, Na Gong, Wai Luen Chan, Rahul Karyappa, and Yuefan Wei

Abstract Global climate changes and worldwide sustainability are pushing the requirement for advanced manufacturing and remanufacturing in various areas, including aerospace, sea- and land-transport, infrastructure, low-carbon energy, etc. Surface engineering and protective coating (SEPC) are a typical area that deals with the surface finishing and enhancement of metallic components. SEPC has long been developed in traditional manufacturing but attracted increasing interest in recent years along with the rapid advancing in additive manufacturing and remanufacturing. For example, surface enhancement process, when introduced into interpass additive manufacturing, can significantly enhance the whole components with reduced necessity for post-manufacturing treatment. Here, we present and discuss a few surface engineering techniques as well as their potential applications in protective coatings produced by direct energy depositions.

Keywords Additive manufacturing · Remanufacturing · Surface engineering · Protective coatings

1 Introduction

To achieve zero emissions and sustainability in ecosystem is pushing the requirement for advanced manufacturing and remanufacturing in worldwide various areas, including aerospace, sea- and land-transport, infrastructure, low-carbon energy, etc. The rapid progress in computer-assisted automation and precision control along with

H. Liu (✉) · T. L. Meng · N. Gong · R. Karyappa
Institute of Materials Research and Engineering (IMRE), A*STAR (Agency for Science, Technology and Research), 2 Fusionopolis Way, Singapore 138634, Singapore
e-mail: liuhf@imre.a-star.edu.sg

W. L. Chan · Y. Wei
Advanced Remanufacturing and Technology Center (ARTC), A*STAR (Agency for Science, Technology and Research), 3 CleanTech Loop, Singapore 637143, Singapore

N. Maharjan and W. He (eds.), *Proceedings of the 3rd International Conference on Advanced Surface Enhancement (INCASE) 2023*, Lecture Notes in Mechanical Engineering, https://doi.org/10.1007/978-981-99-8643-9_43

the great advances in materials science and engineering in the past decade have been bringing up additive manufacturing (AM), also known as 3D printing, entering industrial applications [1]. The capability of being able to produce and/or restore structural components with complex shape and/or geometry makes AM essential for remanufacturing with high value-add, e.g., when repairing turbine engines by direct energy depositions (DEDs) [2–5]. Different from AM of structural components by DEDs, protective coatings can also be produced by DEDs, e.g., with a corrosion resistant layer on structural components in a cost-effective design. It is commonly accepted that such kinds of processes, i.e., AM process for advanced manufacturing and/or remanufacturing are typically fit-for-purpose designed, where surface engineering and protective coating (SEPC) play an important role [6].

From surface integrity enhancement point of view, both surface engineering, e.g., through cold-working and/or laser-treatment [7–12], and protective coating, e.g., by DED such as laser-cladding [5, 13–15], wire-arc AM (WAAM) [16], spray coating [17], etc., can provide fit-for-purpose functions. Unlike traditional manufacturing of metal alloys, where compositional uniformities and/or structural defects can be removed by various treatments, including heat treatments and hot/cold plastic deformations, post-AM process (e.g., forging and rolling) might be limited for AM components. Such process limitations, especially for complex AM components, lead to increased requirements for quality control in AM process. In this light, understanding the effect of SPEC on the process-microstructure-performance relationship of AM alloys, as well as its assistance by theoretical modelling and simulation, machine learning, deep learning, artificial intelligence, etc., has been attracting increasing research interest in recent years [5, 18].

In this work, we present a few cases of cold-working, laser-cladding, and effect of surface engineering on protective coating deposited by laser-cladding, addressing their potential applications towards advanced manufacturing and remanufacturing.

2 Surface Enhancement by Cold-Working

Cold-working process of metal alloys, including those manufactured by traditional and AM methods, has been extensively studied, driven by their effect on surface integrity enhancement. By applying mechanical impacts on the surface of metals, plastic deformation is induced, leading to the generation of dislocations as well as their migrations and interactions, crystallographic shear and stacking faults, and eventually grain refinements. Along with the plastic deformation and grain refinement, compressive residual stress (CRS) can be stored in the surface and near-surface regions while mechanical strength and surface hardness can be enhanced, which in turn result in fatigue life enhancement and increased wear-resistance of the metallic components [19, 20].

Plastic deformation-induced austenite-to-martensite phase transformations have been observed in austenitic steels with relatively low stacking fault energies (SFEs).

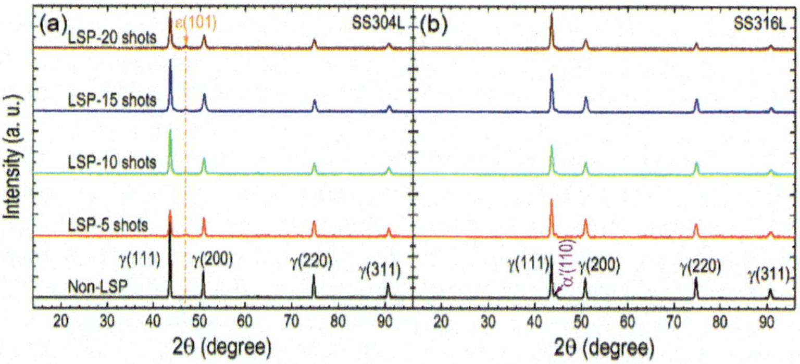

Fig. 1 Comparisons of laser-shock peening (LSP) induced phase transformations in **a** SS304L and **b** SS316L coupons measured by X-ray diffraction (CuKα) at the LSP-induced craters

The phase transformation mechanism, i.e., γ-to-ε and/or γ-to-α′, is greatly dependent on the cold-working method and processing parameters [21–23]. It is generally known that γ-to-ε transformation usually occurs in 'simple plastic deformation' while γ-to-α′ transformation typically occurs in 'complex plastic deformation'. Our recent studies by comparing laser-shock peening (LSP, see Fig. 1) and robotic hammer peening (RHP) on stainless steel (SS) 304L and 316L revealed apparent phase transformations in SS304L rather than in SS316L due to their significantly different lower SFEs [8, 10]. More interesting observation is that the phase transformation induced in SS304L by LSP is dominated by γ-to-ε while that by RHP is dominated by γ-to-α′ under the surface of the craters with the same depth [10].

The effect of cold-working on resistance of metal alloys against corrosion and stress corrosion cracking is also largely dependent on processing method and parameter [24–26]. In general, the CRS induced by cold-working tends to suppress the pitting corrosion and mitigate the initiation and propagation of stress corrosion cracking. However, careful attention has to be paid on the surface roughness since the surface roughness might play even more important role than that of CRS in corrosion and stress corrosion cracking of metallic alloys [7]. It is worth mentioning that crystallographic reorientation and material migrations could occur in cold-working process [8, 10, 27], which tend to densify the material and modify the resistance to corrosion and stress corrosion cracking.

3 Protective Coating by Laser-Cladding

Laser-cladding is a typical laser-based DED method for surface protection as well as AM of metal alloys. Recent advances in materials and feedstock engineering techniques made laser-cladding with various variations in processing configurations [5]. Both wire- and powder-feedstock have been developed for laser-cladding and for the powder-feedstock mode, both pre-lay and blow configurations have been developed [28, 29]. By optimizing the processing parameters, crack- and pore-free layers of steel- and Ni-based alloys can be achieved by laser-cladding, making it entering practical applications, for example, in repair and restoration of turbine blades [30–32].

To protect the combustion chamber waterwall of energy plants from hot corrosion and erosion, Ni-based superalloys have been deposited by laser-cladding as the protective coating on steel-based heat exchanger waterpipes [13, 14, 33]. For this purpose, we have deposited Inconel 625 coatings by laser-cladding on low-alloy steel (LAS) substrate using powder feedstock carried in argon flow. It has been found that a single pass cladding gave rise to a layer of ~1.5 mm and the substrate dilution can be effectively suppressed in the second layer. Heat affected zone (HAZ) was formed in the LAS substrate at the interface region, which apparently degraded the corrosion resistance, i.e., by nearly half, of the substrate [13]. In this care, crack/pore-free coating of the Inconel 625 layer is essential for the guarantee of the corrosion protections. The initial/ending cladding tracks must be designed to mitigate potential electrolyte penetrations.

Within the hot corrosion environment formed by Na_2SO_4 and $MgSO_4$ mixture and sulfur vapor at 900 °C, the Inconel 625 coating exhibited significantly reduced corrosion rate as compared with that of the LAS substrate (see Fig. 2). The dominant anticorrosion mechanism, as revealed by transmission electron microscopy (TEM) based microstructural and energy dispersive spectroscopy (EDS) characterization and analysis, is related to the thermal-induced growth of chromium oxide scales, i.e., TCO, on the surface along with the outward diffusion of Cr through the grain boundaries from the base material of the coating. Also observed that the growth and accumulation of the TCO scales on the surface of the Inconel 625 coating led to thermal stress and eventually cracking and spallation at the TCO/Inconel 625 interface [14].

From materials design point of view, along with the developments in theoretical modelling and simulations, accelerated prototyping is highly desired towards developing advanced alloys with improved functions and/or performance. In this regard, compositional gradient deposition of metallic alloys by DED methods has been proposed in recent years. In this case, laser-cladding assisted by feedstock blender, which can mix the feedstocks from different reservoirs, might play important roles [34, 35].

4 Surface Engineering on Laser-Cladded Coatings

The rapid heating and cooling characteristics of the materials in the melt pool of the laser-cladding-based AM process give rise to unavoidable thermal stress remaining in the coatings and/or AM components. Too large thermal stress could induce cracks of the coating layer while unproperly engineered feedstock might cause porous structures involved in the laser-DED components. More important is that that most of

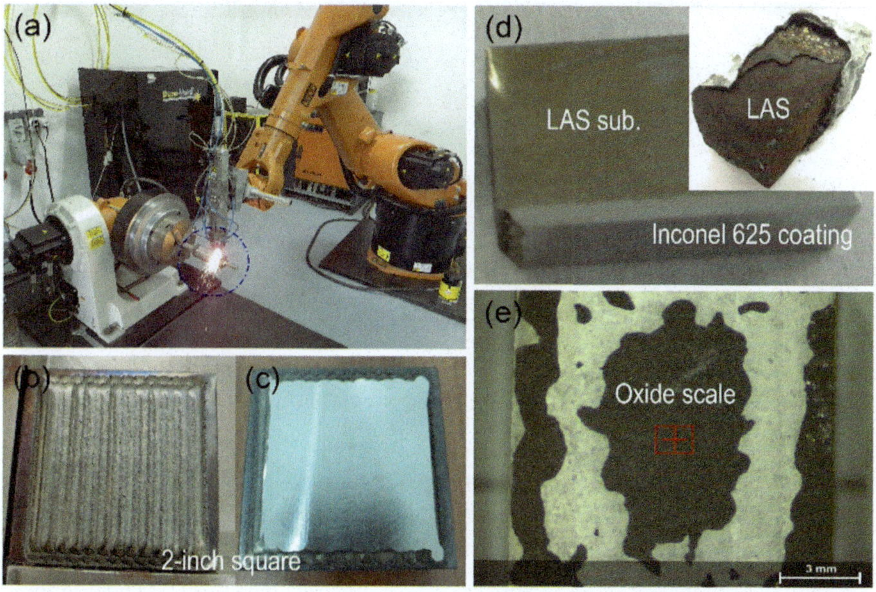

Fig. 2 Protective coatings deposited by laser-cladding (LC) against hot corrosion in molten salt. **a** Typical setup of LC based on an industrial robot platform, **b** Inconel 625 coating deposited by LC on low-alloy steel (LAS) substrate, **c** LC-deposited Inconel 625 coating after grinding, **d** Prepared sample for hot corrosion test in molten salt (mixture of Na_2SO_4 and $MgSO_4$), and **e** the surface of Inconel 625 coating (bright areas) with remaining oxide scales (dark areas, dominated by MgO) after hot corrosion test. The inset shows the corrosion of the LAS substrate without the Inconel 625 protective coating

the laser-DED processes might lack or have limitation to access post-DED treatments, e.g., solution treatment, hot isostatic pressing (HIP), etc., which put much more requirements for high-quality laser-DED process. In this regard, various interpass treatments have been studied in recent years, aiming at compatible integration with laser-DED process towards advanced manufacturing and remanufacturing. For example, laser-induced heat treatment, cold-rolling, machine hammer peening, etc. have been induced as an interpass process into DED processes, including laser-cladding and WAAM treatments. The general mechanism induced by the interpass process is to (i) release the tensile thermal stress immediately after the complete of the individual layers; (ii) removal of potential cracks and/or pores (see a typical example shown in Fig. 3); (iii) disrupt the epitaxial growth from the deposited layer into its consequent layer; (iv) refine the grains in the deposited layer; and (v) modify the shape and/or fine tune the configuration of the AM components.

In an integration of WAAM and interpass rolling process for Inconel 718 [36], it has been revealed that the mechanical force must be optimized and adjusted along with the height of the materials deposition, so as to avoid the crack at the interface between the Inconel 718 wall structure and the substrate. Pore closures have been

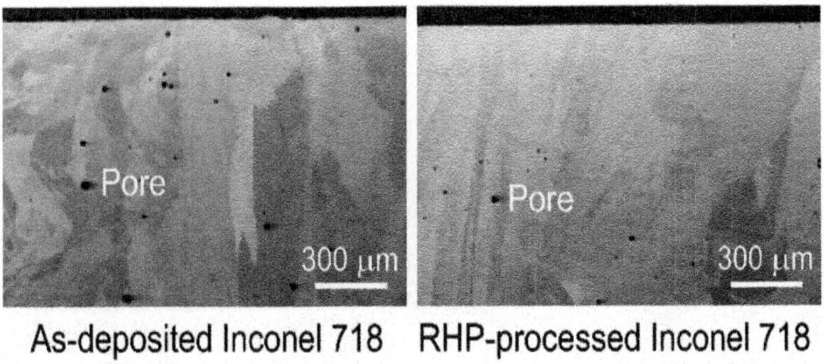

Fig. 3 Microscopic images taken from the cross-section of Inconel 718 coupons deposited by laser-cladding (LC) before (left) and after (right) robotic hammer peening (RHP)

apparently induced by the interpass rolling. We have also carried out RHP on laser-cladded Inconel 718 coatings. Based on the hardness enhancement measured from the cross-section of the coating layer near the surface regions by nanoindentation, we found that the effective thickness of the RHP process is over 1000 μm, which is close to the coating layer thickness deposited by a single pass laser-cladding. Pore closures have also been observed by introducing the RHP process, especially at the regions near the surface of the coating. Besides, the tensile thermal stress has been effectively converted to compressive stress after the RHP process. By introducing interpass machine hammer peening (MHP) into WAAM of Ti-6Al-4 V alloys [37], Neto et al. have observed that the texture can be eliminated while the size of the β-grains can be reduced from centimeters long to 1–2 mm. Both the yield strength and the ultimate tensile strength have been increased by the interpass MHP, i.e., to 907 MPa and 993 MPa, respectively, without degrading the ductility. The MHP-induced grain size reduction exhibited absence of considerable changes in thin and thick wall structures deposited by WAAM, suggesting its potential applications for large components.

5 Conclusion and Outlook

Recent development in AM exhibits promising potential for advanced manufacturing and remanufacturing of metallic alloys towards industrial applications, especially in aerospace, civil infrastructure, and clean energy, driven by zero emission and sustainable ecosystem requirements in response to global climate changes. Surface engineering and protective coating play important roles in advanced manufacturing and remanufacturing. Surface engineering techniques have been developed not only on traditionally manufactured alloy components but also on AM ones to enhance their surface integrities and thus the service lifetimes. The physical mechanism behind cold-working process is plastic deformation upon external mechanical impacts while that for laser treatments is related to melting and resolidification. Compared to hammer peening and shot peening, LSP has a larger depth of working effect while the plastic deformation introduced by LSP is relatively less intense than those by hammer peening or shot peening.

Some of the surface engineering techniques, e.g., laser treatments, RHP, rolling, etc., can be compatibly integrated with laser-DED for enhanced integrity of metal alloys and components with reduced necessity for post-DED treatments. Recent studies have shown that by introducing interpass cold-working process, the mechanical properties such as the yield and ultimate tensile strength can be effectively increased without apparently degrading the ductility of Ti-alloys. Effective depth of industrial hammer peening on AM alloys is quite close to the coating thickness deposited by a single pass laser-cladding, providing a promising strategy for AM towards advanced manufacturing and remanufacturing.

References

1. Lam HKS, Ding L, Cheng TCE, Zhou H (2019) The impact of 3D printing implementation on stock returns. Int J Oper Prod Manage 39(6/7/8):935–961
2. Kanishka K, Acherjee B (2023) A systematic review of additive manufacturing-based remanufacturing techniques for component repair and restoration. J Manuf Process 89:220–283
3. Barke A, Thies C, Melo SP et al (2023) Maintenance, repair, and overhaul of aircraft with novel propulsion concepts—analysis of environmental and economic impacts. Procedia CIRP 116:221–226
4. Velu R et al (2023) Processing challenges in additively manufactured single crystal alloys: a process–structure–property relationship approach. In: Kumar A et al (eds) Advances in additive manufacturing. Elsevier, pp 253–264
5. Lim WYS, Cao J, Suwardi A et al (2022) Recent advances in laser-cladding of metal alloys for protective coating and additive manufacturing. J Adhes Sci Technol 36(23–24):2482–2504
6. Liu H, Meng TL, Cao J, Tan CKI (2022) Advanced surface engineering and protective coating. Springer Singapore, Singapore, pp 138–141
7. Liu H, Ivan Tan CK, Wei Y et al (2020) Robotic hammer peening-induced martensite in austenitic steels: spatial distributions of plastic deformation and phase transformation. Procedia CIRP 87:297–301
8. Liu H, Meng TL, Cao J et al (2022) Comparisons on localized surface modifications of stainless steels induced by laser shock peening and robotic hammer peening. Procedia CIRP 108:118–122
9. Kalentics N, de Seijas MOV, Griffiths S, Leinenbach C, Logé RE (2020) 3D laser shock peening—a new method for improving fatigue properties of selective laser melted parts. Addit Manuf 33:101112
10. Gong N et al (2023) High energy laser-shock induced phase transformation and micro-spallation on surface of SS: the effect of stacking fault energy and deformation mechanisms. Appl Surf Sci 613:156013
11. Liu H, Wei Y, Tan CKI et al (2021) Laser-treatment-induced morphology and structure modifications of stainless steel: element segregations and phase evolutions. Mater Chem Phys 266:124570
12. Gong N, Wei Y, Meng TL, Karyappa R, Cao J, Tan CKI, Suwardi A, Zhu Q, Liu H (2023) Laser-treatment-induced surface integrity modifications of stainless steel. Mater Res Express 10(3):034002

13. Liu H, Tan CKI, Wei Y et al (2020) Laser-cladding and interface evolutions of inconel 625 alloy on low alloy steel substrate upon heat and chemical treatments. Surf Coat Technol 404:126607
14. Liu H, Tan CKI, Meng TL et al (2021) Hot corrosion and internal spallation of laser-cladded inconel 625 superalloy coatings in molten sulfate salts. Corros Sci 193:109869
15. Gong N, Meng TL, Cao J, Liu H et al (2023) Laser-cladding of high entropy alloy coatings: an overview. Mater Technol 38(1):2151696
16. Cunningham CR, Flynn JM, Shokrani A, Dhokia V, Newman ST (2018) Invited review article: strategies and processes for high quality wire arc additive manufacturing. Addit Manuf 22:672–686
17. Liu H, Tan CKI, Meng TL et al (2022) Direct deposition of low-cost carbon fiber reinforced stainless steel composites by twin-wire arc spray. J Mater Process Technol 301:117440
18. Gong N et al (2022) Machine learning guided understanding of microstructural entropy decay in bimodal grain-size distributed alloys with combination of high-strength and high-ductility. Materialia 26:101630
19. Tekeli S (2002) Enhancement of fatigue strength of SAE 9245 steel by shot peening. Mater Lett 57:604–608
20. Çakir FH, Öteyaka MÖ, Er Ü, Bozkurt F (2021) Enhancing wear resistance of AISI 304 alloy with shot peening and investigation of corrosion behaviour in marine water. Trans IMF 99(4):194–202
21. Wang JL, Huang MH, Xi XH, Wang CC, Xu W (2020) Characteristics of nucleation and transformation sequence in deformation-induced martensitic transformation. Mater Charact 163:110234
22. Liu H et al (2020) XRD and EBSD studies of severe shot peening induced martensite transformation and grain refinements in austenitic stainless steel. Mater Charact 168:110574
23. Misra RDK, Challa VSA, Injeti VSY (2022) Phase reversion-induced nanostructured austenitic alloys: an overview. Mater Technol 37(7):437–449
24. Sun Q, Han Q, Xu R, Zhao K, Li J (2018) Localized corrosion behaviour of AA7150 after ultrasonic shot peening: corrosion depth vs. impact energy. Corros Sci 130:218–230
25. Prabhakaran S et al (2018) Laser shock peening without coating induced residual stress distribution, wettability characteristics and enhanced pitting corrosion resistance of austenitic steel. Appl Surf Sci 428:17–30
26. Liu H, Tan CKI et al (2022) Laser-cladding and robotic hammer peening of stainless steel 431 on low alloy steel 4140 for surface enhancement and corrosion protections. J Adhes Sci Technol 36:2313–2327
27. Liu H et al (2020) Effects of hammer peening on structural properties of ni-based single-crystal superalloy: dislocation slip traces and crystallographic reorientations. Metallu Mater Trans A 51(6):3180–3193
28. Zuo P, Chen S, Wei M, Liang J, Liu C, Wang M (2019) Thermal behavior and grain evolution of 24CrNiMoY alloy steel prepared by pre-laid laser cladding technology. Opt Laser Technol 119:105613
29. Heigel JC, Gouge MF, Michaleris P, Palmer TA (2016) Selection of powder or wire feedstock material for the laser cladding of Inconel® 625. J Mater Process Technol 231:357–365
30. Brandt M, Sun S, Alam N et al (2009) Laser cladding repair of turbine blades in power plants: from research to commercialisation. Int Heat Treatment Surf Eng 3(3):105–114
31. Vaníček O et al (2022) Low-pressure turbine blade leading edge protection using robotic laser cladding technology. Int J Adv Manuf Technol 122(5):2543–2559
32. Shrivastava A, Mukherjee S, Chakraborty SS (2021) Addressing the challenges in remanufacturing by laser-based material deposition techniques. Opt Laser Technol 144:107404
33. Uusitalo MA, Vuoristo PMJ, Mäntylä TA (2004) High temperature corrosion of coatings and boiler steels below chlorine-containing salt deposits. Corros Sci 46(6):1311–1331
34. Joshi A, Patnaik A, Gangil B, Kumar S (2012) Laser assisted rapid manufacturing technique for the manufacturing of functionally graded materials. Students Conf Eng Syst 2012:1–3
35. Amado JM et al (2014) Laser cladding of Ni-WC layers with graded WC content. Phys Procedia 56:269–275

36. Hönnige J et al (2021) Study of residual stress and microstructural evolution in as-deposited and interpass rolled wire arc additively manufactured In718 alloy after ageing treatment. Mater Sci Eng A 801:140368
37. Neto L, Williams S, Ding J, Hönnige J, Martina F (2020) Mechanical properties enhancement of additive manufactured Ti-6Al-4V by machine hammer peening. Springer Singapore, Singapore, pp 121–132

Emerging Trends in Surface Engineering

Study of Palladium Reduction and Its Effect on the Electroless Nickel Coating Adhesion on Carbon-Fibre Reinforced Polyetherimide Composite

Zhaohong Huang, Yujie Zhou, and Hong Xie

Abstract Polymer composite is frequently coated with metals by various metallization processes, one of which is electroless nickel (EN) plating process. EN plating has been accepted by industry as a reliable method to produce the high quality, bright reflective surfaces. Aesthetically, the well-established bright metal finishes also offer plastic components with important functional advantages, such as wear resistance and hygienic properties. To improve EN coating adhesion on the polymer composite, chemical etching is used as the pre-treatment method prior to EN coating. However, the immediate consequence of the chemical etching is the exposure of the composite carbon fibre, leading to poor surface finish. To overcome the challenges, chemical-etching-free pre-treatment methods are desirable. In this paper, the key factors controlling the adhesion of EN coating on carbon fibre reinforced polyetherimide (CF-PEI) were investigated. Surface chemistry at different treatment stages was studied by X-ray photoelectron spectroscopy (XPS) analysis. It is found that reducing palladium (Pd) ions into Pd atoms on the composite surface takes a critical role to develop a good adhesion of EN coating to CF-PEI composite surface. When hypophosphite is used as a reducing agent, elevated temperature increases the density of catalytic Pd metal sites. With a stronger reducing agent sodium borohydride ($NaBH_4$), more adsorbed Pd ions can be reduced at room temperature than the high temperature hypophosphite. Even though $NaBH_4$ is used, there is still more than 30% adsorbed Pd ions yet to be reduced.

Z. Huang (✉) · Y. Zhou · H. Xie
Singapore Institute of Manufacturing Technology (SIMTech), Agency for Science, Technology and Research (A*STAR), 2 Fusionopolis Way, Innovis #08-04, Singapore 138634, Singapore
e-mail: zhhuang@simtech.a-star.edu.sg

Y. Zhou
e-mail: yjzhou@simtech.a-star.edu.sg

H. Xie
e-mail: hxie@simtech.a-star.edu.sg

Z. Huang · Y. Zhou
Advanced Remanufacturing & Technology Centre (ARTC), Agency for Science, Technology and Research (A*STAR), 3 Cleantech Loop, #01-01 CleanTech Two, Singapore 637143, Singapore

N. Maharjan and W. He (eds.), *Proceedings of the 3rd International Conference on Advanced Surface Enhancement (INCASE) 2023*, Lecture Notes in Mechanical Engineering, https://doi.org/10.1007/978-981-99-8643-9_44

Keywords Plastic plating · Metallization of polymer · Surface activation · Electroless nickel plating · Coating adhesion

1 Introduction

Components made of polymer and polymer composite are frequently coated with metals by various metallization processes, one of which is EN plating process. EN plating has been accepted by industry as a reliable method to produce the high quality, bright reflective surfaces. Aesthetically, the well-established bright metal finishes also offer plastic components with important functional advantages, such as wear resistance and hygienic properties. As well-known, polymer surfaces lack catalytic properties and therefore require activating treatments that render them catalytic surface for initial nickel deposition [1]. In general, this activation is done by seeding the surface with a catalytically active metal such as Pd. In this process, surface sensitizing is carried out at the first in a solution containing reducing ions such as stannous ions (Sn^{2+}). The sensitizing Sn^{2+} ions adhered on the polymer substrate can reduce the Pd metal ions (Pd^{2+}) into Pd metal (Pd^0) from a catalyst solution containing palladium chloride ($PdCl_2$) according to the follow chemical reaction [2, 3].

$$Pd^{2+} + Sn^{2+} \rightarrow Sn^{4+} + Pd^0$$

The alternate method of catalyzing and activating makes use of a mixed colloidal catalyst bath, so-called one-step activation. The colloid is reduced Pd metal nano particles stabilized by Sn^{2+} and Sn^{4+} ions [4].

The surface of an injection-molded plastic article normally is glossy and quite hydrophobic. Consequently, this surface is unreceptive to aqueous solutions used in electroless metal deposition. Since the sensitizing and activating solutions will not wet the surface, the metal ions are not adsorbed onto the surface and deposition of the metal cannot proceed. Many approaches rendering the polymer surface with hydrophilicity have been proposed, such as solvent conditioning [5, 6], mechanical surface roughening [7], and chemical etching [8].

All the pretreatments of polymer substrate will produce coarse surface, leading to adhesion improvement of EN coating by ether increasing surface area or forming microscale physical anchor layer being beneficial to the Pd catalyst adsorption. However, surface roughening techniques are not suitable for EN coating on the high precision components made of polymer composites. One example is the metallized carbon-fiber (CF) reinforced polyetherimide (PEI) composite, which has been widely used for electronics applications such as components for hard disk drives. The above-mentioned pretreatments prior to Pd activation and EN plating will strip off PEI polymer from the composite surface and then produce CF protrusion, as shown in Fig. 1. Besides the high precision requirements in electronics-related plating fields, plastic plating methods for circuit formation where resin-metal interface roughness

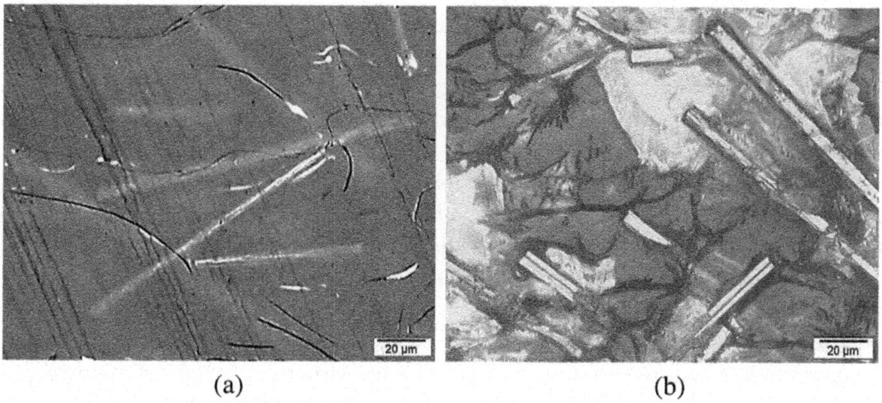

Fig. 1 Optical microscopic images of injection-molded PEI-30%CF composite substrates: **a** pristine, **b** CrO_3 etching in a solution containing CrO_3 (415 g/L) and H_2SO_4 (200 ml/L) at 70 °C for 3.5 min

must be minimized to avoid signal delay due to the skin effect have also been desired. For these reasons, many reports on new resin surface modification methods have been published [9–14].

The surfaces after modifications without chemical/mechanical etching/roughening, however, may lead to less physical/mechanical contribution to the coating adhesion. Therefore, effective bonding of Pd catalyst and the electroless plating may become more influential to the overall coating adhesion. Understanding of the mechanisms behind the effect of modification on the surface chemistry and thereby the adsorption of Pd catalyst is essential to explore new processes for pretreatment of EN plating without chemical/mechanical etching. In this study, by using X-ray photoelectron spectroscopy (XPS) the effect of surface modifications on adhesion of EN coating on PEI-CF composite was investigated. The XPS surface chemistry analysis reveals that Pd reduction takes an important role in adhesion of EN coating on polymer substrate.

2 Experimental Procedures

Testing coupons were made of 30% CF reinforced PEI (PEI-30%CF, State-Kon) by injection molding process. The composite coupons were first cleaned with isopropyl alcohol (IPA) in an ultrasonic bath for 5 min and then nitrogen (N_2) spray dried. The cleaned coupons were immersed in an aqueous adhesion promoter containing 5% polyethyleneimine (Sigma-Aldrich, Mw750,000) at 80 °C for 60 min, followed by deionized (DI) water rinsing in an ultrasonic bath for 5 min to remove the excess polyethyleneimine from the PEI-CF composite surface, and immersion in an aqueous solution containing $PdCl_2$ (0.5 g/L) and HCl (10 ml/L) at room temperature for

5 min for Pd ions adsorption onto the chemically modified surface. After DI water rinsing for 5 min to remove the free Pd ions, the composite coupons were soaked in a water-based solution containing either NaH_2PO_2 or $NaBH_4$ at room or elevated temperatures for 5 min to reduce the adsorbed Pd ions to Pd atoms for surface catalyzation.

The Pd catalyzed coupons were then immersed in an EN plating solution (RONAMAX™ SMT*, R&H) at 90 °C for 40 min, followed by DI water rinsing for 5 min and hot air dry by a hair drier prior to characterization. The sample preparation by following the above conditions is summarized in Table 1, denoted as #1, #2, #3, #4, and #5, respectively.

The characterization of the CF-PEI composite substrates with different surface treatments includes surface appearance and microstructure by optical microscopes, chemical analysis by XPS (Thermo Scientific), and adhesion by standard crosshatch tape test (ASTM D3359). XPS measurements were carried out with a monochromatized Al Kα X-ray source (1486.6 eV photons) at a power of 150 W (15 kV and 10 mA). The core-level spectra were obtained at the photoelectron take-off angle (α, with respect to the sample surface) of 90°. A pass-energy of 150 eV was used for the survey spectra and 40 eV for high-resolution scans. The pressure in the analysis chamber was maintained at 10^{-8} Torr or lower during each measurement. To compensate for surface charging effects, all binding energies (BEs) were referenced to the C 1 s hydrocarbon peak at 284.6 eV. A computer program was employed to perform de-convolution with which one experimental peak could be fit precisely by more than one Gaussian curves, and the relative surface concentration of the specific chemical species in question was obtained from the product of the peak area and the corresponding atomic sensitivity factor. In peak synthesis, the line width (full width at half maximum or FWHM) of Gaussian peaks was maintained constant for all components in a particular spectrum. Surface elemental stoichiometry was determined from the peak area ratios and were accurate to within ± 10%.

Table 1 Sample preparation with different treatments

Treatment conditions	Treatment time (min)				
	#1	#2	#3	#4	#5
Polyethyleneimine (5 g/L, H_2O), 80 °C	0	60	60	60	60
NaH_2PO_2 (30 g/L, H_2O), 25 °C	0	0	5	0	0
NaH_2PO_2 (30 g/L, H_2O), 98 °C	0	0	0	5	0
$NaBH_4$ (19 g/L, H_2O), 25 °C	0	0	0	0	5

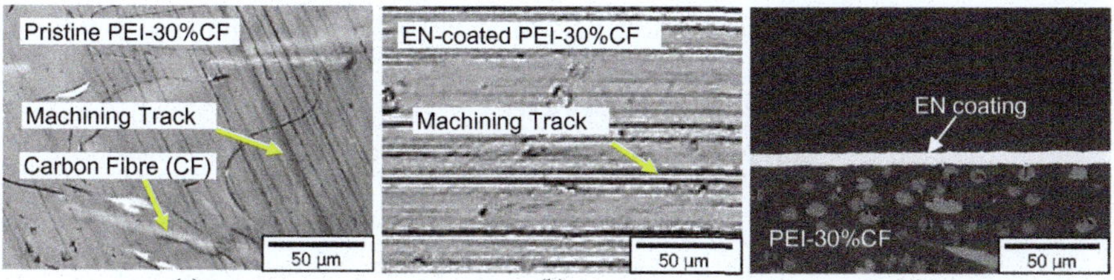

(a) (b) (c)

Fig. 2 Optical microscopic images of the injection-molded PEI-30%CF composite: **a** as-molded pristine surface, **b** the EN coated surface, and **c** cross-sectioned observation of the EN coated PEI-30%CF composite

3 Results and Discussion

3.1 Morphological Observation on the PEI-30%CF Composite with and Without EN Coating

Figure 2 shows the optical microscopic images of (a) the pristine PEI-30%CF composite molded by injection molding process, (b) the EN coated surface, and (c) cross-sectioned observation of the EN coated PEI-30%CF composite. It is noted that the straight dark lines on the surface of pristine PEI-30%CF (#1 in Table 1) as shown in Fig. 2a were replicated during injection molding from the conventional milling of ejector pin surface. The EN coating process didn't change the surface morphology significantly as shown in Fig. 2b, which can be further verified by microscopic observation of the cross-sectioned EN coated PEI-30%CF composite samples, as shown in Fig. 2c. The carbon-reinforced PEI composite surface is fully covered by about the 12 µm thick uniform EN coating without interface defect and carbon fiber protrusion.

3.2 Effect of Surface Treatment Conditions on EN Coating Adhesion

Crosshatch tape tests were carried out by following ASTM D3359 after EN coating on Pd catalyzed PEI-CF composite coupons with different reduction conditions (refer to Table 1) and summarized in Table 2. It can be seen that surface treatment conditions, in particular, the type of reducing agent and the process parameters, have significant effect on metal/composite adhesion. When the Pd reduction is carried out in an aqueous solution containing 30 g/L hypophosphite (NaH_2PO_2) at room temperature (25 °C), the EN coating adhesion is ranked as 2B (#3 in Table 2), which is evaluated as "Failed" according to industry standards for hard disk drive. When the reduction process is conducted in the same solution at 98 °C, a good adhesion ranked as 5B (#4 in Table 2) is obtained, which is the highest scale in the adhesion test (ASTM D3359).

Table 2 Summary on crosshatch tests of EN coated PEI-30%CF with different pretreatment processes

Treatment sample	EN coating and tape test		
	EN coating (min)	Tape test ranking	Evaluation
#3	40	2B	Failed
#4	40	5B	Pass
#5	40	5B	Pass

For benchmarking, a stronger reducing agent, $NaBH_4$, was used for reduction of the adsorbed Pd on the PEI-30%CF composite substrate at 25 °C (#5 in Table 2), followed by EN coating. The same adhesion testing result (ranked as 5B) as sample #4 was achieved. To understand the key factors controlling adhesion, XPS was employed to analyze the changes in surface chemistry of the composite substrate with various surface treatments.

3.3 *XPS Analysis of PEI-30%CF Surface with Different Treatment Conditions*

Effect of Polyethyleneimine Treatment on Surface Chemistry of PEI-30%CF

A basic requirement for substrate to be electrolessly plated is catalytic surface that provides active sites for reduction of metal ions. For EN plating, Pd metal is frequently used as its deposition catalyst. In an etching-free surface preparation of a polymer substrate for EN plating, the challenge is how to activate a polymer surface on which Pd metal catalytic particles are firmly fixed. One approach is modification of polymer surface with amine rich chemicals. In this study, a treatment in an aqueous solution containing branched polyethyleneimine is applied on the PEI-30%CF (#2 in Table 1).

XPS wide scan of PEI-30%CF substrate treated with polyethyleneimine is shown in Fig. 3b, in comparison with that of pristine sample (Fig. 3a). After polyethyleneimine treatment, N content is significantly increased from 4.32 at.% of the pristine sample to 16.37 at.%. The high N on the modified surface is clearly originated from the primary, secondary and ternary amines in the polyethyleneimine molecules, which are attached onto the PEI-30%CF composite surface.

In Fig. 4, C1s core-level spectra of polyethyleneimine treated PEI-30%CF can be curve-fitted with 3 peak components with different BEs at about 284.5, 285.5, and 288.5 eV, attributed to the species of C–C/C–H, C–N, and C=O, respectively [15–17]. Compared with the pristine sample, C–N species are significantly increased from 20 to 75.6 at.% due to the contribution of the attached polyethyleneimine. Meanwhile, detected C–C/C-H species are decreased from 72.9 at.% of pristine surface down to only 18.2 at.% of the polyethyleneimine treated surface, which implies that the density of amine groups are the control species after the polyethyleneimine treatment.

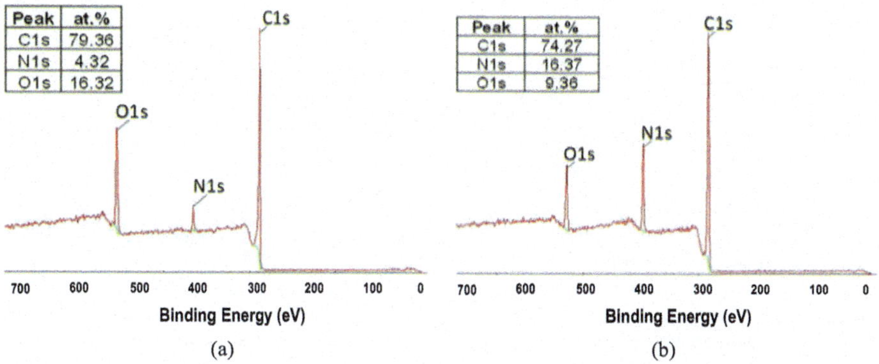

Fig. 3 XPS wide scan and elemental contents on the surface of PEI-30%CF composite before (**a**) and after (**b**) polyethyleneimine treatment

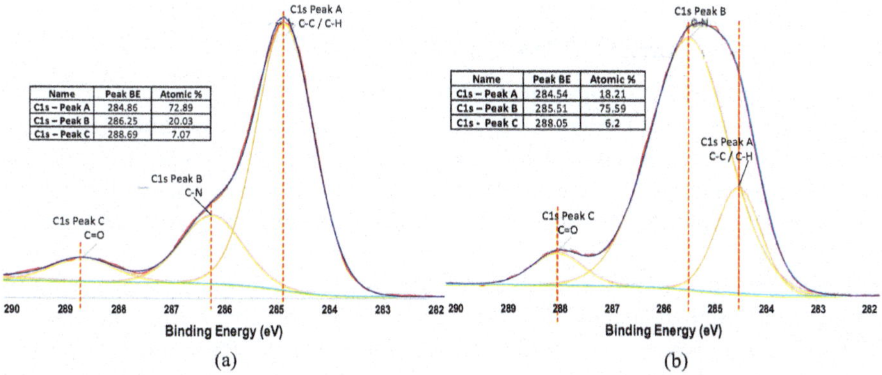

Fig. 4 Deconvoluted C1s core-level spectra of PEI-30%CF composite samples: **a** pristine, **b** treated with polyethyleneimine

N1s core-level spectra before polyethyleneimine treatment (pristine) and after polyethyleneimine treatment (a) and deconvolution of the treated sample (b) are given in Fig. 5. As shown in Fig. 5a, polyethyleneimine treatment leads to N1s peak right shift. The N1s core-level spectra of the polyethyleneimine treated PEI sample can be curve-fitted two peak components at BEs of 400.6 and 399.2 eV as shown in Fig. 5b, which is attributable to the –N–(C = O)$_2$ and NH/NH_2 species, respectively [15]. Although the imide nitrogen is still detected from the XPS analysis, its percentage in the total N is only 27.6 at.%, much lower than the amine (72.4 at.%) that are basically contributed by the attached polyethyleneimine. This result is consistent with the above XPS analysis based on C1s, where the content of the imide carbon (N–C = O) species is 6.2 at.% out of the total carbon, as shown in Fig. 4b.

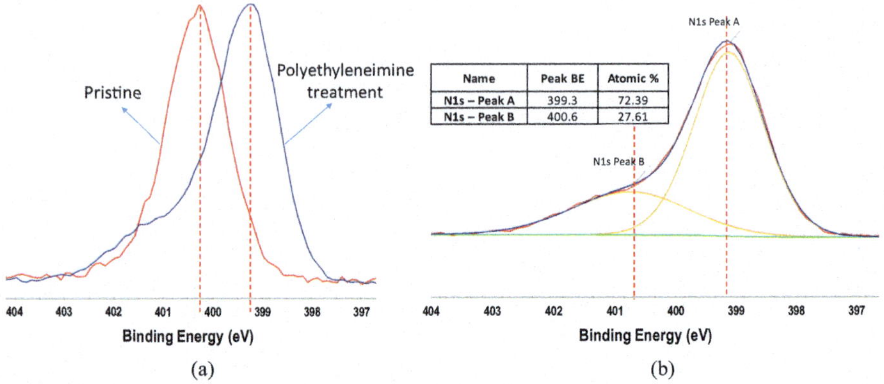

Fig. 5 N1s core-level spectra of PEI-30%CF composite treated with polyethyleneimine: **a** peak shift, **b** deconvoluted peaks

Pd Adsorption on the PEI-30%CF Modified With Polyethyleneimine

As mentioned above, in the chemical-etching-free activation process, no Sn^{2+} sensitization is needed, instead, the polyethyleneimine treated surface is directly immersed into the Pd^{2+} acidic solution, where the Pd^{2+} ions are chemically adsorbed onto the amine rich surface. To activate the surface, a reduction process is still needed to reduce the Pd^{2+} ions into the catalytic Pd^0 metal by reducing agents such as NaH_2PO_2 and $NaBH_4$.

Figure 6a shows the XPS wide scan on PEI-30%CF composite surfaces after the polyethyleneimine treatment and immersion in $PdCl_2$ solution. It is shown that after Pd immersion treatment, Pd and Cl are detectable and the Cl/Pd atomic ratio is about 2, which is identical to that of theoretical value of $PdCl_2$ and implies that Pd exists on the polymer surface in ionic state (Pd^{2+}).

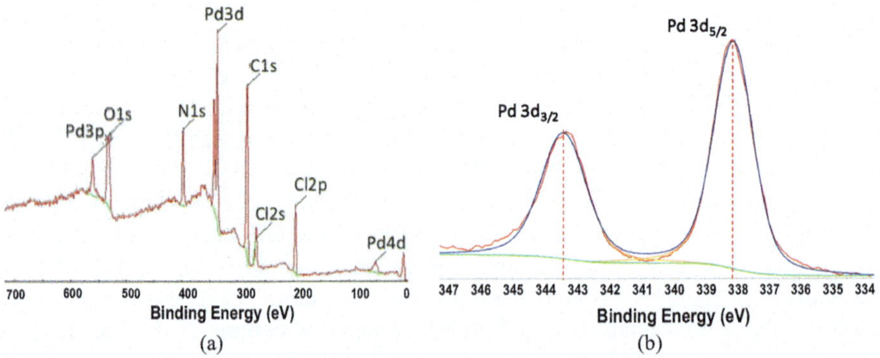

Fig. 6 XPS wide scan on the polyethyleneimine treated PEI-30%CF composite surface after Pd immersion treatment (**a**); Pd 3d core-level spectra (**b**)

It is noted that DI water rinsing was applied for 5 min after Pd immersion treatment in order to remove any water-soluble species from the polymer surface, including Pd^{2+} and Cl^-. Therefore, the presence of a high concentration of Pd and Cl on the amine rich surface is probably due to the formation of Pd–N complex on PEI polymer surface. As the primary and secondary amine groups of the attached polyethyleneimine molecules provide strong electron donors [15] in the formation of the Pd–N complexes.

This speculation on the Cl-Pd–N complex is supported by the Cl $2p_{3/2}$ peak shift to the low binding energy (197.9 eV), which is usually in the range of 198.5–199 eV for metal chloride, and the lower BE shift is attributed to the formation of Pd–N coordination bond. XPS Pd 3d core-level spectra in Fig. 6(b) with the BEs of the spin–orbit-split doublet (Pd $3d_{5/2}$ and Pd $3d_{3/2}$) lying at about 338 and 343 eV, respectively, suggest the presence of predominantly the Pd^{2+} species [16].

Reduction of Pd Ions with Different Reducing Agents and Working Conditions

The Pd at the Pd^0 state has been considered as the catalyst for the electroless metal deposition process [18]. Therefore, reduction of Pd^{2+} adsorbed on the amine modified PEI surface to be Pd metal (Pd^0) is essential and important step in the pretreatment of polymer substrate for EN plating. A first trial of the reduction is carried out in a water solution containing 3 wt% NaH_2PO_2 at 25 °C for 5 min.

XPS wide scan spectra (Fig. 7) on the above sample reveal that the reduction removes the adsorbed Cl but P peak appears. It is found there is no Na element detectable and the P/Pd ratio is 1.1:1, implying that some insoluble P compounds are likely formed and adsorbed onto the polymer surface during the reduction process by using hypophosphite water solution at room temperature, by following below chemical reactions:

$$Pd^{2+} + 2H_2PO_2^- + 2H_2O \rightarrow Pd^0 + 2H_2PO_3^- + 2H^+ + H_2$$

$$Pd^{2+} + 2H_2PO_3 \rightarrow Pd(H_2PO_3)_2 \downarrow$$

In this alkaline based reduction solution, part of Pd^{2+} may not be reduced into Pd^0 and possibly exist in the forms of insoluble $Pd(H_2PO_3)_2$. Another possibility causing the existence of P is formation of a complex compound in combination with amine, leading to high P/Pd ratio. Theoretically, P/Pd ratio in between 0:1 (100% reduction of the adsorbed Pd^{2+}) and 2:1 (0% reduction of adsorbed Pd^{2+}). The measured 1.1:1 of P/Pd ratio implies that there are substantial Pd ions were not reduced into Pd metal.

This speculation can be further verified from the high resolution XPS Pd 3d core-level spectra after reduction in hypophosphite solution at 25 °C, as shown in Fig. 8a. After Pd immersion and reduction, the spin–orbit-split doublet (Pd $3d_{5/2}$ and Pd $3d_{3/2}$) at BEs of 338 and 343 eV are still the main peaks, implying the presence of predominant Pd^{2+} species. Besides the Pd^{2+} doublet, a new doublet (Pd $3d_{5/2}$ and Pd $3d_{3/2}$) at lower BEs of 335 and 340 eV is observed, which is assigned to the Pd^0 species [16]. The fraction of Pd^0, however, is only about 30%, implying an

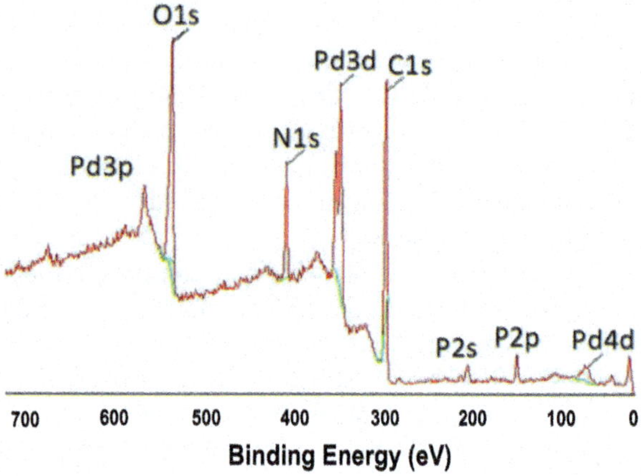

Fig. 7 XPS wide scan on the polyethyleneimin treated PEI-30%CF composite surface after reduction treatment in water solution containing 3 wt% NaH_2PO_2 at 25 °C for 5 min

insufficient reduction. In other words, 70% of Pd element is in ionic status, which may be responsible to the poor EN adhesion as indicated in Table 2.

In order to improve redution efficiency from Pd^{2+} to Pd^0, the reduction treatment in hypophosphite solution is conducted at an elevated temperated, 98 °C. The high resolution XPS Pd 3d core-level spectra after reduction is shown in Fig. 8b. Pd^0 is a predominant species after the treatment temperature is elevated, which increases the Pd^0 portion from 30% up to 58%. Therefore, the resulting high density of Pd^0 plays an important role in improving the EN coating adhesion.

The above results show that the reduction of Pd^{2+} to Pd^0 has remarkable effect on the adhesion of EN coating, which relies on the density of Pd^0 metal particles as catalytic sites on the polymer surface. To further prove this speculation, a strong reducing agent $NaBH_4$ is used to perform the reduction of adsorbed Pd^{2+}. The high resolution XPS Pd 3d core-level spectra after reduction in $NaBH_4$ water solution is shown in Fig. 8(c). This strong reducing agent reduces about 62% of adsorbed Pd^{2+} into Pd^0 metal, higher than that of hypophosphite at high temperature (98 °C). After peak decomposition, a minor doublet with the Pd $3d_{5/2}$ and Pd $3d_{3/2}$ peak components around 336 and 341 eV, respectively, indicates the existence of a third Pd species [19, 20]. The exact nature of this state of Pd is yet to be determined but is probably associated with an NH_2-Pd complexes [15, 21].

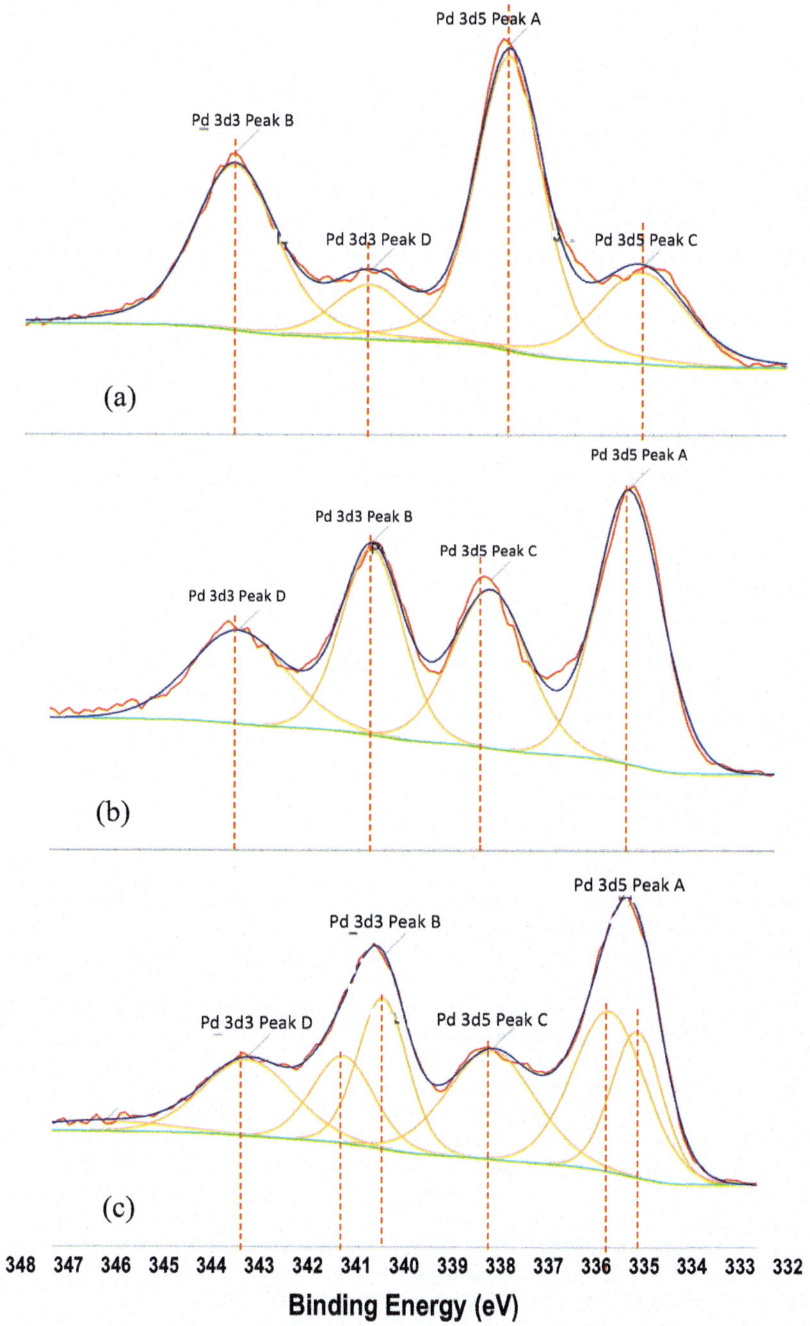

Fig. 8 Pd 3d core-level spectra of PEI-30%CF composite after Pd activation in $PdCl_2$ acidic solution and redution in NaH_2PO_2 water solution at 25 °C (**a**) and 98 °C (**b**), and $NaBH_4$ water solution at 25 °C (**c**)

4 Conclusion

To understand the key factors controlling the adhesion of EN coating on to PEI-30%CF composite substrate, surface chemistry at different treatment conditions was studied by XPS analysis. It was found that amine species can be grafted through a treatment with polyethyleneimine at elevated temperature, which is essential for Pd activation prior to EN coating on a PEI composite without chemical etching. A further finding is that the reduction of adsorbed Pd ions on the amine-grafted composite takes a critical role in developing a good adhesion of EN coating to the polymer. When hypophosphite is used as a reducing agent, elevated temperature should be applied for increasing the density of catalytic Pd metal particles. When a strong reducing agent is used, e.g., $NaBH_4$, more adsorbed Pd ions can be reduced at room temperature than the high temperature hypophosphite. Even the strong reducing agent is used, there is still more than 30% adsorbed Pd ions yet to be reduced. The mechanism behind the finding is still not clear, which is worth to further investigate.

References

1. Marton JP, Schlesinger M (1968) J Electrochem Soc 115:16
2. Kuzmik JJ (2010) Plating on plastics. In: Schlesinger M, Paunovic M (eds) Modern electroplating, 5th ed. Wiley
3. Furness RW (1968) The practice of plating on plastics. D.L. Robert, U. K. Surrey, p 41 (Chapter 8)
4. US Patent No. 3,011,920
5. Karas BR, Foust DF, Dumas WV, Lamby EJ (1991) Metallized plastics 2 (Mittal KL (ed)). Plenum Press, New York, p 387
6. US Patent No. 4,775,449
7. McCaskie JE (2006) Plating on plastics: a survey of mechanisms for adhering metal films to plastic surfaces. Met Fin 104(5):31–39
8. Mandich NV, Krulik GA (1993) On the mechanisms of plating on plastics. Plat Surf Finish 80:68–72
9. Bessho T, Yoshinaga F, Inoue K, Koiwa I, Honma H (2006) J Surf Finish Soc Jpn 9:162
10. Bessho T, Tashiro K, Sugimoto M, Honma H (2006) J Jpn Inst Electron Packaging 74:472
11. Inoue K, Mathui K, Watanabe M, Honma H (2008) J Surf Finish Soc Jpn 1:47
12. Sugimoto M, Honma H (2008) J Surf Finish Soc Jpn 59:294
13. Yoshikane Y, Nagao T, Yoshikawa J, Okumura H (2008) J Surf Finish Soc Jpn 59:288
14. Baba K, Nishimura N, Watanabe M, Honma H (2010) J Jpn Inst Electron Packaging 113:383
15. Zhang Y, Tan KL, Yang GH, Kang ET, Neoh KG (2001) J Electrochem Soc 148:C574
16. Moulder JF, Stickle WF, Spbol PE, Bomben KD (1992) In: Chastian J (ed) X-ray photoelectron spectroscopy. Perkin-Elmer, Eden Prairie, NN
17. Vuori L, Hannula M, Lahtonen K, Jussila P, Löytty HA, Hirsimäki M, Pärna R, Nõmmiste E, Valden M (2014) Appl Sur Sci 317:856–866
18. Wang WC, Kang ET, Neoh KG (2002) Appl Surf Sci 199:52–66
19. Ma ZH, Tan KL, Kang ET (2000) Synth Met 114:17–25
20. Dressick WJ, Dulcey CS, George JH, Calabrese GS, Calvert JM (1994) J Electrochem Soc 141:210
21. Yang GH, Kang ET, Neoh KG, Zhang Y, Tan KL (2001) Langmuir 17:211

Two-Step Post-processing Treatment to Improve Additive Manufactured AlSi10Mg Surface Finish

Yinsong Huang and Yong Teck Tan

Abstract Additive manufactured (AM) aluminum alloys show good performance as lightweight critical parts for different industrial applications. However, the poor surface finish of as-printed parts due to the "Staircase Effect" is an obstacle to the implementation of AM for industry practice. In this study, a two-step post processing technique was developed to improve surface finish of 3D-printed AlSi10Mg. In the first step, the as-printed AlSi10Mg sample was chemically etched to remove the outer skin oxide layer and the dangling loose powder particles; in the second step, the chemically treated sample was electrochemically polished in an alkaline electrolyte to further smoothen the surface. The surface morphology of treated and reference parts were observed with SEM; the surface roughness (Ra) was measured with stylus profilometer to understand the smoothening mechanism and the effects of the process parameters. After this post processing treatment, the Ra was found to be typically decreased by at least 50%, for a much-improved surface finish.

Keywords Additive manufacturing · Post-processing technique · Chemical polishing · Electrochemical polishing

Y. Huang (✉) · Y. T. Tan
Agency for Science, Technology and Research (A*STAR), Singapore Institute of Manufacturing Technology (SIMTech), 2 Fusionopolis Way, Innovis #08-04, Singapore 138634, Republic of Singapore
e-mail: huangys@simtech.a-star.edu.sg

Advanced Remanufacturing & Technology Centre (ARTC), Agency for Science, Technology and Research (A*STAR), 3 Cleantech Loop, #01-01 CleanTech Two, Singapore 637143, Republic of Singapore

Y. T. Tan
e-mail: tanyongt@simtech.a-star.edu.sg

N. Maharjan and W. He (eds.), *Proceedings of the 3rd International Conference on Advanced Surface Enhancement (INCASE) 2023*, Lecture Notes in Mechanical Engineering, https://doi.org/10.1007/978-981-99-8643-9_45

1 Introduction

Additive manufacturing (AM) has wide applications across a range of industries. However, some disadvantages, such as lower strength, printing accuracy limits, poor surface finish due to the "Staircasing Effect", are obstacles to the implementation of AM in practice [1, 2]. Post-processing is therefore an essential stage of the AM process, to improve the quality of AM parts and ensure that the AM parts meet their design specifications [3].

AlSi10Mg is a common additive manufactured alloy suitable for use in areas such as aerospace engineering and the automotive industry [4]. Although there a lot of technologies and process are reported to improve the AM AlSi10Mg surface quality, including traditional methods such as machining, abrasion, shot peening, etc. and non-traditional methods such as chemical polishing, electrochemical polishing, laser machining, etc., these one-step post processing techniques have their limitations [5–7]

In this study, a two-step post processing technique was introduced to improve surface finish of 3D-printed AlSi10Mg.

2 Experimental

100 × 100 × 10 mm AlSi10Mg plates were fabricated by DMLS (Direct Metal Laser Sintering) using EOS M290 metal 3D printer with process parameter setting of layer thickness 30 um, hatch spacing 0.19 mm, laser power = 370 W, scan speed 1300 mm/s, and with a generic particle size between 25 and 70 μm. Table 1 shows the nominal chemical composition of the AlSi10Mg powder used for printing. No further post-treatment was done after printing. The fabricated plates were cut into smaller pieces for chemical and electrochemical post processing treatment to improve their surface finish.

The surface morphologies of as-printed and post-treated AM AlSi10Mg were observed with JEOL IT300 LV SEM, and the surface compositions were analyzed with Energy-dispersive X-ray spectroscopy (EDX). The surface roughness (Ra) of the upskin was analyzed with Taylor-Hobson Form Talysurf Series 2 profilometer.

The chemical etching solution (composition given in Table 2) was made from nitric acid, hydrofluoric acid, and deionized water. Electrochemically treatment of above chemical etched AlSi10Mg was conducted in an alkaline electrolyte made of sodium citrate and sodium hydroxide, the composition and process parameters listed in Table 3.

Table 1 Powder chemical composition (wt%)

Element	Al	Si	Fe	Cu	Mn	Mg	Ni	Zn	Pb	Sn	Ti
wt%	Balance	9–10	< 0.55	< 0.05	< 0.45	0.25–0.45	< 0.05	< 0.10	< 0.05	< 0.05	< 0.15

Table 2 Chemical etching bath compositions and conditions

HNO_3 (65%) (ml)	HF (48%) (ml)	Deionized water (ml)	Temperature (°C)	Time (min)
50	25	925	80–90	10–30

Table 3 Electrochemical treatment electrolyte compositions and process parameters

Sodium citrate (g/L)	Sodium hydroxide (g/L)	Temperature (°C)	Voltage (V)	Time (mins)
20–30	5–10	45–60	6–15	10–30

3 Results and Discussion

Figure 1 shows the as-printed 3D AlSi10Mg upskin morphologies. It can be observed that there are many partially melted powder particles dangling on the surface, melting pools and holes if the dangling powder particles are torn away, which becomes increasingly visible under high magnification. Such defects result in high surface roughness of Ra ~ 7.18 μm, and Rz ~ 48.10 μm which is corresponding the average as-received AlSi10Mg powder particle size.

During chemical etching, nitric acid first oxidized dissolving Al and Si and weaken the link of dangling particles on the 3D printed substrate surface, hydrofluoric acid dissolved away the outer surface skin oxide layers and etching product of Al_2O_3 and SiO_2 to maintain a uniform whole surface etching. In processing, nitric acid is contributes to etch off the dangling loose particles, and hydrofluoric acid's function is to dissolve the passivated surface oxide layers and etching residues to provide a fresh surface for next electrochemical treatment to further improve the substrate surface finishing. Following reactions contribute to the surface smoothing:

$$2Al + 6HNO_3 = Al_2O_3 + 6NO_2 + 3H_2O$$

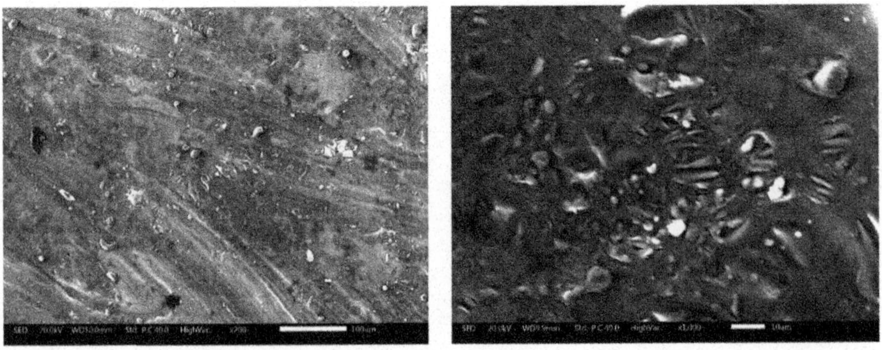

Fig. 1 SEM morphologies of as-printed 3D AlSi10Mg upskin (left 200X magnification, right 1000X magnification)

$$Si + 4HNO_3 = SiO_2 + 4NO_2 + 2H_2O$$

$$Al_2O_3 + 6HF = 2AlF_3 + 3H_2O$$

$$SiO_2 + 4HF = SiF_4 + 2H_2O$$

As AlF_3 is an insoluble product and adheres to the sample surface forming a smut layer, after every 5 min of chemical etching treatment, the sample was ultrasonic cleaned with DI water for 5 min.

Figure 2 shows the surface morphology of 3D AlSi10Mg upskin after 20 min chemical etching treatment. The dangling partially melted particles were all removed, and the outer skin oxide layer was dissolved away. Overlapping, segregated melt pools with distinct boundaries are clearly visible and void defects were observed [8]. The surface roughness of Ra and Rz are reduced to ~4.50 μm and ~23.76 μm, respectively. Table 4 gives the surface roughness after different durations of chemical etching.

The current–voltage relationship of 3D AlSi10Mg electropolishing in alkaline electrolyte was monitored and is shown in Fig. 3. According to Jacquet's study [9], when a positive voltage is applied, etching, passivating, polishing and pitting effects is happening with the increased voltage. However, despite the well-established current–voltage relationship, optimization of electrochemical polishing parameters is still

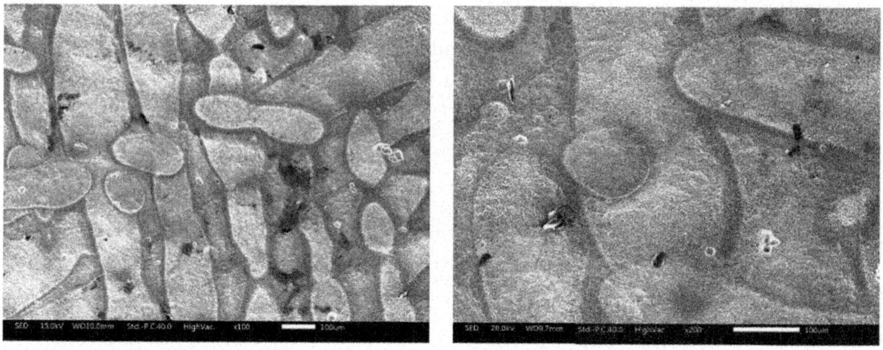

Fig. 2 SEM images of 3D AlSi10Mg upskin after 20 min chemical etching (left 100X magnification, right 200X magnification)

Table 4 Surface roughness (Ra, Rz) of AlSi10Mg after chemical etching

	As-printed	After chemical etching			
		5 min	10 min	15 min	20 min
Ra (μm)	7.18	5.44	5.22	4.70	4.50
Rz (μm)	48.10	30.92	28.02	24.20	23.76

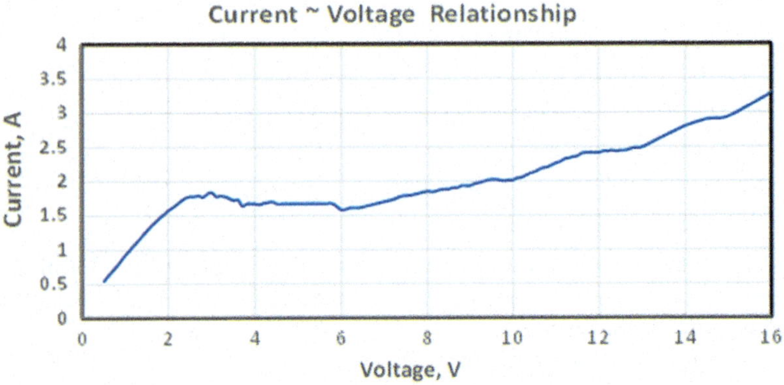

Fig. 3 Current–voltage relationship of 3D AlSi10Mg electropolishing in alkaline electrolyte

necessary to achieve best polishing results. The surface roughness of samples treated at different voltages for 10 min are summarized in Table 5. The effective polishing voltage is located between 8 and 14 V and the most effective is 12 V. The surface roughness increased when the applied voltage increased to 14 V, which may be an indication of pitting.

Figure 4 shows the AlSi10Mg upskin surface morphologies after the 2nd post-processing step of 10 min alkaline polishing at 12 V. It can be seen that most void defects in the melting pools have disappeared and the surface is smoothed although the segregated melt pool boundaries are clear. It can be inferred that a post-printing heat treatment and/or further optimization of the printing parameters are necessary.

Figure 5 shows visual images of as-printed 3D AlSI10Mg, after 1st step of 20 min chemical etching and 2nd step of 10 min of 12 V electrochemical polishing in alkaline electrolyte.

Table 5 Surface roughness (Ra, Rz) of AlSi10Mg after chemical etching and electrochemical polishing

	After chemical etching (20 min)	Electrochemical polishing after chemical etching			
		8 V/10 min	10 V/10 min	12 V/10 min	14 V/ 10 min
Ra (μm)	4.50	4.45	3.46	3.14	5.60
Rz (μm)	23.76	24.25	17.71	15.58	25.86

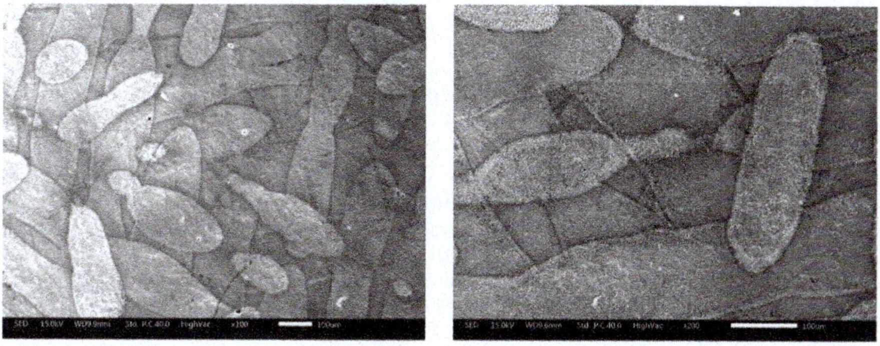

Fig. 4 SEM images of 3D AlSi10Mg upskin after 2-step post processing (left: 100X magnification, right 200X magnification)

Fig. 5 Visual images of 3D AlSi10Mg at different post-processing stages

4 Conclusion

A two-step post-processing treatment was applied to additive manufactured AlSi10Mg part to improve the surface finishing. The 1st step of chemical etching treatment contributed to remove dangling or partially melted particles, and also to dissolve the outer skin oxide layer. During the 2nd step of alkaline electrochemical polishing treatment, the void defects of melting pools are smoothed. After the two-step post-processing treatment, the upskin surface roughness was found to be decreased from 7.18 μm Ra (as-printed) to 3.14 μm Ra (chemical etched and electropolished), with the surface finishing much improved by more than 50%.

Acknowledgements The authors are grateful for the financial support from Singapore Institute of Manufacturing Technology (SIMTech), Agency for Science, Technology and Research (A*STAR), under in-house project C20-S2-030.

References

1. Townsend A, Senin N, Blunt L et al (2016) Surface texture metrology for metal additive manufacturing: a review. Precis Eng 46:34–47
2. Abdulhameed O, Al-Ahmari A et al (2019) Additive manufacturing: challenges, trends, and applications. Adv Mech Eng 11(2):1–27
3. Kumbhar NN, Mulay AV et al (2018) Post processing methods used to improve surface finish of products which are manufactured by additive manufacturing technologies: a review. J Inst Eng India Ser C. https://doi.org/10.1007/s40032-016-0340-z
4. Aversa A, Marchese G et al (2019) New aluminum alloys specifically designed for laser powder bed fusion: a review. Materials 12:1007. https://doi.org/10.3390/ma12071007
5. Hassanin H et al (2018) Surface finish improvement of additive manufactured metal parts. https://doi.org/10.1007/978-3-319-68801-5_7
6. Chang S, Liu A et al (2019) Highly effective smoothening of 3D-printed metal structures via overpotential electrochemical polishing. Mater Res Lett 7(7):282–289
7. Mu J, Sun T et al (2023) Application of electrochemical polishing in surface treatment of additively manufactured structures: a review. Prog Mater Sci 136:10110
8. Ng GKL, Jarfors AEW et al (2009) Porosity formation and gas bubble retention in laser metal deposition. Appl Phys A 97:641
9. Jacquet P (1936) On the anodic behavior of copper in aqueous solutions of orthophosphoric acid. Trans Electrochem Soc 69:629–655

Improvement of All Solid-State Lithium Metal Battery Performance with Sulfide and Oxide Electrolyte by Shot Peening

Manabu Kodama, Kanta Miyamoto, Kai Takashima, and Shuichiro Hirai

Abstract The all solid-state lithium metal battery is considered a promising technology for electric vehicles due to its high theoretical capacity. However, during fast charging, the growth of lithium dendrites causes internal short circuits, which limits the charging rate. This study investigates the fast charging performance of sulfide and oxide solid electrolytes, which are representative solid electrolytes, by suppressing crack formation in the electrolyte through shot peening. The results show that shot peening did not increase the charging rate of sulfide solid electrolytes, but increased it by 10 times in the case of oxide solid electrolytes. Shot peening increased the fracture toughness of the oxide solid electrolyte by about 2 times, which is believed to have contributed to the increase in charging rate. The difference in the results for the sulfide and oxide solid electrolytes is due to the difference in their pellet production process. The oxide solid electrolytes are produced by high-temperature sintering, which results in compressive residual stresses that increase fracture toughness when shot peening is performed. However, the sulfide solid electrolytes are formed into pellets by cold pressing, and sintering is not performed after pressing, so there is residual stress by the pressing, resulting in no increase in fracture toughness due to shot peening.

Keywords Shot peening · All solid-state battery · Lithium-metal anode

1 Introduction

The solid-state lithium-ion battery is a type of battery that replaces the liquid electrolyte used in existing lithium-ion batteries with a solid electrolyte. While the liquid electrolyte used in existing lithium-ion batteries has high ionic conductivity, it is flammable due to the use of organic solvents and is subject to severe temperature

---

M. Kodama (✉) · K. Miyamoto · K. Takashima · S. Hirai
Tokyo Institute of Technology, 2-12-1 Ookayama, Meguro, Tokyo 152-8550, Japan
e-mail: tanaka.m.ay@m.titech.ac.jp

N. Maharjan and W. He (eds.), *Proceedings of the 3rd International Conference on Advanced Surface Enhancement (INCASE) 2023*, Lecture Notes in Mechanical Engineering, https://doi.org/10.1007/978-981-99-8643-9_46

limitations due to freezing at low temperatures. On the other hand, solid-state lithium-ion batteries do not use liquid electrolytes and instead use inorganic solid electrolytes as the ion transport media between the positive and negative electrodes. Because inorganic solid electrolytes are made of inorganic materials, they are non-flammable and can operate at temperatures as low as −20 °C and as high as over 100 °C, with some variation in ionic conductivity [1]. In addition, since the transport number of solid electrolytes is 1, concentration polarization, which is one of the causes of voltage loss (overvoltage) during charging and discharging of lithium-ion batteries, does not occur. In addition, ionic conductivity does not decrease due to changes in lithium-ion concentration (diffusion coefficient reduction), making it suitable for high-speed charging and discharging. As described above, solid-state lithium-ion batteries have significant advantages over existing lithium-ion batteries that use liquid electrolytes. However, they have not yet been commercialized as automotive batteries, in part due to the ionic conductivity problem of solid electrolytes. Until the early 2000s, only an ionic conductivity about 2–3 orders of magnitude lower (~0.1 mS/cm) than that of organic electrolytes (~10 mS/cm) had been achieved.

Through the efforts of researchers in the field of materials science, solid electrolytes based on sulfide compounds have been developed, especially in Japan, resulting in improved ionic conductivity. In 2002, Hayashi and Tatsumisago investigated the mechanical milling method and achieved 0.72 mS/cm using Li_2S-P_2S_5 glass, and subsequently achieved 5 mS/cm using Li_2S-P_2S_5 glass ceramic by sintering the material [2, 3]. In 2011, Kanno et al. achieved 12 mS/cm using $Li_{10}GeP_2S_{12}$ [4], and in 2016, they achieved 25 mS/cm using $Li_{9.54}Si_{1.74}P_{1.44}S_{11.7}Cl_{0.3}$, surpassing the ionic conductivity of liquid electrolytes [5]. Research has also been conducted on solid oxide electrolytes, which are ceramic materials that exhibit lithium ionic conductivity. In 2007, Weppner et al. achieved 0.77 mS/cm with $Li_7La_3Zr_2O_{12}$, which has a garnet structure [6], and research to increase ionic conductivity by doping with various metals have been conducted. Recently, Huan et al. [7] achieved 1.2 mS/cm with $Li_{6.25}Ga_{0.25}La_3Zr_2O_{12}$. Compared to sulfide solid electrolytes, oxide solid electrolytes have about an order of magnitude lower ionic conductivity, but are superior in terms of electrochemical stability (i.e., they have a wider potential window). In addition, sulfide solid electrolytes are easily degraded by moisture in the air, requiring handling in a dry room or glove box, whereas oxide solid electrolytes have high resistance to water vapor and can be handled in ambient air. Various other solid electrolytes have been developed, such as polymer-based solid electrolytes [8], halide-based solid electrolytes [9], and antiperovskite-type solid electrolytes [10], but to date, the above-mentioned sulfide and oxide solid electrolytes are the only ones that have achieved high ionic conductivity using readily available starting materials and are suitable for mass production.

As described above, the all-solid-state battery, which has been a challenge for practical use as an automotive battery due to insufficient ionic conductivity, is now being solved. The solid-state battery is expected to provide high safety and fast charging, as well as a slight increase in energy density due to the reduction of the battery cooling system. On the other hand, no significant improvement in energy density can be expected from lithium-ion batteries using existing electrolytes, even with the use of new solid electrolytes. This is because the solid-state battery is only a modification of the electrolyte of existing lithium-ion batteries, and as long as metal oxides such as lithium cobalt oxide are used as the active material of the positive electrode and graphite is used as the active material of the negative electrode, the energy density per unit volume and per unit weight can only be equivalent because the storage materials for lithium in each electrode are the same. In today's electric vehicles, the driving range is shorter than that of gasoline vehicles, and this is one of the challenges for popularization. Considering that the driving range depends on the energy density of the secondary battery installed in the vehicle, further improvement of the energy density of all-solid-state batteries is essential for the future popularization of electric vehicles.

To achieve high energy density, a lithium metal negative electrode using lithium metal itself instead of graphite has been proposed. Figure 1 shows the conventional lithium-ion battery with metal oxide anode, liquid electrolyte and graphite anode, and the all solid-state battery with lithium metal oxide anode, solid electrolyte and lithium metal anode. While the theoretical energy density of graphite is 372 mAh/g, the energy density of lithium metal is approximately 10 times higher at 3860 mAh/g, making high capacity possible. The use of lithium metal in the negative electrode has been considered for lithium-ion batteries using electrolytes, but practical implementation has not been achieved. When lithium metal is used in the negative electrode, lithium is transported from the positive electrode to the negative electrode during charging, and metallic lithium is deposited on the surface of the lithium metal. However, it is known that during this process, dendrites (tree-like crystals) form and grow from the negative electrode toward the positive electrode, piercing the porous separator filled with electrolyte and causing a short circuit between the positive and negative electrodes, resulting in loss of battery function. In liquid electrolyte lithium batteries, it is easy for dendrites to penetrate the separator and cause loss of battery function because the separator is a porous membrane filled with electrolyte. However, in solid-state batteries, dendrites are believed to be less likely to penetrate the separator, which is a dense solid electrolyte, compared to lithium batteries using existing electrolytes. Therefore, the use of lithium metal as the negative electrode is being investigated. However, it has been found that even in all-solid-state lithium batteries using a solid electrolyte as a separator, dendrites precipitate within the separator under high load conditions or long cycle times, leading to loss of battery function.

Figure 2 illustrates the mechanism of dendrite growth in an all-solid-state battery. The following mechanisms are believed to be involved in the growth of dendrites in an all-solid-state battery using a lithium metal anode. First, due to the uneven contact between the solid electrolyte and the lithium metal, or the distribution of grain boundaries in the electrolyte, the deposition of lithium becomes uneven. Then, due to the uneven deposition, stress concentration occurs in the electrolyte, leading to the formation of cracks. Lithium is then deposited inside the cracks. As lithium is deposited inside the cracks, the cracks progress by applying stress to them, and lithium continues to advance toward the counter electrode. This process is repeated until the lithium reaches the positive electrode, causing the battery to short circuit internally.

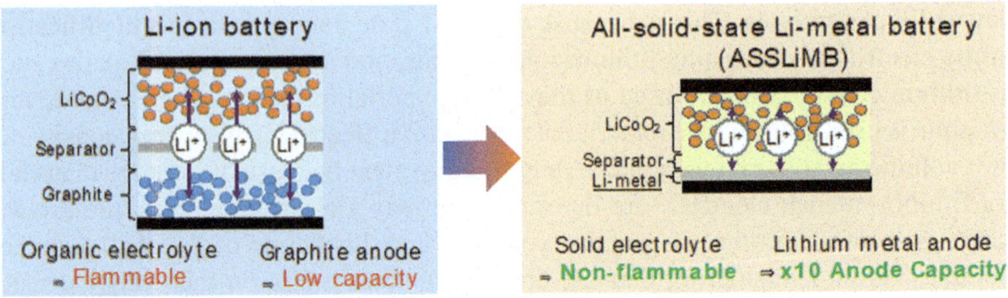

Fig. 1 Comparison of conventional lithium ion battery with liquid electrolyte and all solid battery with lithium metal anode

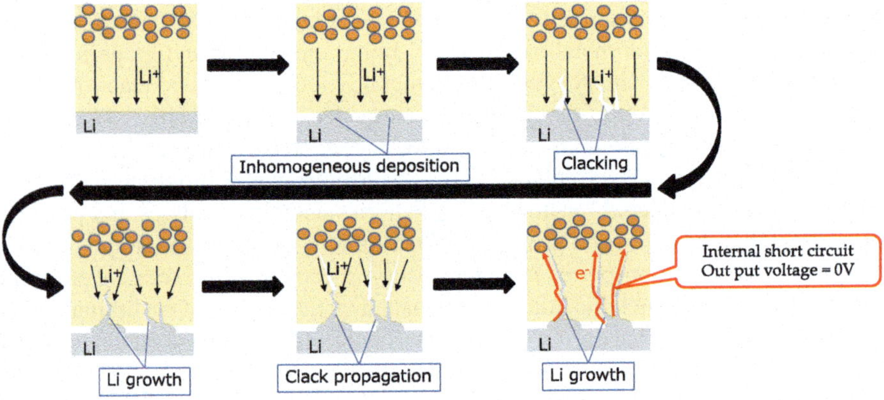

Fig. 2 Dendrite growth mechanism in all solid battery

Considering this mechanism, if the crack formation in the electrolyte can be suppressed, the dendrite growth can be reduced and the charging rate can be improved. To this end, the authors' research group proposed and demonstrated a method of shot peening the oxide-based solid electrolyte of an all-solid-state lithium metal anode battery. Shot peening is a process in which abrasive particles are blasted onto a surface using high-pressure gas, and the impact of the particles imparts residual compressive stress to the surface, thereby improving fracture toughness and suppressing crack formation. Experimental results showed that after shot peening the surface of the solid electrolyte in contact with the lithium metal anode, batteries were constructed and charged and discharged. As a result, the charging rate was shown to increase compared to batteries without shot peening [11].

As mentioned earlier, there are several types of solid electrolytes that can be used in all-solid-state batteries, not just the oxide-based solid electrolyte used for the shot peening test. In particular, sulfide-based solid electrolytes are attractive materials for use in high-power batteries due to their high ionic conductivity. In this study, we compared and examined the effect of shot peening on the high-speed charging of lithium metal anode all-solid-state batteries using oxide-based and sulfide-based solid electrolytes, and investigated the effect of shot peening on the electrolyte type.

2 Experimental Procedures

2.1 *Synthesize of Oxide Solid Electrolyte*

In this study, $Li_{6.25}Ga_{0.25}La_3Zr_2O_{12}$, which has the highest reported ionic conductivity, was used as the oxide-based solid electrolyte [7]. Figure 3 shows the process of preparing the oxide-based solid electrolyte separator in this study. The starting materials were La_2O_3, ZrO_2, Ga_2O_3, and LiOH. The materials are weighed in a glove box to avoid errors due to moisture and carbon dioxide adsorption in lanthanum oxide and weighing errors due to deliquescence of lithium hydroxide. After weighing, the powders are placed in a mortar and mixed for 5 min with a strong magnetic stirrer (Fig. 3①) outside the glove box. Then, the resulting mixture was placed in two zirconia pots, each containing 10 Φ10 mm zirconia balls, 10 Φ5 mm zirconia balls, and a grinding aid (hexane), and mechanical milling was performed for 2 h at a rotational speed of 300 rpm using a planetary mill (Fig. 3②).After mechanical milling, the resulting slurry were vacuum-dried at 80°C to obtain a mixed powder (Fig. 3③). We then placed the resulting powder in a platinum crucible (25 ml), covered it, and fired it at 800 °C for 12 h on an alumina setter in an electric furnace (Fig. 3④). The resulting solid was ground with a stainless steel mortar to obtain $Li_{6.25}Ga_{0.25}La_3Zr_2O_{12}$ powder (Fig. 3⑤). The powder was filled with zirconia balls and hexane in a zirconia pot and ground for 2 h (Fig. 3⑥). The obtained slurry was dried with the same vacuum drying (Fig. 3⑦). The obtained powder was filled into a carbide pellet press, and a solid electrolyte pellet was prepared by cold pressing with a hydraulic press (Fig. 3⑧). The pellet produced by cold pressing was placed to cover the powder before cold pressing (powder bed method). The use of the powder bed method suppressed the reaction between the platinum in the crucible and the pellet, and also suppressed the loss of lithium. Sintering was carried out at 1050 °C for 12 h (Fig. 3⑨). After sintering, the pellets were buried in the weakly sintered powder bed and removed from the powder bed using a milling cutter. The surface of the removed pellets was further smoothed with grit paper #1500 (Fig. 3⑩).

2.2 *Synthesize of Sulfide Solid Electrolyte*

In this study, the easily synthesizable and highly ion-conductive $Li_{5.35}Ca_{0.1}PS_{4.5}Cl_{1.55}$ was used as the sulfide-based solid electrolyte [12]. Figure 4 shows the process of preparing the sulfide-based solid electrolyte separator in this study. Li_2S, P_2S_5, CaS and anhydrous LiCl were used as the starting materials. Each powder was placed in

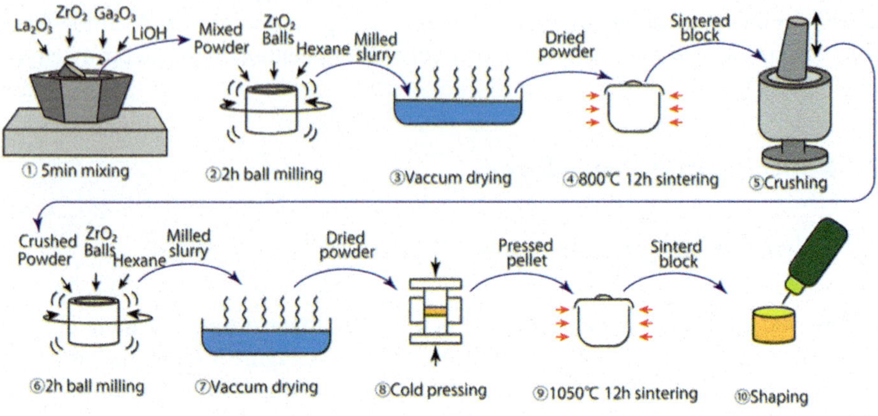

Fig. 3 Synthesize procedure of oxide solid electrolyte

a mortar and mixed with a strong magnetic stirrer for 5 min (Fig. 4①). The mixed powder was divided in half and placed in two 50 mL zirconia pots, with 16 Φ10 mm zirconia balls and a grinding aid (diethyl ether) added to each pot. These processes were performed in a glove box with a low dew point to prevent deterioration of the material due to water vapor. The zirconia pots were mounted on a high energy vibratory mill (Retch, Emax) and mechanically milled at 1200 rpm with a cooling water circulation temperature of 20 °C (Fig. 4②). After processing, the pots were returned to the glove box and the resulting $Li_{5.35}Ca_{0.1}PS_{4.5}Cl_{1.55}$ was synthesized by calcining the powder in an alumina crucible at 450 °C for 5 h using an electric furnace. The synthesized $Li_{5.35}Ca_{0.1}PS_{4.5}Cl_{1.55}$ was then crushed using the mortar to adjust the particle size (Fig. 4④). The solid electrolyte powder was then loaded into a carbide pellet press (Φ5 mm) and cold-pressed by a hydraulic press to form a solid electrolyte pellet, which was used as a separator (Fig. 4⑤).

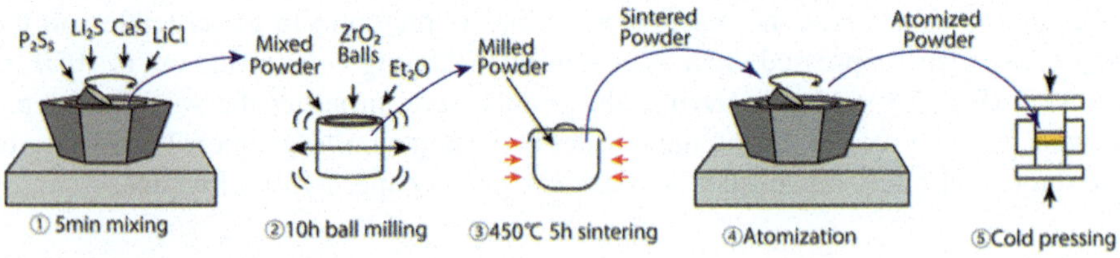

Fig. 4 Synthesize procedure of sulfide solid electrolyte

2.3 Shot Peening and Performance Evaluation

Pellets of oxide-based and sulfide-based solid electrolytes were processed by shot peening. Since sulfide-based solid electrolytes easily decompose and deteriorate in the presence of moisture in the atmosphere, the shot peening process was carried out in a glove box with a dew point temperature of −60 °C or lower. A gun for shot peening was prepared inside the glove box, and dry alumina with a particle size of 100 μm was used as the media. Shot peening was performed by blowing the media onto the solid electrolyte using high purity dried nitrogen at 0.4 MPa.

The lithium-symmetric cells are constructed with the shot-peened and non-shot-peened samples. The lithium-symmetric cell is a battery in which the electrolyte is sandwiched between lithium metal, and although no electromotive force is generated because both electrodes are lithium metal, the phenomenon occurring inside the cell is derived only from the lithium metal electrode. Therefore, it is a battery commonly used in the study of lithium metal negative electrode. Figure 5 shows a list of samples. Sample 1 is a battery that uses a solid oxide electrolyte and has not been shot-peened, with lithium metal attached. Sample 2 is a battery in which both sides of the oxide solid electrolyte have been shot-peened and lithium metal has been applied. Samples 3 and 4 are batteries that use a sulfide solid electrolyte for the electrolyte, and sample 4 is a battery that has been shot-peened similar to sample 2.

The Critical Current Density (CCD) test was performed on the manufactured lithium-symmetric cells. In this test, the battery was connected to a charge–discharge tester and the positive and negative currents were alternately increased to increase the current density. When a certain current density was reached, dendrites grew and reached the opposite electrode, causing a sharp drop in cell voltage. The CCD, which is the upper limit of the chargeable rate, was measured by detecting this voltage drop.

To study the increase in fracture toughness by shot peening, Vickers hardness tests are performed. A diamond indenter makes a square indentation in the SE pellet.

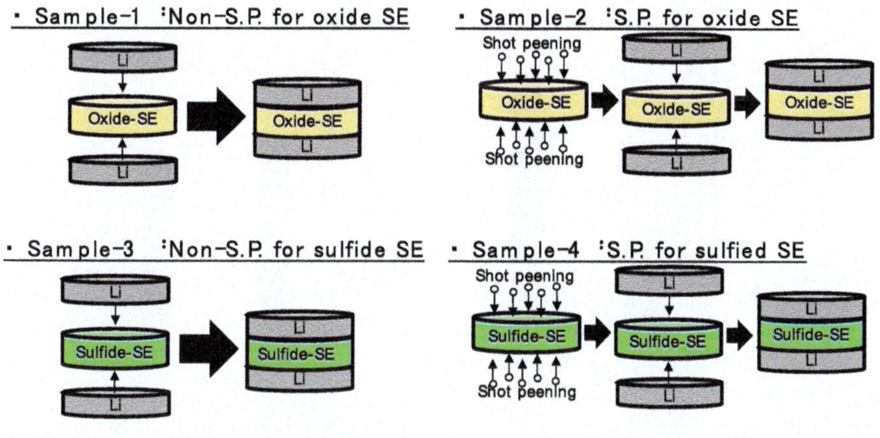

Fig. 5 Samples for performance evaluation

Cracks form from the edges of the square indentation. The fracture toughness is calculated from the crack length and the indentation length [13]. The indenter load was 1 kgf.

In order to examine the variation of each sample, both the CCD tests and the Vickers hardness tests were conducted at least three times for the same conditions. Although no significant variations were observed, the average values of these measurements are presented in this paper.

3 Results and Discussion

Figure 6 shows the maximum charge current (CCD, critical current density) of each sample. Figure 6a shows the results of a lithium-symmetric cell with oxide-based solid electrolyte without and with shot peening, while Fig. 6b shows the results of a lithium-symmetric cell with sulfide-based solid electrolyte without and with shot peening. From Fig. 6a, it can be seen that for the case of using oxide-based solid electrolyte, without shot peening, the CCD is low, but with shot peening, it increases by about 10 times. On the other hand, as shown in Fig. 6b, for the case of using a sulfide-based solid electrolyte, there is almost no difference in CCD, regardless of whether shot peening is applied or not. Therefore, it can be concluded that the performance improvement of all-solid-state lithium metal batteries by shot peening can only be achieved with oxide-based solid electrolytes and not with sulfide-based solid electrolytes. Note that in Fig. 6, although a higher CCD was obtained with the sulfide solid electrolyte compared to the oxide solid electrolyte, this is believed to be due to the higher ionic conductivity of the sulfide solid electrolyte, which resulted in the suppression of lithium segregation.

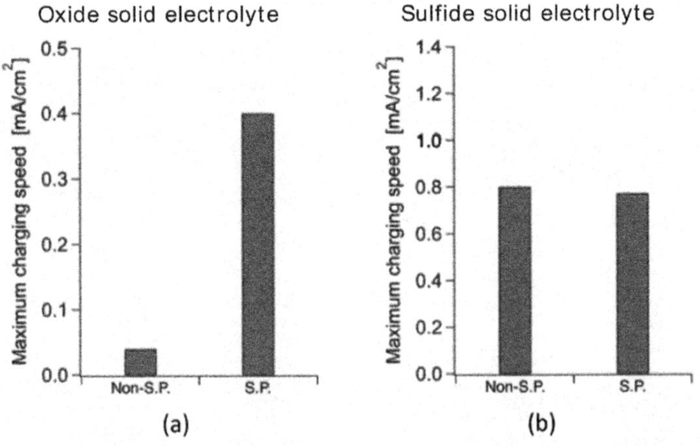

Fig. 6 Maximum charging speed (CCD: Critical current density) of **a** oxide solid electrolyte and **b** sulfide solid electrolyte

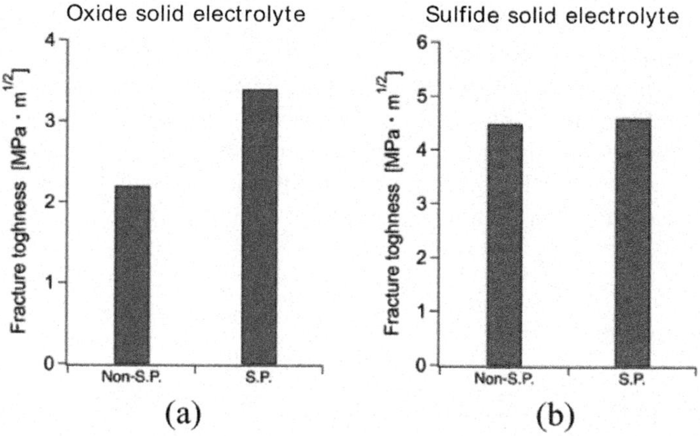

Fig. 7 Fracture toughness of **a** oxide solid electrolyte and **b** sulfide solid electrolyte

The fracture toughness obtained from Vickers hardness tests is shown in Fig. 7. Figure 7a shows that the fracture toughness of the oxide-based solid electrolyte increased about 1.5 times with shot peening. It is believed that this increase in fracture toughness suppressed crack formation within the electrolyte and hindered dendrite growth, resulting in the CCD increase by shot peening as shown in Fig. 6a. On the other hand, no increase in fracture toughness was observed for the sulfide-based solid electrolyte with shot peening, as shown in Fig. 7b. It is considered that the lack of increase in fracture toughness prevented the suppression of dendrite growth, resulting in the absence of CCD increase by shot peening as shown in Fig. 6b.

It is considered that the reason why the fracture toughness did not increase and the CCD did not increase even with shot peening in the case of the sulfide solid electrolyte, while the fracture toughness increased and the CCD increased with shot peening in the case of the oxide solid electrolyte, is due to the pellet preparation process for each electrolyte.

As shown in Fig. 3, for the solid oxide electrolyte, the pellets are pressed and then sintered at a high temperature of 1050 °C. Residual stress is generated in the electrolyte during the pressing process, but the subsequent sintering process releases the residual stress. Therefore, it is assumed that the compressive residual stress is imparted by shot peening and, as a result, the fracture toughness is increased after shot peening compared to before. On the other hand, as shown in Fig. 4, for the sulfide solid electrolyte, there is no sintering after the pellets are pressed. It is assumed that residual stresses from the pressing process remain in the formed pellets. Since the residual stress is already imparted, it is believed that shot peening cannot impart much additional stress, and as a result, there was no difference in fracture toughness due to the presence or absence of shot peening. The residual stress is measured by X-ray diffraction in many studies. Therefore, the authors tried the measurement of the residual stress of the solid electrolyte by X-ray diffraction. However, the grain size of the solid electrolyte is large and Debye ling could not be observed and the residual stress could not be calculated from the X-ray diffraction. So, in this paper, the influence of the shot peening is only discussed with the fracture toughness.

In this paper, shot peening parameters (spraying pressure, grain size, grain shape, etc.) were performed under specific conditions, but there is a possibility that higher fracture toughness and higher CCD can be obtained by optimizing the shot peening parameters, especially when using oxide-based solid electrolytes, where the effect of peening was particularly pronounced. This is a subject for future study.

4 Conclusion

The all solid-state lithium metal battery is expected to be the next generation storage battery for electric vehicles due to its high theoretical capacity. However, during high-speed charging, lithium dendrites grow and penetrate the electrolyte layer, causing internal short circuits and limiting the charging rate. When dendrites penetrate the electrolyte, it is accompanied by the formation of cracks in the electrolyte, and suppressing the formation of cracks in the electrolyte can improve the charging speed. Therefore, in this study, we experimentally investigated the fast charging performance by suppressing crack formation in the electrolyte by shot peening. The study was conducted on sulfide and oxide solid electrolytes, which are representative solid electrolytes. As a result, the following findings were obtained.

1. In the case of using sulfide solid electrolytes, there was little difference in CCD, an index of charging rate, whether shot peening was performed or not. However, when oxide solid electrolytes were used, the charging rate increased by a factor of 10.
2. From the Vickers hardness test, it was confirmed that shot peening did not increase the fracture toughness of the sulfide solid electrolytes, but the fracture toughness of the oxide solid electrolytes increased by about 2 times due to shot peening. It is considered that this increase in fracture toughness occurred only in oxide solid electrolytes and contributed to the increase in charging rate.
3. In the case of solid oxide electrolytes, electrolyte pellets are produced by high-temperature sintering after solid electrolyte powder is formed by pressing. Residual stress is applied to the electrolyte during pressing, but the sintering process anneals it, and as a result, compressive residual stress is applied to the electrolyte by shot peening, resulting in an increase in fracture toughness. On the other hand, sulfide solid electrolytes are formed into pellets by cold pressing and sintering at room temperature, and sintering is not performed after pressing. Therefore, residual stress is applied to the electrolyte during pressing, and even if shot peening is performed afterwards, little residual stress is applied, resulting in no increase in fracture toughness due to shot peening.

Acknowledgements This study was supported by the Japan Keirin Autorace Foundation, the Kurata Grants, and JSPS KAKENHI (grant number 22K14761).

References

1. Chaoshan W et al (2021) Nano Energy 87:106081
2. Hayashi A et al (2003) Electrochem Commun 5:111–114
3. Hayashi A et al (2010) J Non-Cryst Solids 356:2670–2673
4. Kamaya N et al (2011) Nat Mater 10:682–686
5. Kato Y et al (2016) Nat Energy 1:16030
6. Ramaswamy M et al (2007) Angew Chem Int Ed 46:7778–7781
7. Huang X et al (2021) Ceram Int 47:2123–2130
8. Zhigang X et al (2015) J Mater Chem A 3:19218–19253
9. Asano T et al (2018) Adv Mater 30:1803075
10. Alexandra E et al (2013) Chem Mater 25:4663–4670
11. Kodama M et al (2022) J Power Sources 537:231556
12. Adeli P et al (2021) Chem Mater 33:146–157
13. Ponton CB et al (2013) Mater Sci Technol 5:865–872

Predicting the Effect of Deep Cold Rolling on the CO_2-Footprint of Dynamically Loaded Parts

Oliver Maiss, Karsten Röttger, and Stefan Zenk

Abstract Mechanical surface treatments, as for example machine hammer peening or deep cold rolling, are well known to increase the service life of dynamically loaded parts. Using this increased strength, a part can be designed more resource- or weight efficient. Both parameters are to reduce the carbon-footprint of the part. Within this presentation the effect of deep cold rolling on the overall CO_2-footprint will be pointed out. The aim of this project is to identify the main effects on the carbon footprint. Using a life cycle assessment in- and output values are analyzed for different processes and are evaluated for their importance. Additionally, a method will be presented to quickly evaluate the benefits of mechanical surface treatments for any dynamically loaded part within the design phase for the whole life cycle. Within this framework different assumptions can be drawn to increase the quality of the prediction and it will be possible to quickly evaluate if a mechanical surface treatment will be beneficial for the CO_2-footprint of a part.

Keywords Deep cold rolling · Surface integrity · LCA · CO_2

1 Introduction

Climate change is progressing faster and faster. The OEMs are setting ambitious goals for a climate-neutral future, putting many of their suppliers under pressure. In industrial companies there are many different levers to achieve a climate-neutral future. One possibility is to consider the CO_2-footprint directly in the design stage, to

O. Maiss (✉) · K. Röttger · S. Zenk
ECOROLL AG Werkzeugtechnik, H.-H.-Warnke-Street 8, 29227 Celle, Germany
e-mail: oliver.maiss@ecoroll.de

K. Röttger
e-mail: karsten.roettger@ecoroll.de

S. Zenk
e-mail: stefan.zenk@ecoroll.de

N. Maharjan and W. He (eds.), *Proceedings of the 3rd International Conference on Advanced Surface Enhancement (INCASE) 2023*, Lecture Notes in Mechanical Engineering, https://doi.org/10.1007/978-981-99-8643-9_47

reduce the effects of the machine or the individual component. Measurements such as reducing power losses, lightweight construction using new materials, or the use of high-strength materials are just a few examples. For this, however, the greatest influencing variables must be known in the design. In various research projects, for example the "Antriebsstrang 2025" project funded by the German Government (BMWK), it has been shown that one of the biggest influencing factors on the CO_2-footprint of a component is the proportion of material required [1], i.e., the amount of raw material. Subsequent manufacturing steps, such as machining, forming or heat treatment, have a significantly lower share in comparison. This is also where the challenge lies in the lightweight construction measures mentioned above. The use of new or even higher-strength materials usually requires compromises and also higher energy consumption for production or disposal.

A simpler way to improve the climate balance is to exploit the so-called surface and subsurface properties. These properties can significantly affect the performance of a component. For example, by producing a specifically ground or honed surface, the proportion of friction can be significantly reduced when the component is used. Furthermore, the service life of dynamically loaded components can be reduced by the targeted exploitation of residual compressive stresses in the subsurface area. This is known from shot peening of turbine blades or gears. The targeted use of residual compressive stresses can increase the service life by more than 1.2–5 times, which means that the components in the design stage can be designed smaller and lightweight construction effects can be exploited.

If the design specifies the use of residual compressive stresses, the question naturally remains how these residual compressive stresses can be introduced in a targeted and process-reliable manner. Shot peening would be an option, but always requires an additional machine system. The processes of deep rolling or machine hammer peening, on the other hand, can be used directly on the machine tool and do not require any additional plant technology.

During deep rolling, or deep cold rolling, a rolling element is pressed onto the surface of a component with a defined force. The contact between the component and the rolling body creates high stresses and the material is plastically deformed locally on the surface. This deformation smoothed the surface and causes changes in the microstructure in the near-surface area [2], which leads to a significant increased service life.

The aim of this research is, to evaluate the effect of deep rolling regarding the reduction of CO_2-emissions. Therefore, different analysis, experiments and literature reviews are considered.

2 Evaluating the CO_2-Footprint of Deep Cold Rolled Parts

An increased lifetime can be used with different intent. On the one hand to increase the service life for consumable parts. For example, if it is possible to double the service life and the part has to be replaced 10 times within the lifetime of the whole

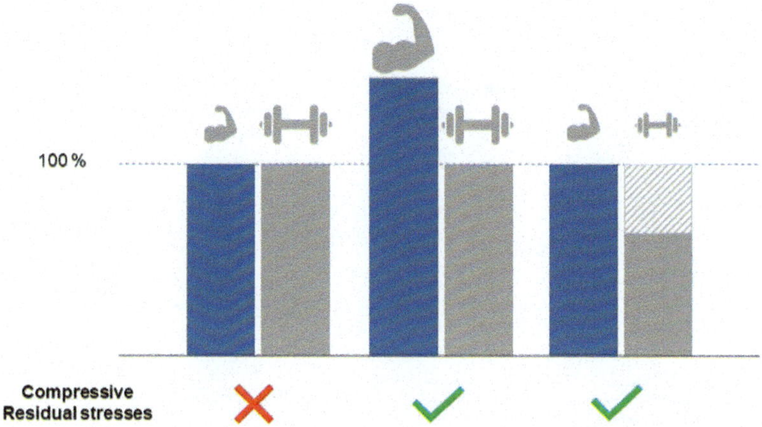

Fig. 1 Use cases for an increased resource efficiency due to an increased service life of cold deep rolled parts

machine, only the resources for 5 parts will be used. On the other hand, there are parts, which are already designed to last for the whole service life of the machine. There is no need to increase the service life. In this case it is more important to reduce the weight of the part to increase the resource efficiency (Fig. 1).

Within this paper, the last situation is considered (Fig. 1). As mentioned in the introduction, so far no valid framework (Fig. 2) exists to evaluate the CO_2-emissions for a deep rolled part. Therefore, the authors came up with a simple and modular framework. It is important to start by defining the part service life and load situation. As a second step the main challenge will be to identify the optimal rolling parameters. However, after many years of research, it is still not possible to simply predict the parameters, to achieve a specific service life for a specific material, without trying. It is possible to use experiments, literature or expert knowledge. A suitable approach should be used, according to the necessary accuracy of the prediction.

In the next step a lifetime increasement factor must be identified. Again, this can be done by estimations, literature review or actual service life tests. This step is followed by a redesign or lightweight design with the goal to decrease the amount of material which is used for the part. In the last step of the framework the actual CO_2-footprint is evaluated by using a Life Cycle Assessment (LCA).

3 Experimental Procedure

To test this framework two kinds of experiments are used. On the one hand lifetime test with rotating bending specimens are conducted. The specimens have a working diameter of d = 13 mm and are made of AISI 4140. Two different specimens are machined. One specimen is produced by turning and the second type is produced by

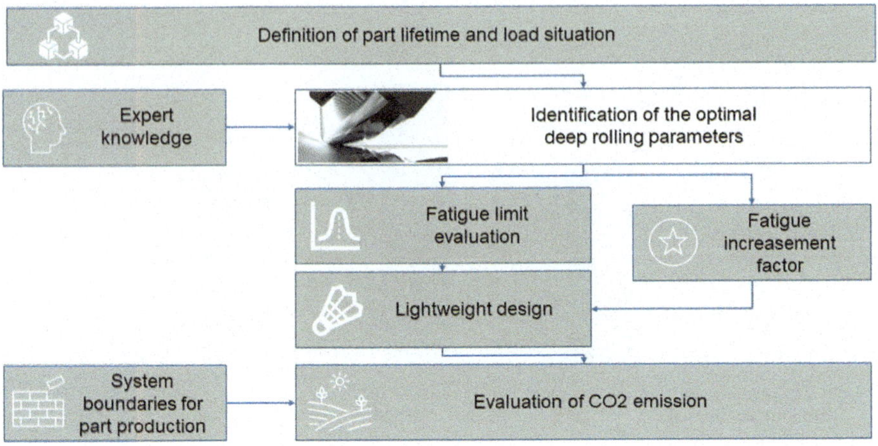

Fig. 2 Framework to evaluate the CO_2-emissions of a cold deep rolled part [3]

turning with an additional deep rolling using a ECOROLL tool type EG5-1-40 M. Both specimen types are tested in rotating bending test machine from SincoTec.

On the other hand literature data is used to evaluate the CO_2-footprint for deep rolled roller bearings. In this case the data is given in References [4, 5]. Within the cited research, the roller bearings are hard turned and deep rolled in comparison to a conventional manufacturing chain of grinding and honing.

4 Measurement of CO_2-Footprint for Cold Deep Rolling

The rotating bending test show an increase of the service life of 10% by using deep rolling. Assuming a lightweight design with a test diameter of $d_{red} = 12.59$ mm, the same lifetime as for the just turned specimen will be achieved. This means about 8.6% less raw material is used for the specimen.

The LCA was conducted within a project by the Mittelstand-Digitalzentrum Hannover. The energy- and material input for the process chain "raw material", "turning" and "deep rolling" was considered. For the deep rolling an energy measurement has been conducted. All information are considered within the tool "Umberto" to use well known CO_2-databases. Umberto is a commercial software and database to evaluate the CO_2-footprint.

The model shows very clearly that the most important factor to consider is the amount of raw material. The data shows, that just by reducing the material amount by 8.6% the equivalent carbon emissions are reduced by 96 g CO_2-eq. per part. To conduct the additional rolling process about 15 g CO_2-eq. is emitted. In total there is a reduction of 7.3%.

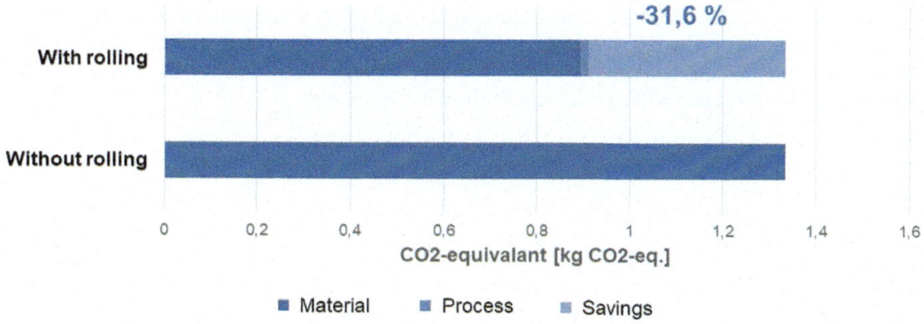

Fig. 3 Resulting CO2-reduction for deep rolled roller bearings

5 Potential of CO_2-Reduction for Different Applications

For the roller bearings the effect is even bigger. By using a bearing type NU205 instead of a NU206 the material can be reduced by 33%. The basis for this conclusion is 2.5 $\times$ L_{10}(small/rolled) = L_{10}(large/not rolled). Assuming the same loads, the dynamic load rating C can be calculated. For the NU206 this calculation leads to the value of a NU205. With the additional rolling process just 18 g CO_2-eq. are emitted, which leads to a CO_2-reduction of about 31.6%. Assuming just 50,000 bearings per year manufactured, results in 21 to CO_2-eq. reduced (Fig. 3).

6 Conclusion and Outlook

The presented results show a very clear effect of the cold deep cold rolling process on the resulting carbon emissions of a dynamically loaded part. If all life cycle parameters of a part are taken int account, the use phase is the most important factor to reduce carbon emissions. However, the definition of the amount of CO_2-emissions is given in the design phase. This is, why a consequent use of subsurface properties is necessary. Therefore, designers must know about the mechanical surface treatments as e. g. cold deep rolling, roller burnishing or machine hammer peening.

References

1. Denkena B, Wichmann M, Kettelmann S, Matthies J, Reuter L (2022) Ecological planning of manufacturing process chains. Sustainability 14(5)
2. https://www.surfacematters.tech
3. Maiß O, Röttger K, Meyer K (2022) Increase the resource efficiency by evaluation of the effects of deep rolling within the design and manufacturing phase. In: Future automotive production conference 2022, Springer Vieweg, pp 86–96

4. Neubauer T (2016) Betriebs- und Lebensdauerverhalten hartgedrehter und festgewalzter Zylinderrollenlager. Dr.-Ing. Dissertation, Gottfried Wilhelm Leibniz Universität Hannover
5. Maiß O (2019) Lebensdauererhöhung von Wälzlagern durch mechanische Bearbeitung. Dr.-Ing. Dissertation, Gottfried Wilhelm Leibniz Universität Hannover

Improving Life Cycle Assessment Accuracy and Efficiency with Transformers

Yang Zhao

Abstract Life Cycle Assessment (LCA) is a methodology employed to evaluate the environmental effects of goods or services over their complete life cycle. LCA reports are key for the industry to assess their contribution on the sustainability. It is a complex and time-consuming process that can be improved with the use of deep learning models due to a large amount of data are involved in prediction. Transformers have been successful in natural language processing and can also be applied to numerical data to predict environmental impacts. By detecting the phases in a product's life cycle that generate the most significant environmental consequences and automating the data compilation and analysis procedures, they can reduce the time and expense connected with LCA. The use of transformers for LCA analysis has the potential to improve the accuracy and efficiency of sustainability assessments, providing more comprehensive information about environmental impacts. Through experimenting with real-world datasets, the proposed transformer framework has been shown to effectively contribute to making informed sustainability-related decisions by providing comprehensive information about environmental impacts. This has the potential to benefit a wide range of industries and sectors, enabling more sustainable development and decision-making.

Keywords Life cycle assessment · LCA · Transformer · Sustainability

1 Introduction

The evaluation of the environmental impact of a product or process is frequently accomplished using the well-established method of life cycle assessment (LCA) [1]. LCA considers the entire life cycle of a product, from the extraction of raw materials to the end-of-life disposal, providing a comprehensive analysis of its environmental

Y. Zhao (✉)
Singapore Institute of Manufacturing Technology, Agency for Science, Technology and Research (A*STAR), INNOVIS, 2 Fusionopolis Way, Singapore 138634, Singapore
e-mail: zhao_yang@simtech.a-star.edu.sg

N. Maharjan and W. He (eds.), *Proceedings of the 3rd International Conference on Advanced Surface Enhancement (INCASE) 2023*, Lecture Notes in Mechanical Engineering, https://doi.org/10.1007/978-981-99-8643-9_48

performance. Given the growing demand for sustainable products and processes, LCA has become an indispensable tool for firms to measure and decrease their ecological footprint. In this article, we examine the use of transformer models in LCA and their potential to enhance the precision and effectiveness of LCA research.

2 Background

LCA studies typically involve a large amount of data, which needs to be collected, processed, and analyzed. Traditional methods for LCA rely on manual data entry and analysis, which can be time-consuming and error-prone. Moreover, LCA studies require the use of complex models and software tools for calculating environmental impact indicators, such as carbon footprint, water footprint, and toxicity potential. These models are often based on mathematical equations and require significant computing power to run. As a result, LCA studies can be expensive and time-consuming, limiting their accessibility to smaller companies and organizations.

Transformer models are a category of deep learning architecture that has proven to be extremely successful in natural language processing duties, such as language modelling, text classification, and machine translation. They leverage a self-attention mechanism to enable the model to concentrate on distinct parts of the input sequence and capture long-range dependencies. Recently, transformer models have been employed in other domains, including computer vision and audio processing, with encouraging outcomes.

3 Methodology

To explore the use of transformer models in LCA, we used a transformer-based architecture for time-series forecasting. Specifically, we used the BERT-based transformer model [2] from the Hugging Face transformers library as the backbone of our architecture. Our dataset consisted of hourly power consumption data for a household over a period of several years, which will be illustrated in detail in the next section. We first pre-processed the data by filling missing values with forward fill and resampling it to hourly frequency. We then normalized the data by subtracting the mean and dividing by the standard deviation. We created a custom PyTorch dataset class to prepare the data for inputting into the transformer, where each input sequence consisted of 24 h of data and the target was the power consumption for the 25th hour. We trained the model using a mean squared error loss function and an AdamW optimizer [3] with a learning rate of 1e-4 and weight decay of 1e-5. We used a ReduceLROnPlateau [2] learning rate scheduler with a factor of 0.5 and a patience of 5 epochs. We trained the model for 150 epochs and saved the best model based on the validation loss. Finally, we evaluated the model on a held-out test set and reported the test loss. Details of

Table 1 Details of the methodology

Step No.	Methodology step	Description
1	Load dataset	Load the hourly power consumption dataset for a household over several years
2	Normalize data	Subtract the mean and divide by the standard deviation to normalize the data
3	Create sequence	Create a custom PyTorch dataset class where each input sequence is 24 h of data and the target is the power consumption for the 25th hour
4	Create transformer model	Use the BERT-based transformer model from the Hugging Face transformers library
5	Loss function	Use a mean squared error loss function for training
6	Optimizer	Use an AdamW optimizer with a learning rate of 1e-4 and weight decay of 1e-5
7	Learning rate scheduler	Use a ReduceLROnPlateau learning rate scheduler with a factor of 0.5 and a patience of 5 epochs
8	Train model	Train the model for 150 epochs, saving the best model based on the validation loss
9	Evaluate model	Evaluate the model on a held-out test set and report the test loss

the methodology is illustrated in the Table 1.The workflow of the methodology is shown in the Fig. 1.

4 Dataset

The "household_power_consumption.txt" dataset [4] contains electricity consumption data from a single household with a time resolution of 1 min over a period of almost four years (2006-12-16 to 2010-11-26). The data were collected using a single-phase energy meter and include measurements of 'Global_active_power', 'Global_reactive_power', 'Voltage', 'Global_intensity', 'Sub_metering_1', 'Sub_metering_2' and 'Sub_metering_3' of the three electrical sub-systems (kitchen, laundry, and climate control). The dataset contains 2075,259 observations and 7 variables. This dataset is commonly used in time-series forecasting and energy consumption analysis tasks. In this project, the dataset is pre-processed to be resampled to hourly intervals and missing values are filled using forward fill method. The 'Global_active_power' variable is used as the target variable, while the other variables are used as features. We compare the transformer algorithm with GRU algorithm, and the results of MSE of two algorithms are shown in the Table 2. The MSE of transformer is smaller than GRU.

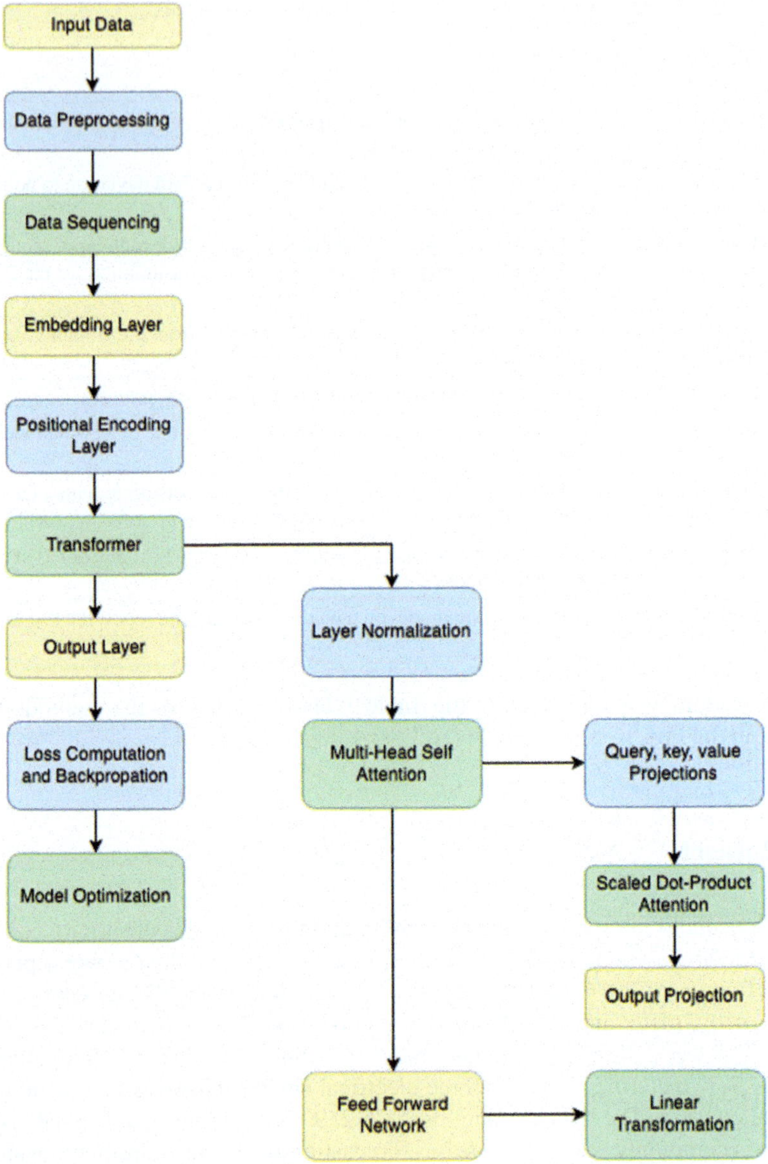

Fig. 1 Workflow of the methodology

Table 2 MSE for GRU and transformer algorithms

Model	GRU	Transformer
MSE	0.045	0.041

5 Discussion

The use of the transformer model can help reduce the manual labour required for data entry and analysis, and provide more accurate predictions of energy consumption. Moreover, the attention weights of the transformer model can provide valuable insights into the most significant variables that affect the environmental performance of a product or process. This can help companies identify areas for improvement and develop more sustainable products and processes.

However, there are also some limitations to our study. First, our dataset was relatively hard to obtain satisfied results, and more extensive datasets may be required to develop more robust models. Second, the model we trained was based on a pretrained model, which means that it requires labelled data to train. LCA studies often involve complex and multi-dimensional data, which may be difficult to label accurately.

6 Conclusion

In conclusion, we try to explore the potential of transformer models for LCA studies. The use of the Transformer model can provide more accurate predictions of the energy consumption and help identify areas for improvement in product and process design. Through the experiments, we obtain that the MSE of transformer is smaller than the GRU. However, there are also limitations to our study, and further research is required to develop more robust and interpretable models that can leverage the complex and multi-dimensional data involved in LCA studies. Overall, we believe that the application of machine learning techniques such as transformer models has the potential to transform the field of LCA and contribute to more sustainable and environmentally responsible practices in industry and society.

References

1. Hauschild MZ, Rosenbaum RK, Olsen SI (2018) Life cycle assessment. Springer International Publishing, Cham. https://doi.org/10.1007/978-3-319-56475-3
2. Acheampong FA, Nunoo-Mensah H, Chen W (2021) Transformer models for text-based emotion detection: a review of BERT-based approaches. Artif Intell Rev 1–41
3. Yao Z, Gholami A, Shen S, Mustafa M, Keutzer K, Mahoney M (2021) Adahessian: an adaptive second order optimizer for machine learning. Proc AAAI Conf Artif Intell 35(12):10665–10673
4. Kim T-Y, Cho S-B (2018) Predicting the household power consumption using CNN-LSTM hybrid networks. In: Intelligent data engineering and automated learning–IDEAL 2018: 19th International conference, Madrid, Spain, November 21–23, 2018, Proceedings, Part I 19. Springer International Publishing, pp 481–490

Development of Photocatalytic Self-cleaning 316 Stainless Steel Surface

Jun Hui Ong, Guangxu Yan, Niroj Maharjan, and Zhong Chen

Abstract Surfaces that can keep themselves clean are described as self-cleaning surfaces. Such surfaces have drawn a lot of attention due to their numerous applications in food, aerospace, manufacturing and other industries. Self-cleaning for metals and alloys is usually achieved by creating hydrophobic textured surfaces. However, such surfaces obtained by texturing are unstable as their wettability changes over time. An alternative to creating self-cleaning surfaces is through the photocatalytic effect where a semiconductor material can generate radicals to remove organic contaminants upon the illumination of UV light. We studied the photocatalytic efficiency of TiO_2 coating deposited on smooth and textured 316 stainless steel surfaces. Thermal oxidation at different temperatures was explored to study how it affects the photocatalytic efficiency of TiO_2 coating. The results showed that the textured surfaces achieved a good bonding of TiO_2 coating and a better self-cleaning effect as compared to the smooth samples. Generally, a better self-cleaning effect could be observed for TiO_2 films that were obtained by thermally oxidising Ti films at around 500 °C.

Keywords Self-cleaning · Photocatalytic coating · Titanium dioxide

J. H. Ong (✉) · G. Yan · Z. Chen
Nanyang Technological University, 50 Nanyang Avenue, Singapore 639798, Singapore
e-mail: ongj0082@e.ntu.edu.sg

Z. Chen
e-mail: ASZChen@ntu.edu.sg

N. Maharjan
Advanced Remanufacturing and Technology Centre, Agency for Science Technology and Research, 3 Cleantech Loop, Singapore 637143, Singapore
e-mail: maharjan_niroj@artc.a-star.edu.sg

N. Maharjan and W. He (eds.), *Proceedings of the 3rd International Conference on Advanced Surface Enhancement (INCASE) 2023*, Lecture Notes in Mechanical Engineering, https://doi.org/10.1007/978-981-99-8643-9_49

1 Introduction

Self-cleaning surfaces are surfaces that can keep themselves clean with the help of water. These surfaces can be classified as hydrophilic or hydrophobic. Generally, if the contact angle is $< 90°$, the surface is considered hydrophilic, which means that they have high surface energy and attract water. Water spreads over the surface to carry dirt away. Whereas for hydrophobic surface, the contact angle is $> 90°$, which implies low surface energy, and they repel water. Water droplets form and they slide and roll over the surface to clean the surface.

The development of hydrophobic self-cleaning surfaces was inspired by nature. A well-known example is the lotus leaf, which remains dirt-free even in muddy waters. The dirt particles are removed when water droplets roll off the leaf. This mechanism is also known as the "lotus effect" [1]. There has been great interest in self-cleaning surfaces due to their diverse applications in food, aerospace, manufacturing industries and more. Benefits of self-cleaning surfaces include reduced maintenance efforts and costs.

For metals and alloys, self-cleaning surfaces are usually produced by creating hydrophobic textured surfaces. One of the most popular techniques to produce micro/nano surface topography is laser surface texturing. This is due to its shorter processing time, precise machining, and good control in generating micro/nano surfaces. The understanding of laser parameters, such as laser power, fluence, hatch distance, scanning speed, and number of scans, is essential for process optimization. The desired surface topography can then be obtained to achieve hydrophobicity [2]. However, such rough superhydrophobic surfaces are not very stable as the wettability changes over time. It is super hydrophilic right after the laser texturing process and transitions to superhydrophobic surfaces over time [3, 4].

An alternative to creating self-cleaning surfaces is through the photocatalytic effect. When light is irradiated on a semiconductor material, radicals are generated to remove organic contaminants from the surface chemically. One of the most used semiconductor materials is titanium dioxide (TiO_2). In 1972, water photoelectrolysis with TiO_2 was discovered by Fujishima and Honda [5]. Subsequently, the photocatalysis and hydrophilicity of TiO_2 films were discovered [6]. The self-cleaning coatings received significant attention due to its many benefits. However, TiO_2 has a wide bandgap which means that it can only absorb ultraviolet light which makes up 4 to 5% of the electromagnetic radiation output from the Sun [7]. Another challenge is finding the right surface roughness and condition for better bonding of TiO_2 coating on metal surface. This is because surface roughness increases interfacial area which results in mechanical interlocking between the coating and the surface [8].

Therefore, this study aims to explore the use of photocatalytic TiO_2 coating on 316 stainless steel surfaces to obtain the self-cleaning effect.

2 Materials and Methods

An overview of the experimental procedures is illustrated in Fig. 1. Detailed experimental procedures and conditions are described in the following sub-sections.

2.1 Sample Preparation

Stainless steel 316 was used as substrate material for the deposition of Ti film. Samples (15 mm by 15 mm by 3 mm) were prepared by cutting the as-received SS316. Manual grinding using P2500 grit size and polishing using 3 μm diamond suspension were carried out to prepare for the final surface finish. Laser surface texturing was then performed on some of the SS316 substrates using a nanosecond laser. The texturing was carried out in air using a nanosecond laser from SPI Lasers (20 W, Air-cooled EP-S Laser). The pulse duration and pulse repetition rate were fixed at 26 ns and 140 kHz respectively. The substrates were textured by scanning the laser beam with a fixed scanning speed of 500 mm/s. Crosshatch pattern was selected. The hatch distance was varied to obtain different surface morphology.

Before the Ti film deposition, the substrates were cleaned using acetone, followed by ethanol, then DI water—10 min each with ultrasonication. Ti thin film was deposited onto non-textured and textured SS316 substrate using PVD 75, manufactured by Kurt J. Lesker. Substrates were cleaned before placing it onto the substrate holder. Ti films were deposited onto SS316 substrates by sputtering a Titanium target with 99.99% purity using DC power. Then, thermal oxidation of Ti films was performed at varying temperatures from 400 to 600 °C for a period of two hours. Thermal oxidation for Ti films deposited on smooth surfaces was done at 500 °C, 550 °C and 600 °C. Based on the characterization of those samples, lower temperatures (400 °C, 450 °C and 500 °C) were chosen for the textured samples. The samples thermally oxidised at 500 °C were used to compare the smooth samples and textured samples.

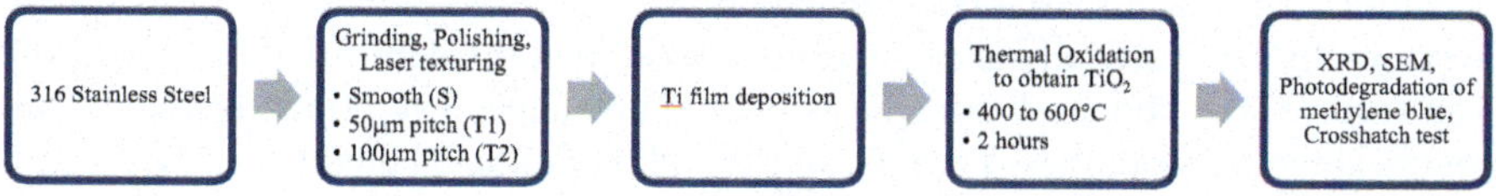

Fig. 1 Flowchart illustrating experimental procedures

2.2 Materials Characterization

The Shimadzu XRD-6000 X-Ray Diffractometer (XRD) was used to identify the phases present in the TiO_2 coating. JEOL JSM-6360 Thermionic Scanning Electron Microscope (SEM) and JEOL JSM-7600F Field Emission Scanning Electron Microscope (FESEM) were used to observe surface morphology.

2.3 Photocatalytic Activity Study

The degradation of methylene blue [9] was used to investigate the photocatalytic activity. It was conducted by placing the sample in a petri dish containing 20 mL of 15 μmol/L aqueous methylene blue solution. To reach adsorption equilibrium, the sample was placed in the dark for 30 min. A 365 nm single wavelength UV-LED light source (Intensity = 15.2 mW/cm^2) was used. The light absorption spectra of the methylene blue solution were captured using a Shimadzu UV-2700 UV–Visible spectrophotometer.

Under the influence of UV or solar radiation, dyes are degraded by TiO_2. Methylene blue has a strong absorption peak in the visible light region (λ_{max} = 664 nm). Based on Beer-Lambert law, $A = \epsilon cL$, the absorbance of methylene blue is directly proportional to its concentration. By measuring the change in concentration of methylene blue upon UV irradiation, the degradation of methylene blue can be determined, which indicates the photocatalytic activity of TiO_2 coating. The absorbance of methylene blue was measured at 0, 1, 2, 4, 6, and 8 h. The greater the photodegradation of methylene blue, the better the photocatalytic activity.

2.4 Mechanical Reliability of Coating

The adhesion of the coating to the substrate was evaluated using the crosshatch test following the ASTM D3359 standard. A cutter was run through the surface horizontally and vertically across the same area, then the debris was brushed away. The surface was observed using an optical microscope. The failure of the coating occurs along the grid lines. The adhesion could be estimated qualitatively on a scale of 0B to 5B.

3 Results and Discussion

3.1 *Heat Treatment of Smooth Steel Surface*

In this set of study, Ti film was deposited on smooth SS316 substrate followed by thermal oxidation at various temperatures to obtain TiO_2. Samples S-500, S-550, and S-600 (the number denotes the Ti film oxidation temperature) are discussed in this section.

Figure 2 shows the XRD pattern of the SS316 substrate and deposited Ti film. For the XRD pattern of deposited Ti film on SS16, hexagonal alpha Ti peaks can be observed.

As shown in Fig. 3, the particulate size increased as temperature increased from 500 °C to 600 °C. This could be due to conglomeration of TiO_2 deposits at higher temperature as well as increase in grain size. Previous studies also showed similar trends where the grain size increased as thermal oxidation temperature increased [10]. The XRD patterns of TiO_2 films obtained by oxidizing Ti films at 500 °C, 550 °C and 600 °C (samples S-500, S-550 and S-600) are shown in Fig. 4. The peaks at 43.9°, 51.1° and 75.2° belong to the SS316 substrate. The relative peak intensities of anatase decreased as temperature increased. Whereas the relative intensity of rutile peaks increased in intensity as temperature increased. This means that there is a lower anatase to rutile ratio at a higher temperature. This is also supported by the larger grain size for the S-600 sample. This is because there is grain growth during the transition from anatase to rutile [11].

Figure 5 shows how the concentration of methylene blue changes with UV irradiation time. It can be seen from Fig. 5 that the concentration of methylene blue remained relatively constant for the control sample which is uncoated SS316 substrate. The slight increase may be attributed to water loss from evaporation during the UV irradiation time. It is evident that the TiO_2 exhibited photocatalytic activity which is

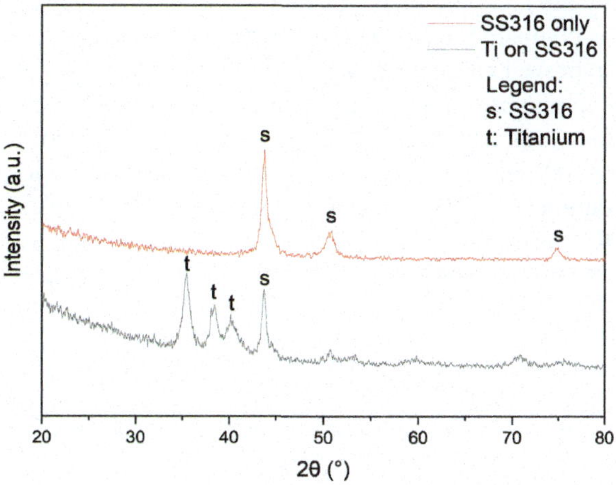

Fig. 2 XRD patterns of SS316 substrate and deposited Ti film

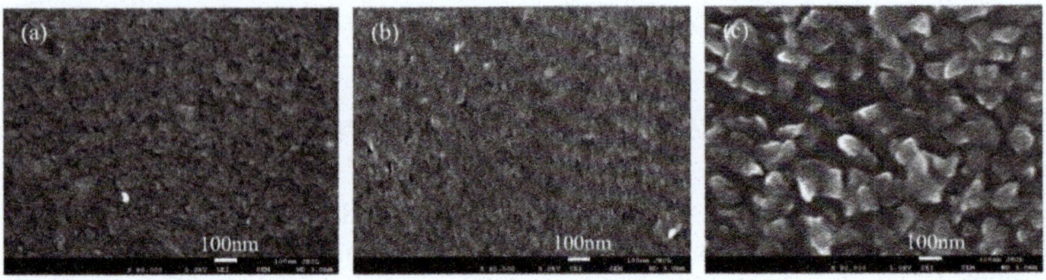

Fig. 3 SEM images of smooth samples heat treated at **a** 500 °C, **b** 550 °C and **c** 600 °C to obtain TiO_2 coating

Fig. 4 Thin film XRD patterns of smooth samples heated at various temperatures

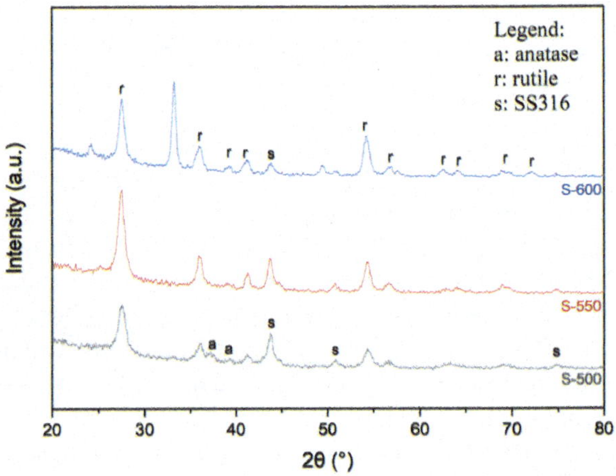

indicated by a significant decrease in methylene blue concentration, as compared to the control sample where the C/C_0 stayed relatively constant. Samples S-500 and S-550 exhibited a better photocatalytic activity as compared to S-600. This is supported by the fact that the S-500 and S-550 have a relatively higher anatase-to-rutile ratio as compared to S-600. The higher photocatalytic activity could be attributed to the improved charged carrier separation because the photoexcited electrons from rutile can migrate to the conduction band of anatase. This reduces the recombination of electrons and holes [12]. S-550 has better photocatalytic activity as compared to S-500. We may conclude that the best anatase/rutile ratio for methylene blue degradation is at 550 °C and thus for the smooth samples, S-550 is the best. However, in this series of samples, only experimental repeats were performed for S-500 sample due to time constraints. Therefore, more repeats could be done for future works.

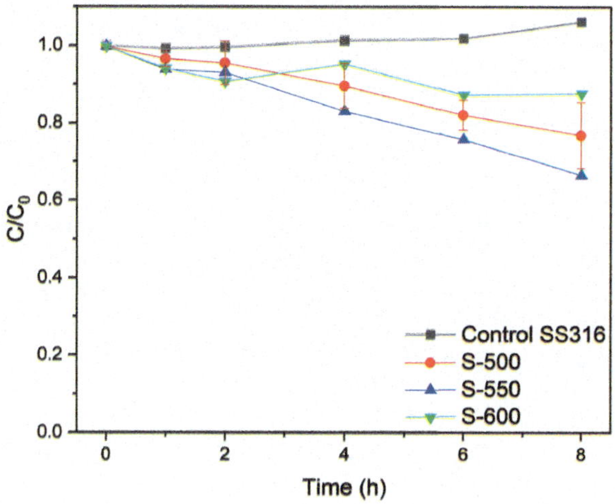

Fig. 5 Methylene blue concentration changes with UV irradiation time on uncoated SS316 sample, S-500, S-550, and S-600

3.2 *Effect of Laser Texturing*

In this set of samples, SS316 substrates were laser textured with varying hatch distances. The surfaces textured with hatch distances of 50 μm and 100 μm are referred to as T1 and T2 respectively. Then, Ti thin films were deposited on laser textured SS316 substrates, followed by thermal oxidation at various temperatures to obtain TiO_2 films.

Figure 6 shows the surface of SS316 after laser texturing. Laser machined grooves were applied in a square grid to form crosshatch textures. The globules were formed because of the laser remelting the surface. When the hatch distance is closer to the laser beam spot size, it is observed that the surface resembles an egg box shape. When the hatch distance is further from the spot size, the surface pattern resembles trenches.

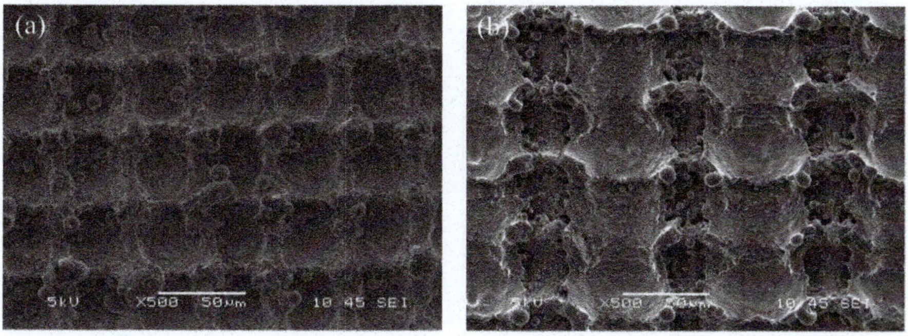

Fig. 6 SEM images of laser textured SS316 surfaces with a hatch distance of **a** 50 μm and **b** 100 μm

Figure 7 shows how the water contact angle changes with time after laser texturing. It was superhydrophilic right after laser texturing and the surface transitions to hydrophobic over time. This shows that the laser textured surfaces take time to become superhydrophobic and thus, the deposition of photocatalytic TiO_2 coating on SS316 is explored. The duration for it to transition to a superhydrophobic surface is dependent on the processing conditions.

The peaks of high intensities in Fig. 8a belong to the SS316 substrate. Zooming in to the 20° to 40° in Fig. 8b, rutile peaks can be observed. The relative intensities of the rutile peaks for both T1-500 and T2-500 are similar. A possible explanation could be that the texturing with varying hatch distances does not affect the phases of the TiO_2 obtained at the same thermal oxidation temperature.

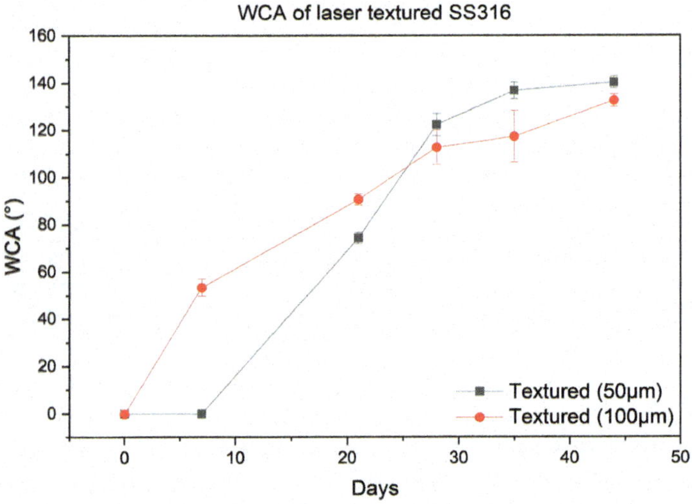

Fig. 7 WCA of SS316 only after laser texturing

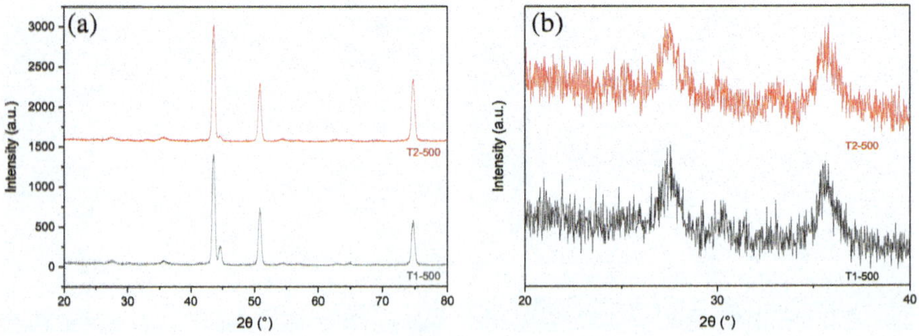

Fig. 8 XRD profiles of textured samples with TiO_2 coating obtained by thermal oxidation at 500 °C using 2θ-θ scan **a** from 20° to 80° and **b** zoomed in to 20° to 40° range

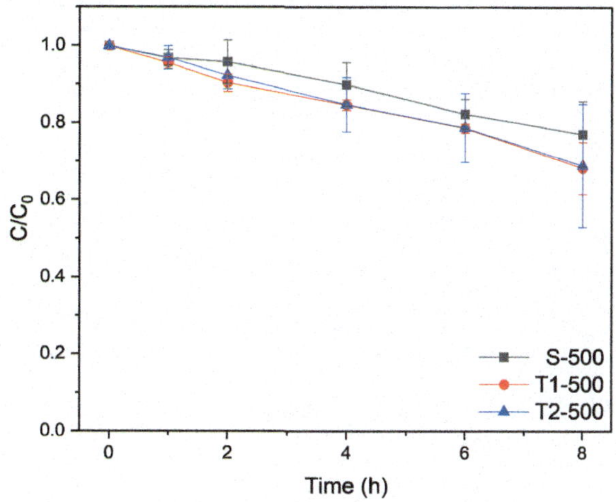

Fig. 9 Methylene blue concentration changes with UV irradiation time on samples with different surface finishes heat treated at 500 °C

From Fig. 9, samples T1-500 and T2-500 show better photocatalytic activity as compared to that for S-500. This shows that textured surfaces could potentially enhance the photocatalytic activity, which could be due to the larger surface area of the textured surfaces. A larger surface would provide more available sites for photocatalysis to occur.

It is observed that the two textured samples T1-500 and T2-500 have similar photocatalytic activity. This could be because the two types of textured surfaces have a rather similar roughness value. Therefore, surface textures with more distinct patterns and roughness values could be explored in future works.

From Fig. 10a, a slightly better photocatalytic activity was observed when the thermal oxidation temperature decreases. The better photocatalytic activity could be a result of the improved charged carrier separation because the photoexcited electrons from rutile can migrate to the conduction band of anatase. This reduces the recombination of electrons and holes.

From Fig. 10b, there is a general decreasing trend for all samples which indicates that the TiO_2 exhibits the self-cleaning effect. However, there are no significant trends observed when comparing the different conditions. More experimental repeats could be performed for future studies.

According to ASTM D3359 [13], the coating on textured surfaces in Fig. 11b and c were graded as 5B whereas the coating on smooth surface in Fig. 11a was graded as 4B.

3.3 Discussion

Textured surfaces could potentially enhance photocatalytic activity as compared to smooth surfaces. This could be attributed to the larger surface area of the textured

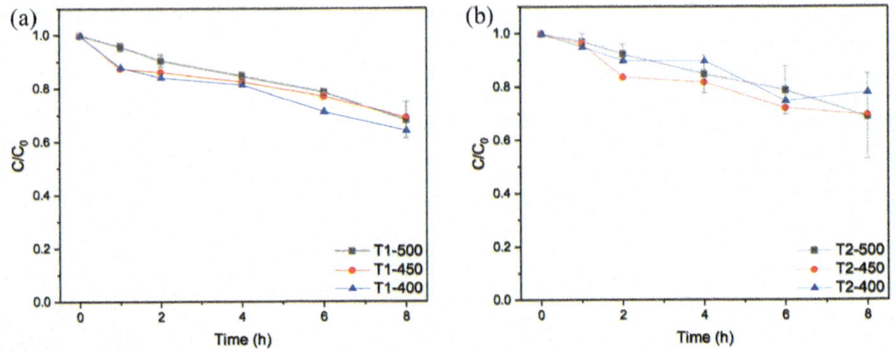

Fig. 10 Methylene blue concentration changes with UV irradiation time on textured samples with TiO_2 coating obtained by thermal oxidation at 400 °C, 450 °C and 500 °C

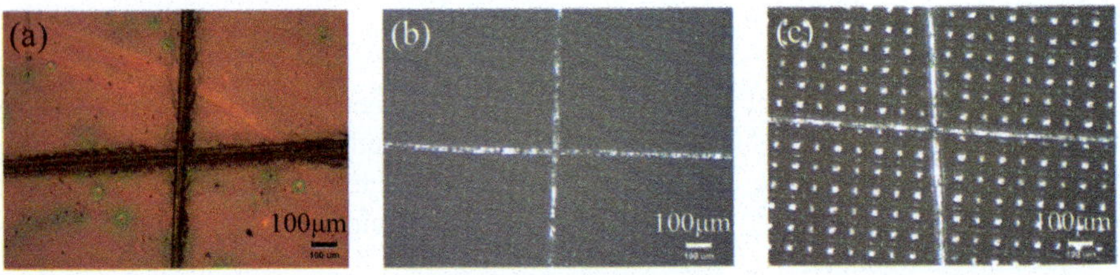

Fig. 11 Crosshatch test for samples **a** S-500, **b** T1-500 and **c** T2-500

surfaces which would provide more available sites for photocatalysis to occur. However, the 2 textures used in this study have similar photocatalytic activity, which could be due to their rather similar roughness values. Surface textures with more distinct patterns and roughness values could be explored in future works.

By conducting thermal oxidation at various temperatures, a change in phase composition was observed. In general, the results suggest that the samples which underwent thermal oxidation at lower temperatures demonstrated better photocatalytic activity. This could be attributed to the improved charged carrier separation in mixed phase TiO_2. The photoexcited electrons from rutile can migrate to the conduction band of anatase which reduces the recombination of electrons and holes [12]. However, more experimental repeats could be performed to better understand the observations.

Laser texturing could potentially be useful in improving the adhesion of the coating to the substrate as seen from Fig. 11. This could be because texturing provides additional mechanical interlocking. Textured surface has a large surface area, which is beneficial to the photocatalytic effect.

4 Conclusions

This study set out to explore the use of photocatalytic TiO_2 coating on 316 stainless steel surfaces to obtain the self-cleaning effect. Thermal oxidation temperature in the range of 400 °C to 600 °C was varied to obtain TiO_2 thin film. The effect of laser texturing was also studied.

The main findings are as follows:

- Based on the results from the photodegradation of methylene blue, there was a greater decrease in the concentration of methylene blue solution for the textured samples with around 32% decrease as compared to the smooth samples with a 23% decrease. This means that the photocatalytic self-cleaning effect is better for textured samples.
- Mixed phase TiO_2 obtained from a lower thermal oxidation temperature around 500 °C generally improves self-cleaning efficiency. For the most part, there was a relatively larger decrease in the concentration of methylene blue solution for the samples thermally oxidized at a lower temperature. This indicates a better photocatalytic activity.
- Texturing could improve adhesion strength of the coating to the substrate. As seen from the crosshatch tests, the coatings on textured surfaces were graded as 5B whereas the coating on smooth surface was graded as 4B. This means that the coatings on textured surfaces had smooth cut edges and no chippings, indicating excellent adhesion of coating to the substrate, while the coating on smooth surface had minor chippings.

References

1. Barthlott W, Neinhuis C (1997) Purity of the sacred lotus, or escape from contamination in biological surfaces. Planta 202:1–8
2. Raja RSS, Selvakumar P, Babu PD, Rubasingh BJ, Suresh K (2021) Influence of laser parameters on superhydrophobicity—a review. Eng Res Express 3:022001
3. Kietzig A-M, Hatzikiriakos SG, Englezos P (2009) Patterned superhydrophobic metallic surfaces. Langmuir 25:4821–4827
4. Gregorčič P, Šetina-Batič B, Hočevar M (2017) Controlling the stainless steel surface wettability by nanosecond direct laser texturing at high fluences. Appl Phys A 123:766
5. Fujishima A, Honda K (1972) Electrochemical photolysis of water at a semiconductor electrode. Nature 238:37–38
6. Hashimoto K, Irie H, Fujishima A (2005) TiO_2 photocatalysis: a historical overview and future prospects. Jpn J Appl Phys 44:8269
7. Malato-Rodríguez S (2004) Solar detoxification and disinfection. In: Cleveland CJ (ed) Encyclopedia of energy. Elsevier, New York, pp 587–596
8. Pizzi A, Mittal KL (2017) Handbook of adhesive technology. CRC Press
9. Sun Y, Singh Rawat R, Chen Z (2022) Mechanically robust multifunctional antifogging coating on transparent plastic substrates. Appl Surf Sci 580:152307
10. Ting C-C, Chen S-Y, Liu D-M (2002) Preferential growth of thin rutile TiO_2 films upon thermal oxidation of sputtered Ti films. Thin Solid Films 402:290–295

11. Hanaor DAH, Sorrell CC (2011) Review of the anatase to rutile phase transformation. J Mater Sci 46:855–874
12. Siah WR, Lintang HO, Shamsuddin M, Yuliati L (2016) High photocatalytic activity of mixed anatase-rutile phases on commercial TiO_2 nanoparticles. IOP Conf Ser Mater Sci Eng 107:012005
13. D01 Committee. Test methods for rating adhesion by tape test. ASTM International

The manufacturer's authorised representative in the EU is Springer Nature Customer Service Centre GmbH, Europaplatz 3, 69115 Heidelberg, Germany. If you have any concerns regarding our products, please contact ProductSafety@springernature.com

Printed and bound by CPI Group (UK) Ltd, Croydon, CR0 4YY

01/10/2026

02225901-0006